PROCEEDINGS OF A WORKSHOP JOINTLY ORGANIZED BY THE COMMISSION OF THE EUROPEAN COMMUNITIES & OECD NUCLEAR ENERGY AGENCY/ BRUSSELS/15-17 MAY 1984

Design and Instrumentation of In Situ Experiments in Underground Laboratories for Radioactive Waste Disposal

Edited by
B.CÔME
Commission of the European Communities, Brussels

P.JOHNSTON & A.MÜLLER
OECD-NEA, Paris

Published for the Commission of the European Communities by

A.A.BALKEMA / ROTTERDAM / BOSTON / 1985

JOINT SECRETARIAT/SECRETARIAT CONJOINT

CEC   B.CÔME
Nuclear Fuel Cycle Division
DG XII/D/1
Rue de la Loi, 200
B-1049 Brussels

OECD-NEA   P.JOHNSTON
Division of Radiation Protection and Waste Management
Bd. Suchet, 38
F-75016 Paris

Now at:
UK Department of the Environment
Romney House
43 Marsham St.
UK-London, SW1P 3PY

A.MÜLLER
Division of Radiation Protection and Waste Management
Bd. Suchet, 38
F-75016 Paris

Cataloguing data can be found at the end of this publication.
Une fiche bibliographique figure à la fin de l'ouvrage.

Published for the Commission of the European Communities,
Directorate-General Information Market & Innovation, Luxembourg

EUR 9575 EN-FR

ISBN 90 6191 594 5

Published by A.A.Balkema, P.O.Box 1675, 3000 BR Rotterdam, Netherlands
Distributed in USA & Canada by: A.A.Balkema Publishers, P.O.Box 230, Accord, MA 02018
Printed in the Netherlands

OECD

Pursuant to article 1 of the Convention signed in Paris on 14th December, 1960, and which came into force on 30th September, 1961, the Organisation for Economic Co-operation and Development (OECD) shall promote policies designed:

- to achieve the highest sustainable economic growth and employment and a rising standard of living in Member countries, while maintaining financial stability, and thus to contribute to the development of the world economy;
- to contribute to sound economic expansion in Member as well as non-member countries in the process of economic development; and
- to contribute to the expansion of world trade on a multilateral, non-discriminatory basis in accordance with international obligations.

The Signatories of the Convention on the OECD are Austria, Belgium, Canada, Denmark, France, the Federal Republic of Germany, Greece, Iceland, Ireland, Italy, Luxembourg, the Netherlands, Norway, Portugal, Spain, Sweden, Switzerland, Turkey, the United Kingdom and the United States. The following countries acceded subsequently to this Convention (the dates are those on which the instruments of accession were deposited): Japan (28th April, 1964), Finland (28th January, 1969), Australia (7th June, 1971) and New Zealand (29th May, 1973).

The Socialist Federal Republic of Yugoslavia takes part in certain work of the OECD (agreement of 28th October, 1961).

The OECD Nuclear Energy Agency (NEA) was established on 20th April 1972, replacing OECD's European Nuclear Energy Agency (ENEA) on the adhesion of Japan as a full Member.

NEA now groups all the European Member countries of OECD and Australia, Canada, Japan, and the United States. The Commission of the European Communities takes part in the work of the Agency.

The primary objectives of NEA are to promote co-operation between its Member governments on the safety and regulatory aspects of nuclear development, and on assessing the future role of nuclear energy as a contributor to economic progress.

This is achieved by:

- *encouraging harmonisation of governements' regulatory policies and practices in the nuclear field, with particular reference to the safety of nuclear installations, protection of man against ionising radiation and preservation of the environment, radioactive waste management, and nuclear third party liability and insurance;*
- *keeping under review the technical and economic characteristics of nuclear power growth and of the nuclear fuel cycle, and assessing demand and supply for the different phases of the nuclear fuel cycle and the potential future contribution of nuclear power to overall energy demand;*
- *developing exchanges of scientific and technical information on nuclear energy, particularly through participation in common services;*
- *setting up international research and development programmes and undertakings jointly organised and operated by OECD countries.*

In these and related tasks, NEA works in close collaboration with the International Atomic Energy Agency in Vienna, with which it has concluded a Co-operation Agreement, as well as with other international organisations in the nuclear field.

OCDE

En vertu de l'article 1er de la Convention signée le 14 décembre 1960, à Paris, et entrée en vigueur le 30 septembre 1961, l'Organisation de Coopération et de Développement Économiques (OCDE) a pour objectif de promouvoir des politiques visant :

- à réaliser la plus forte expansion de l'économie et de l'emploi et une progression du niveau de vie dans les pays Membres, tout en maintenant la stabilité financière, et à contribuer ainsi au développement de l'économie mondiale ;
- à contribuer à une saine expansion économique dans les pays Membres, ainsi que non membres, en voie de développement économique ;
- à contribuer à l'expansion du commerce mondial sur une base multilatérale et non discriminatoire conformément aux obligations internationales.

Les signataires de la Convention relative à l'OCDE sont : la République Fédérale d'Allemagne, l'Autriche, la Belgique, le Canada, le Danemark, l'Espagne, les Etats-Unis, la France, la Grèce, l'Irlande, l'Islande, l'Italie, le Luxembourg, la Norvège, les Pays-Bas, le Portugal, le Royaume-Uni, la Suède, la Suisse et la Turquie. Les pays suivants ont adhéré ultérieurement à cette Convention (les dates sont celles du dépôt des instruments d'adhésion) : le Japon (28 avril 1964), la Finlande (28 janvier 1969), l'Australie (7 juin 1971) et la Nouvelle-Zélande (29 mai 1973).

La République socialiste fédérative de Yougoslavie prend part à certains travaux de l'OCDE (accord du 28 octobre 1961).

L'Agence de l'OCDE pour l'Énergie Nucléaire (AEN) a été créée le 20 avril 1972, en remplacement de l'Agence Européenne pour l'Énergie Nucléaire de l'OCDE (ENEA) lors de l'adhésion du Japon à titre de Membre de plein exercice.

L'AEN groupe désormais tous les pays Membres européens de l'OCDE ainsi que l'Australie, le Canada, les États-Unis et le Japon. La Commission des Communautés Européennes participe à ses travaux.

L'AEN a pour principaux objectifs de promouvoir, entre les gouvernements qui en sont Membres, la coopération dans le domaine de la sécurité et de la réglementation nucléaires, ainsi que l'évaluation de la contribution de l'énergie nucléaire au progrès économique.

Pour atteindre ces objectifs, l'AEN :

- *encourage l'harmonisation des politiques et pratiques réglementaires dans le domaine nucléaire, en ce qui concerne notamment la sûreté des installations nucléaires, la protection de l'homme contre les radiations ionisantes et la préservation de l'environnement, la gestion des déchets radioactifs, ainsi que la responsabilité civile et les assurances en matière nucléaire ;*
- *examine régulièrement les aspects économiques et techniques de la croissance de l'énergie nucléaire et du cycle du combustible nucléaire, et évalue la demande et les capacités disponibles pour les différentes phases du cycle du combustible nucléaire, ainsi que le rôle que l'énergie nucléaire jouera dans l'avenir pour satisfaire la demande énergétique totale ;*
- *développe les échanges d'informations scientifiques et techniques concernant l'énergie nucléaire, notamment par l'intermédiaire de services communs ;*
- *met sur pied des programmes internationaux de recherche et développement, ainsi que des activités organisées et gérées en commun par les pays de l'OCDE.*

Pour ces activités, ainsi que pour d'autres travaux connexes, l'AEN collabore étroitement avec l'Agence Internationale de l'Énergie Atomique de Vienne, avec laquelle elle a conclu un Accord de coopération, ainsi qu'avec d'autres organisations internationales opérant dans le domaine nucléaire.

Synopsis

A joint CEC/NEA workshop was held in Brussels on 15-17th May, 1984 on the theme "Design and Instrumentation of in situ Experiments in Underground Laboratories for Radioactive Waste Disposal". About 100 specialists attended this meeting, in which a review of the current development of such underground facilities was made.

Several of them (e.g. the Asse mine, FRG, in salt, or the Stripa mine, Sweden, or the Underground Research Laboratory at Pinawa, Canada, both in granite) are developing testing and instrumentation methodologies, to be transferred later to repository sites; others, such as the Mol excavation in Belgian clay, will also serve to qualify the site in which actual disposal of waste could take place later.

It appeared from the discussions that underground laboratories are well suited for rock-mechanical experiments and testing of multicomponent systems in a realistic environment (e.g. heater tests).

Visits to the Mol facility were arranged for the participants.

The proceedings of this workshop are edited jointly by the CEC and the NEA.

Synopsis

Une séance de travail conjointe AEN-CCE s'est tenue à Bruxelles du 15 au 17 mai 1984 sur le thème "Conception et instrumentation d'expériences in situ en laboratoires souterrains pour l'évacuation de déchets radioactifs". Cette séance a réuni environ cent spécialistes; on y a procédé à une revue du développement actuel de ces installations souterraines.

Parmi celles-ci, certaines servent à mettre au point des méthodes d'essai et d'instrumentation (par exemple la mine de Asse, RFA, ou la mine de Stripa, en Suède, ou encore, le Laboratoire Souterrain de Recherche à Pinawa, Canada, ces deux derniers dans le granite); ces méthodes seront transposées plus tard à des sites de dépôts. D'autres, telles la cavité de Mol dans l'argile belge, serviront aussi à caractériser le site dans lequel on devrait réellement évacuer plus tard des déchets radioactifs.

Les débats ont montré que les laboratoires souterrains se prêtent bien aux expériences de mécanique des roches, et à des essais faisant intervenir plusieurs composants dans un milieu rocheux représentatif (par exemple essais de chauffage en place).

Des visites aux installations de Mol furent également organisées.

Le compte-rendu de cette séance de travail a été préparé conjointement par l'AEN et la CCE.

Table of contents
Table des matières

Session 1. *Underground laboratories: Experience and objectives*
Séance 1. *Les laboratoires souterrains: Expérience et objectifs*

Session 2. *Reviews of experiments in the major rock types*
Séance 2. *Revue des expériences dans les principaux types de formations*

Session 5. *Design and instrumentation of in situ experiments in salt*
Séance 5. *Conception et instrumentation d'expériences in situ dans le sel*

XI

Session 6. *Overviews and conclusions*
Séance 6. *Résumés et conclusions*

Opening address

M.S.FINZI
Commission of the European Communities, Brussels, Belgium

Ladies and Gentlemen,

In the name of the Commission of the European Communities and the Nuclear Energy Agency of the O.E.C.D., I have pleasure in welcoming you to Brussels and thanking you for participating in this workshop.

The problem of radioactive waste is generally regarded with apprehension by the public in our countries. And yet, the large countries involved in the peaceful use of nuclear energy, especially including the production of electricity from nuclear sources, are now all undertaking work whose aim is to find safe solutions to the management of this waste and to its disposal.

At present, one can consider disposal in deep geological formations as the most realistic option. A consensus from the scientific, international Community on the merits of this option and the demonstration of the feasability of this disposal should contribute to removing the reticence of the public and allowing the realisation of underground facilities for disposal.

There are great advantages to allowing the research phase and the demonstration preceding the operational phase of disposal to be co-ordinated on an international level. The Nuclear Energy Agency of the O.E.C.D. and the Commission of the European Communities have been working in this direction for several years, each in their own way.

As regards the European Communities, the Commission proposed activities relating to three experimental, underground facilities, in Belgium, France and the Federal Republic of Germany, for the new Community programme 1985-1989 which it has just sent to the Council of European Ministers. These facilities will be open to Community co-operation in the framework of the programme. Taking into account their experience in this field, our two organisations took the initiative at the end of last year, to undertake a joint report of work carried out, or to be carried out, in what are called underground laboratories.

Why harp on this theme of underground laboratories ? Above all because these laboratories represent an indispensable stage in order to be able to construct deep repositories for radioactive waste in the future. Actually, they constitute the link of a chain which starts at research sites and their qualification by surface and laboratory tests, to culminate in the realisation of demonstration facilities, then in the construction and putting into operation of final repositories.

1

Another reason for having chosen this theme is its present interest : indeed, a certain number of such underground laboratories have been constructed or are in the process of being constructed in various types of potential host rocks, and it is already possible to draw useful instruction from them. Others are foreseen or are about to be designed amongst the countries involved in the peaceful use of nuclear energy.

The scientists are now fully aware of the limits set by the laboratory tests on samples, or futhermore in deep bore-holes, when it is a question of characterising the properties of a deep formation. These limits result in the difficulty of reproducing natural conditions in laboratory, and in the slight volume of rock which the bore-hole measurement techniques allow to be reached. It is only on the site, in depth and on a real scale, that the true properties of the rock mass, that is to say its physical, mechanical, chemical and hydrogeological characteristics can be obtained. Tests can also be carried out in underground cavities, on great scale, on multiple component systems, such as on site heating tests. In this framework, the temperature conditions, fluid movements, ground pressure, etc., are truly representative of the rock mass.

From these tests, it will also be possible to verify if the design of a repository, to be implanted on the same test site, has been adapted. Experience leads us to believe that modifications will have to be made in due time.

Finally, the duration of these tests undertaken, for several years in some cases, will improve our confidence in the forecasts of the rock mass behaviour which one can derive from numerical simulations.

As you can ascertain from title of the programme, the retained theme concern the design and instrumentation of these experiences to be carried out underground with a view to geological disposal. Previous to the numerous detailed technical presentations, the people responsible for the projects on a national level will have the possibility to review the interest and the present situation of underground laboratories. Then it will be up to the scientists and technicians to collate their concepts on the tests in progress or in preparation.

At the end of this 3-day meeting, a general discussion, which, we hope, will be lively, will allow the present evaluations to be put in concrete form and perhaps bring out possible courses of action for follow-up work.

Finally, I would like to thank our colleagues at the Centre d'Etude de l'Energie Nucléaire at MOL (CEN/SCK) who kindly accepted to arrange visits of their surface and underground facilities as part of this Workshop.

Before requesting the Chairman of this first session to take this meeting in hand, I would like to remind you that the Nuclear Energy Agency and the Commission have jointly established a document presenting the general state of knowledge in the field of geological disposal.

Ladies and Gentlemen, allow me to finish this speech by expressing my wishes for the success of your meeting. The level of the presentations and the quality of the participants alone are guarantees for its interest.

Allocution d'ouverture

M.S.FINZI

Commission des Communautés Européennes, Bruxelles, Belgique

Mesdames, Messieurs,

Au nom de la Commission des Communautés Européennes et de l'Agence pour l'Energie Nucléaire de l'O.C.D.E., j'ai le plaisir de vous souhaiter la bienvenue à Bruxelles et de vous remercier d'avoir bien voulu participer à cette séance de travail.

Le problème des déchets radioactifs est généralement perçu avec appréhension par le public de nos pays. Et pourtant, les grands pays impliqués dans l'utilisation pacifique du nucléaire, sont tous maintenant engagés dans des travaux dont le but est de trouver des solutions sûres à la gestion de ces déchets et à leur stockage définitif.

Actuellement, on peut considérer le stockage définitif en formations géologiques profondes comme l'option la plus réaliste. Un consensus de la communauté scientifique internationale sur le bien-fondé de cette option et la démonstration de la faisabilité de cette évacuation devrait contribuer à lever les réticences du public et permettre de réaliser des installations souterraines de stockage définitif.

Les avantages sont grands pour que la phase de recherche et de démonstration précédant la phase opérationnelle du stockage définitif puisse être réalisée de façon coordonnée au niveau international. L'Agence pour l'Energie Nucléaire de l'O.C.D.E. et la Commission des Communautés Européennes oeuvrent dans cette direction depuis de nombreuses années, chacune à leur manière.

Pour ce qui concerne les Communautés Européennes, la Commission a proposé pour le nouveau programme communautaire 1985-1989 qu'elle vient d'envoyer au Conseil des Ministres européens, des activités relatives à trois installations expérimentales souterraines, en Belgique, en France et en République Fédérale d'Allemagne. Ces installations seront, dans le cadre du programme, ouvertes à la coopération communautaire. Compte tenu des acquis dans ce domaine, nos deux organismes ont pris, à la fin de l'année dernière, l'initiative de procéder à un bilan conjoint des travaux réalisés, ou à réaliser, dans ce que nous appelons les laboratoires souterrains.

Pourquoi ce thème des laboratoires souterrains ? Avant tout parce que ces laboratoires représentent une étape indispensable pour pouvoir construire, dans le futur, des installations de dépôts profonds de déchets radioactifs.

En effet, ils constituent la maillon d'une chaîne qui part de la recherche des sites et de leur qualification par des essais de surface et de laboratoire, pour aboutir à la réalisation d'installations de démonstration, puis à la construction et à la mise en service de dépôts définitifs.

Une autre raison d'avoir choisi ce thème est son actualité : en effet, un certain nombre de tels laboratoires souterrains ont été construits ou sont en cours de construction dans divers types de roches-hôtes potentielles, et il est déjà possible d'en tirer d'utiles enseignements. D'autres sont prévus ou en sont au niveau de la conception parmi les pays impliqués dans l'utilisation pacifique de l'énergie nucléaire.

Les scientifiques sont maintenant bien conscients des limites que présentent les essais de laboratoires sur échantillons, ou encore les mesures en forage profonds, lorsqu'il s'agit de caractériser convenablement les propriétés d'une formation profonde. Ces limites tiennent à la difficulté de reproduire en laboratoire les conditions naturelles, et au faible volume de roche que les techniques de mesures en forages permettent d'atteindre. C'est en place, en profondeur et à l'échelle réelle qu'on peut obtenir les véritables propriétés des massifs rocheux, à savoir leurs caractéristiques physiques, mécaniques, chimiques, et hydrogéologiques. C'est aussi dans des cavités souterraines qu'on peut procéder à des essais, en vraie grandeur, de systèmes à composantes multiples, tels des essais de chauffage en place. Dans ce cadre, les conditions de température, de mouvements des fluides, de pression des terrains, etc., sont véritablement représentatives du milieu rocheux.

A partir de ces essais, il sera possible de vérifier si la conception d'un dépôt, à implanter sur le même site d'essai, est adaptée. L'expérience porte à croire que des modifications devront y être apportées, auxquelles il sera procédé en temps utile.

Enfin, la durée des essais entrepris, de l'ordre de plusieurs années dans certains cas, améliora notre confiance dans les prédictions du comportement du massif rocheux que l'on peut faire à partir de simultations numériques.

Comme vous pouvez le constater d'après l'intitulé du programme, le thème retenu concerne la conception et l'instrumentation de ces expériences à réaliser en souterrain en vue de l'évacuation géologique. En préalable aux nombreuses présentations techniques détaillées, les responsables des projets au niveau national auront la possibilité de faire le bilan sur l'intérêt et la situation actuelle des laboratoires souterrains. Ce sera ensuite au tour des scientifiques et techniciens de confronter leurs concepts sur les essais en cours ou en préparation.

A l'issue de ces trois jours de réunion, une discussion générale, que nous espérons animée, permettra de concrétiser les bilans actuels et peut-être de dégager d'éventuelles lignes d'actions pour la suite des travaux.

Enfin, je tiens à remercier ici nos collègues du Centre d'Etude de l'Energie Nucléaire à MOL (CEN/SCK) qui ont bien voulu organiser des visites de leurs installations de surface et de leur laboratoire souterrain à l'occasion de cette réunion de travail.

Avant de demander au président de la première session de prendre en main cette réunion, j'aimerais vous rappeler que l'Agence de l'Energie Nucléaire et la Commission ont établi en commun un document présentant l'état général des connaissances dans le domaine de l'évacuation géologique.

Mesdames et Messieurs, permettez-moi de clore ici cette allocution en formulant mes voeux pour le succès de votre réunion. A eux seuls, le niveau des présentations et la qualité des participants sont garants de son intérêt.

Session 1/Séance 1
Underground laboratories: Experience and objectives
Les laboratoires souterrains: Expérience et objectifs

Chairman/Président:
A.BARTHOUX
ANDRA, Paris, France

Les laboratoires souterrains de mécanique des roches avant les déchets nucléaires

P.DUFFAUT
Pierre Londe & Associés, Paris, France

RESUME

Beaucoup d'essais de mécanique des roches sont effectués *in situ*, et la plupart en souterrain, depuis le barrage de Béni Bahdel en 1936. Les principaux essais destinés à comprendre le comportement du massif rocheux sont les essais de déformabilité (essai à la plaque, essai en caverne, y compris les mesures de fluage) et les essais de résistance (compression de pilier de mine, cisaillement du terrain ou d'un joint). L'influence de l'humidité, de la chaleur, du froid et de la congélation sont d'autres domaines qui méritent des laboratoires souterrains. Le comportement de galeries expérimentales, sans soutènement ou avec divers types de soutènement, est souvent étudié en fonction du temps et aussi en fonction des travaux, vitesse d'avancement, élargissement ou approfondissement de la section. Les exemples donnés aident à préciser la notion de laboratoire souterrain malgré la variété de ses objectifs.

ABSTRACT

Underground laboratories for rock mechanics before radioactive waste
Many rock mechanics tests are performed *in situ*, most of them underground since 1936 at the Beni Bahdel dam. The chief tests for understanding the rock mass behaviour are deformability tests (plate test and pressure cavern test, including creep experiments) and strength tests (compression of a mine pillar, shear test on rock mass or joint). Influence of moisture, heat, cold and freeze are other fields of investigation which deserve underground laboratories. Behaviour of test galleries, either unsupported or with various kinds of support, often is studied along time, and along the work progression, tunnel face advance, enlargement or deepening of the cross section. The examples given here help to clarify the concept of underground laboratory in spite of its many different objectives.

INTRODUCTION

L'échantillon de roche ne suffit pas pour représenter le terrain; banale pour les mineurs, cette constation fait justement l'originalité de la mécanique des roches et du même coup elle limite l'efficacité des essais de laboratoire. Un grand nombre des essais de mécanique des roches sont des essais de terrain, donc des essais *in situ*; c'est-à-dire qu'au lieu de transporter les échantillons dans un laboratoire, on transporte le laboratoire sur le terrain, et dans bien des cas en souterrain. En outre, beaucoup de chantiers d'ouvrages souterrains et d'exploitations minières sont précédés par des galeries ou chantiers d'essai, où l'on choisit et met au point les méthodes qui vont être employées. Enfin, dans les ouvrages eux-mêmes, des mesures sont souvent poursuivies jusqu'à la fin des travaux, et parfois au-delà. Dans tous ces essais et mesures, il entre des préoccupations immédiates relatives au projet, à la construction et à l'exploitation des ouvrages, mais aussi, en plus ou moins grande proportion, un souci de meilleure connaissance du comportement du terrain.

C'est cette étude expérimentale du comportement du terrain, placé dans des conditions nouvelles bien définies, qui fonde le concept de laboratoire souterrain de mécanique des roches. Encore doit-il répondre à quelques critères supplémentaires de durée, d'échelle, de confinement et d'équipement qui seront esquissés au fur et à mesure des exemples donnés ci-dessous.

En mécanique des roches, les conditions expérimentales que l'on fait varier sont le plus souvent les forces appliquées, mais quelquefois l'écoulement de l'eau, l'hygroscopie de l'air, et la température. Ces conditions s'appliquent au massif entourant la cavité-laboratoire, ou seulement à une portion de sa surface, voire à des échantillons préparés et conservés sur place. Il y a lieu d'écarter les essais qui sont faits à la surface, et aussi en forages ou entre forages: le laboratoire souterrain implique une cavité pénétrable où l'homme de laboratoire installe ses dispositifs de mesure et d'essai, même s'il doit l'abandonner au cours de certains essais.

Il y a lieu d'écarter aussi, évidemment, tous les laboratoires souterrains qui sont consacrés à autre chose que la mécanique des roches, physique des particules dans les anneaux du CERN, ou plus profondément dans les tunnels routiers du Mont-Blanc, du Fréjus et du Gran Sasso, climatologie et biologie souterraines dans la grotte de Moulis, Ariège, et probablement bien d'autres consacrés à des domaines d'études très variés.

Malgré ces restrictions évidentes, il reste un grand nombre d'expériences qui méritent plus ou moins d'être qualifiées de laboratoires souterrains et les chapitres ci-dessous s'attachent surtout à en illustrer la grande variété qui s'explique par les problèmes à résoudre, par les méthodes employées, et bien entendu par la variété des terrains. Le tableau 1 donne une idée schématique des principaux emplois des essais *in situ* en fonction des problèmes à étudier.

1. MESURES DE CONTRAINTES

Beaucoup de comportements mécaniques dépendent des contraintes. Si la mesure de contrainte *in situ* ne suffit pas à justifier le mot de laboratoire, elle est un outil essentiel de beaucoup de laboratoires souterrains et ne peut donc être méconnue. La méthode la plus employée est la libération partielle par saignée suivie d'une recompression au vérin plat, mise au point dans les carrières souterraines de la région parisienne (Habib et Marchand 1952), largement utilisée dans les mines de fer de Lorraine (Tincelin 1952) puis dans les ouvrages hydroélectriques. Après une évolution vers des vérins de grande surface (1 m2 ou davantage, notamment en Yougoslavie) c'est au contraire une miniaturisation qui a prévalu après que Rocha (et al 1966) ait utilisé une scie circulaire. De nombreuses

	Ouvrages miniers	Galeries en charge	Grands barrages	Grandes cavernes	Stockages froids	Stockages chauds
mesure de contraintes	xx	xx	x	xx	xx	xx
essai à la plaque		x	xx			x
essai en caverne		xx	x			x
fluage	x	xx	xx	x		xx
compression de pilier	xx			x		
cisaillement	x		xx	x		x
traction				x	xx	
hygromécanique			(suivant nature des roches)			
thermomécanique				xx	xx	xx
comportement d'une cavité	xx	x		xx	xx	xx
choix du soutènement	xx	xx		xx	x	x

TABLEAU 1

Essais à engager dans un laboratoire souterrain en fonction de la nature des ouvrages projetés

XX indique un essai très souhaitable
X indique un essai éventuel
(l'absence de signe n'exclut pas totalement l'essai)

équipes sont aujourd'hui équipées pour ces mesures utilisées dans plusieurs des exemples cités ci-dessous.

L'expression laboratoire souterrain paraît s'appliquer à la comparaison dans un même site souterrain de plusieurs méthodes de mesure de contraintes, c'est le cas de la mine d'uranium de Fanay-Augères, Haute-Vienne, où le B.R.G.M. a constaté que les mesures à la paroi des galeries étaient peu fiables alors que les mesures en forages par surcarottage ou par fracturation hydraulique donnaient des résultats plus cohérents (Bertrand et Durand 1983).

2. ESSAIS DE DEFORMABILITE

Les mesures de déformabilité et de résistance *in situ* ont probablement commencé dans les mines longtemps avant l'individualisation de la mécanique des roches, mais c'est l'industrie de la Houille Blanche qui leur a donné un caractère systématique et qui a contribué à fixer les premiers modes opératoires, grâce en particulier à la diffusion par les Congrès des grands barrages.

L'essai à la plaque, emprunté à la mécanique des sols, a été adopté d'abord pour des fondations de barrages sur des roches relativement tendres, calcaires marneux en Algérie (Drouhin 1936 - fig. 1), mollasse en Suisse au barrage de Rossens. Pour obtenir les réactions nécessaires pour un chargement au vérin, ces premiers essais de mécanique des roches ont été faits dans des excavations (ce n'est que beaucoup plus tard que la réaction a pu être fournie par des tirants et l'essai appliqué à la surface du sol) et l'usage s'est rapidement établi d'employer à cet effet les galeries de reconnaissance qui sont nécessaires à l'étude des sites de barrages, de centrales souterraines et surtout de conduites forcées (Habib 1950, Talobre 1959 et 1967, Londe 1973).

C'est en effet l'économie sur l'épaisseur des blindages en acier qui a justifié les conduites forcées souterraines tout en exigeant une meilleure connaissance de la déformabilité du terrain à l'échelle de la conduite. A l'essai à la plaque s'est ajouté l'essai en caverne, véritable tronçon de galerie en charge avec ou sans blindage, où la force appliquée est la pression de l'eau à l'intérieur. Parti probablement d'Italie (Oberti 1949), cet essai est employé par E.D.F. dès 1951 à Montpezat, et régulièrement depuis pour tous les projets importants de galerie en charge. Le dernier exemple en date est celui de Super-Bissorte, Savoie, en 1981-1982 (Doucerain 1983).

Pour une haute chute exploitée avec pompage, E.D.F. a soumis à une pression de 25 MPa un tronçon d'essai bétonné (longueur 20 mètres, diamètre intérieur 2 mètres) revêtu d'un blindage de 12 mm d'épaisseur au lieu des 50 exigés par la pratique habituelle. Les essais ont duré 3 mois et demi par cycles progressifs et ont confirmé que le paramètre important est la déformation admissible de l'acier. Avant de bétonner et de blinder cette caverne d'essai, des essais à la plaque avaient défini la déformabilité locale du terrain; pendant les essais, des extensomètres en forages (Distofor de Télémac) ont précisé les déformations en fonction de la profondeur. L'influence de cet essai permet de dépasser la zone superficielle dont le module est affaibli par le creusement du tunnel, et

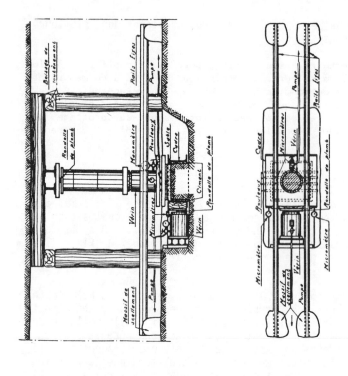

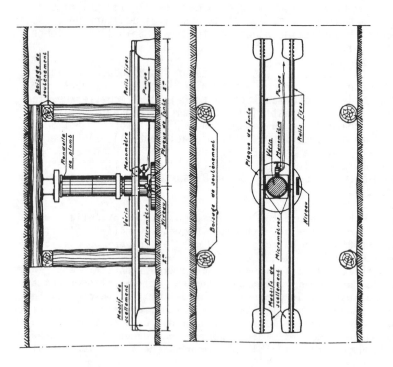

Figure 1 - Essai à la plaque et essai de cisaillement dans une galerie de reconnaissance du barrage de Béni Bahdel, Algérie (figures extraites de Drouhin, 1936)

d'atteindre le module du massif au large, supérieur à 50 GPa (contre 5 à 10 au voisinage immédiat de la galerie).

Ce même essai est employé dans tous les pays du monde, parfois même sans la justification d'une conduite forcée (Rossi 1984), pour fournir une meilleure évaluation de la déformabilité que l'essai à la plaque. Un tableau de comparaison portant sur plus de cent sites est donné par Hoek et Brown (1980) qui mentionne 48 essais en caverne contre 113 essais à la plaque. Il donne aussi des mesures en forages avec divers dilatomètres qui sont en quelque sorte des modèles réduits de caverne (Kujundzic 1958). Et il mentionne en bonne place la variante autrichienne de l'essai en caverne (Lauffer et Seeber 1961): le vérin radial Tiwag (de la Tiroler Wasserkraft A.G.) où la mise en charge du revêtement est obtenue par des vérins plats prenant appui sur une charpente circulaire centrale. Cette méthode plus légère a connu beaucoup de succès en Europe de l'Est, par exemple dans les dernières années au barrage de Nurek (Rukin et al 1983), et elle a été également introduite en Chine (Yang et Je 1983).

Les mêmes méthodes sont applicables aux essais de déformabilité de longue durée à charge constante, c'est-à-dire aux essais de fluage, qui prennent une importance particulière dans certains types de roches. En outre pour tous les essais de déformabilité se pose le problème d'interpréter les graphiques obtenus pour déduire le (ou les) module utile (groupe de travail 1964, Duffaut 1968, Bernaix 1974).

3. ESSAIS DE RESISTANCE

La compression uniaxiale est appliquée à des piliers de mine, soit de façon non mesurable rigoureusement, en augmentant le dépilage de la couche autour des piliers, soit en intercalant des vérins entre le toit et le pilier. Cette méthode a été employée notamment par Bieniawski (1967) dans les mines de charbon d'Afrique du Sud, ainsi que par une "action concertée" française dans une mine de fer de Lorraine (Mancieulles, Tincelin 1968). Dans les deux cas, l'effet d'échelle était le principal problème à résoudre. Dans les deux cas des blocs de charbon ou de minerai de différentes dimensions ont été aussi écrasés à la presse.

Le pilier expérimental de Mancieulles a été découpé au fil sur quatre faces verticales formant un prisme carré de deux mètres de côté, puis la tête a été sciée horizontalement pour ménager la place des vérins plats scellés au mortier de ciment. Pendant la mise en charge, la déformation était suivie par des extensomètres à corde vibrante. En réalité, la rupture en compression n'a pu être obtenue que de façon locale, au voisinage d'un angle de la section, et elle s'accompagnait de la mise hors service du vérin plat (deux ruptures successives à 19,6 et 22,5 MPa). Toutefois des fissures verticales se sont manifestées à 16,5 et 21,5 MPa au cours de ces essais. La déformation est apparue linéaire mais une déformation permanente importante s'est manifestée à la décharge.

Le cisaillement est utilisé surtout pour l'étude des fondations de barrages, qu'il s'agisse de la résistance de la matrice rocheuse, de joints de stratification ou de diaclases, ou encore de la surface de contact entre le béton et la roche. On le trouve déjà en 1936 (Drouhin) en parallèle avec l'essai à la plaque. Un exemple remarquable a été donné par le grand barrage-poids espagnol de Mequinenza sur l'Ebre (Jimenez Salas 1968), où des lignites tendres étaient interstratifiées dans les calcaires horizontaux des versants.

Un essai tout à fait comparable a été fait au barrage-voûte de Vouglans, Ain (Groupe de travail 1967) où la série calcaire horizontale montrait quelques joints argileux minces. Le plus épais, qui a été soumis à l'essai, avait 10 à 20 mm d'épaisseur. A partir d'une galerie de reconnaissance préexistante, les faces d'un "pilier" ont été dégagées à l'explosif puis le tiers supérieur a été enlevé également à l'explosif, ce qui n'a pas manqué de remanier quelque peu le banc calcaire au-dessus du joint. La surface essayée était 5,5 m2 alors que beaucoup d'essais *in situ* sont faits sur des surfaces de 0,25 à 1 m2. Les efforts normaux et tangentiels étaient fournis par des vérins plats.

En fait, l'effet d'échelle est peu important sur le cisaillement des joints. En conclusion de cet essai Coyne et Bellier a construit une machine de cisaillement permettant d'essayer en laboratoire une surface de 0,1 m2 environ (les échantillons étant prélevés par carottage, diamètre 250 mm). Une autre retombée de cet essai a été la décision de découper au fil le pilier de Mancieulles.

La traction est mentionnée pour mémoire. Elle peut être appliquée par arrachement de boulons. La torsion a été parfois recommandée, et aussi diverses formes d'essais plus ou moins triaxiaux. L'essai à la plaque se transforme en essai de poinçonnement lorsqu'il est poussé à la rupture (Groupe de travail 1964).

4. ESSAIS HYDRAULIQUES

Les conditions d'écoulement de l'eau dans le massif rocheux sont étudiées par des essais dérivés de l'essai Lugeon à partir de forages (Louis et Maini 1970), mais aussi à partir de cavernes en pression, cavernes cette fois non revêtues. Le reste de l'essai est assez analogue à ce qui a été décrit ci-dessus.

La galerie d'amenée en charge de la chute de Roselend, Savoie (longueur 13 km, débit 50 m3/s, diamètre intérieur 4 m), était, lors de sa construction, la plus audacieuse de sa catégorie (charge de service statique de 120 à 160 mètres). Aussi de nombreux tronçons ont été mis en pression jusqu'à 200 mètres d'eau (Pousse et Jacquin 1961) avec ou sans revêtement bétonné. Et avant la mise en service, la galerie entière a encore été maintenue en pression pour l'évaluation de fuites éventuelles.

5. COMPORTEMENT HYGROMECANIQUE

Les roches "sensibles" à l'eau forment une catégorie à part qui pose des problèmes spécifiques. On sait qu'elles peuvent être détruites ou leur résistance fortement altérée par la première dessication, d'où l'idée de les essayer en souterrain, avant tout retrait. Ainsi Bieniawski avait-il descendu une presse dans une mine de charbon (en parallèle avec les essais de piliers, chapitre 3). A Sisteron, Alpes de Provence, l'étude d'une centrale souterraine dans des calcaires marneux a conduit de même à conserver des échantillons en souterrain, et à les essayer sur place.

Mais le laboratoire souterrain de Sisteron a été plus ambitieux:

quatre rameaux de galerie ont été climatisés pendant six mois afin d'étudier l'effet de l'hygroscopie non seulement sur des échantillons (Diernat et Duffaut 1969), mais aussi sur le massif (Diernat et al 1970). Comme il fallait s'y attendre, ce sont les ambiances les plus sèches qui ont conduit à des éboulements après progression de fissures de retrait, et l'arrêt de la climatisation a permis une stabilisation totale.

Dans d'autres roches, le gonflement est plus critique que le retrait et son étude a donné lieu à quelques laboratoires souterrains (métro de Marseille, et plus récemment tunnel autoroutier de Chamoise, Jura) mais elle rentre plutôt dans le chapitre 8 ci-dessous.

6.   COMPORTEMENT THERMOMECANIQUE

Il est difficile de qualifier de laboratoire souterrain les observations sur l'écaillage par échauffement brutal d'un massif rocheux granitique: il s'agit de l'échappement de moteurs diesel dans un poste de commande du NORAD à North Bay, Ontario. Du moins ces observations se poursuivent-elles depuis plus de vingt ans.

Par contre l'effet du froid a donné lieu à quelques expériences de Gaz de France dès 1960: trois cavités expérimentales d'environ trois mètres de diamètre ont été creusées à 20-25 mètres de profondeur dans trois roches différentes (craie à Denain, marnes à Thionville, granite à Nantes) et ont été remplies d'azote liquide. Il semble bien que ce type de laboratoire souterrain ait été à l'époque une première mondiale (Toche 1961). Les développements plus récents sont l'objet d'une autre communication à la même réunion, comme aussi les essais de chauffage de Géovexin (de Laguérie).

7.   COMPORTEMENT DE CAVITES SOUTERRAINES

Le premier comportement est celui d'une cavité abandonnée à elle-même. L'observation de galeries expérimentales creusées dans la glace dans les années 40 à l'occasion d'une prise d'eau sous-glaciaire (glacier de Trélatête, Haute Savoie) a montré qu'elles se refermaient lentement de façon très dissymétrique. Le cas des mines et carrières souterraines abandonnées qui sont auscultées pour vérifier leur stabilité et signaler toute évolution dangereuse échappe au concept de laboratoire souterrain.

Par contre les comportements des ouvrages souterrains en cours de creusement peuvent avoir une portée générale. Ainsi de nombreuses centrales souterraines ont été auscultées au cours des phases successives de leur creusement, et plusieurs grands tunnels permettent aussi d'étudier l'influence de l'avancement (l'effet des soutènements étant renvoyé au chapitre suivant: 8). Toutes les méthodes modernes de conception des tunnels reposent sur la mesure des déformations (Lane, Panet 1975; Londe 1977)

Sur une section de tunnel proche du front, les déformations dépendent de la distance au front mais dans beaucoup de terrains elles dépendent aussi du temps et lorsque le front avance ces deux effets se superposent. Les mesures de convergence dans plusieurs tunnels et notamment le tunnel routier transalpin du Fréjus ont permis de séparer ces deux effets et de proposer des lois générales (Sulem 1983).

Pour l'étude de cavernes de "grande" portée, des essais

d'élargissement progressif ont été faits sur divers sites, avec une auscultation appropriée du massif (de même en mine pour choisir une largeur de chambre). Quant aux cavernes de grande hauteur, elles sont parfois auscultées au fur et à mesure de leur approfondissement, à partir d'une excavation initiale au niveau de la voûte.

Un cas remarquable de laboratoire souterrain est Venteuil-Tincourt, Marne, pour l'étude d'un projet qui n'a pas eu de suite (Dessenne et Duffaut 1970). En l'absence de toute expérience à cette époque sur des ouvrages souterrains dans la craie, un programme d'essais très important a eu lieu avec mesures de contraintes, essais à la plaque, et essais de laboratoire. Mais l'essai le plus important était un modèle réduit au cinquième de la caverne projetée, à la profondeur prévue de 90 mètres. Comme pour une caverne véritable, les phases de creusement ont été respectées: la voûte d'abord par la méthode allemande (à partir de deux galeries latérales) puis le stross par tranches successives. Des mesures de contraintes dans la voûte et des mesures de convergence horizontale ont été poursuivies.

Parallèlement, un calcul élastique par éléments finis (en 1967 c'était un des tout premiers de ce genre) permettait de suivre aussi le creusement phase par phase. Compte tenu de la bonne connaissance des contraintes et du module, on a pu s'étonner de ce que les convergences soient restées deux ou trois fois plus faibles que ne le prédisait le calcul. Cette différence paraît explicable par une adaptation plastique, avec augmentation de compacité de la craie. Des mesures sismiques ont confirmé l'élévation de la vitesse du son (Dagnaux et al 1970).

8. COMPORTEMENT DE TUNNELS EN FONCTION DES SOUTENEMENTS

Les exemples sont très nombreux où des tronçons successifs d'un même tunnel dans un même terrain reçoivent des soutènements différents, et il serait abusif de les accepter tous comme autant de laboratoires souterrains. Toutefois, beaucoup de ces études ont effectivement des objectifs généraux en sus des besoins immédiats de l'ouvrage projeté, par exemple lors des études pour le tunnel ferroviaire sous la Manche (Plouviez et al 1974). Un bon exemple classique (Ward et al 1976) est celui du projet anglais Four Fathom. Un exemple français récent concerne un tunnel routier à l'étude près de Nice, Alpes maritimes, le tunnel des 4-chemins (Gaudin et al 1981).

Dans une galerie de reconnaissance de 780 mètres, un rameau perpendiculaire de 38 mètres présente quatre zones successives de 6 mètres de longueur dans des marnes gonflantes (section circulaire de diamètre égal à 4 mètres): béton projeté et boulons – béton projeté sans boulons – béton coffré – résine à fibres de verre et boulons. Treize profils de convergence, trois profils de mesure de contraintes radiales et deux profils d'extensométrie en sondage. Le béton projeté mince est trop raide pour encaisser le gonflement, les boulons à ancrage réparti sont peu efficaces et derrière l'anneau de béton coffré la pression du terrain augmente en fonction du temps.

9. AUTRES ESSAIS

Il y aurait lieu de mentionner encore les essais d'abattage et de

16

creusement, que ce soit à l'explosif (mise au point de plans de tir), avec diverses machines, ou avec ces procédés appelés "exotiques" en américain. Ainsi le laboratoire souterrain d'Atlas Copco à Stockholm qui est plutôt un chantier de démonstration, situé quelques mètres au-dessous des bureaux et ateliers.

Quant aux essais de fracturation hydraulique, ils échappent au concept de laboratoire pénétrable, sinon peut-être à Rainier Mesa, Nevada (Teufel et Warpinski 1982) où une autopsie du massif a permis de suivre et de repérer la direction et l'extension des fractures. Cette même autopsie avait été utilisée précédemment pour contrôler les effets d'injections de ciment. Dans les études de gazéification souterraine du charbon où la fracturation a été menée à partir d'étages profonds de mines en exploitation, le concept s'applique aussi partiellement.

Il y a certainement bien d'autres cas particuliers qui ont échappé à la revue ci-dessus qui ne pouvait être exhaustive dans un cadre limité d'avance; l'auteur appréciera toute information complémentaire.

CONCLUSION

La définition du laboratoire souterrain s'est précisée au fur et à mesure. La mécanique des roches a fait un progrès décisif il y a une trentaine d'années en reconnaissant les différences entre le massif rocheux et l'échantillon. C'est vrai aussi en socio-économie où l'étude de l'individu ne suffit pas à définir les comportements des groupes sociaux.

Grâce aux laboratoires souterrains, un pas significatif est franchi entre l'échelle de l'échantillon et celle de l'ouvrage, mais ce pas n'est que rarement suffisant et la tendance vers des essais à plus grande échelle qui s'est manifestée dans les années 60 n'a pas eu de suite, sinon l'auscultation des ouvrages eux-mêmes. L'échelle raisonnable est celle qui permet de faire avec l'argent disponible plusieurs essais au lieu d'un seul à plus grande échelle, qui pourrait être biaisé (ce que Sharp, 1983, a appelé "the unique large scale in situ test" pour en souligner le danger).

Pour beaucoup d'essais de déformabilité et de résistance, et singulièrement pour les essais de fluage, c'est le confinement naturel plus encore que l'échelle qui justifie le laboratoire souterrain. Pour tous les cas où un gradient thermique est ajouté, l'échelle reprend toute son importance, mais le confinement reste essentiel.

Il reste un problème d'autant plus difficile que le nombre d'essais est plus limité, c'est le choix du site représentatif. Pour mettre les chances de son côté, le mécanicien doit être aussi naturaliste.

REFERENCES

Bernaix, J.: "Rapport général, th. 1", 3ème Congrès international de mécanique des roches, Denver, 1974.

Bertrand, L. et Durand, E.: "Mesure de contrainte in situ; comparaison de différentes méthodes", Symposium international de Paris, vol. 2, p. 449, BRGM Orléans, 1983.

Bieniawski, Z.T.: "In situ large testing of coal", Conference in situ investigations on rock and soils, British Geotechn. Soc. Londres, 1969.

Dagnaux, J.P., Lakshmanan, J. et Garnier, J.C.: "Auscultation sismique de la craie soumise à des contraintes au laboratoire et in situ", Congrès international de mécanique des roches, Thème 4, Comm. 57, Belgrade, 1970.

Dessenne, J.L. et Duffaut, P.: "Les propriétés rhéologiques de la craie et leur influence sur le percement de galeries", La Houille Blanche, pp. 477-488, Grenoble, 1970.

Diernat, F., Comes, G. et Rivoirard, R.: "Etude en souterrain des déformations hygroscopiques des marnes calcaires de Sisteron", Congrès international de mécanique des roches, Thème 4, Comm. 55, Belgrade, 1970.

Diernat, F. et Duffaut, P.: "Contrôle de l'altération de marnes par écrasement d'échantillons de forme irrégulière", Colloque de géotechnique de Toulouse, Thème 2, pp. 96 et suivantes, INSA Toulouse, 1969.

Doucerain, T.: "Essai de conduite forcée bloquée au rocher, étude du comportement du massif rocheux", Symposium international d'Aachen, vol. 2, p. 493, Balkema ed., 1983.

Drouhin, G.: "Essais géotechniques des terrains de fondation", 2ème Congrès international des grands barrages (Washington), CIGB, Paris, 1936.

Duffaut, P.: "Les déformations en mécanique des roches", Symposium international de Madrid, Editorial Blume, Madrid, 1968.

Gaudin, B., Folacci, J.P., Panet, M. et Salva, L.: "Soutènement d'une galerie dans les marnes du Cénomanien", Congrès international de mécanique des sols de Stockholm, Comm 2/4, Balkema ed., 1981.

Groupe de travail CFGB: "La déformabilité des massifs rocheux, analyse et comparaison des résultats", Congrès international des grands barrages d'Edimbourg, Q28 R15, CIGB, Paris, 1964.

Groupe de travail CFGB: "Essais et calculs de mécanique des roches appliqués à l'étude de la sécurité des appuis d'un barrage-voûte, exemple de Vouglans", Congrès international des grands barrages d'Istamboul, Q32 R49, CIGB, Paris, 1967.

Habib, P.: "Détermination du module d'élasticité des roches en place", Annales de l'ITBTP, septembre, Paris, 1950.

Habib, P. et Marchand, R.: "Mesure des pressions de terrain par l'essai de vérin plat", Annales de l'ITBTP, octobre, Paris, 1952.

Hoek, E. et Brown, E.T.: "Underground excavations in rock", Institute of Mining and Metallurgy, Londres, 1980.

Jimenez Salas, J.: "Mechanical resistances", Symposium international de Madrid, p. 115, Editorial Blume Madrid, 1968.

Kujundzic, B.: "Mesure des caractéristiques des roches en place", Annales de l'ITBTP, No 125, Paris, 1958.

Lauffer H. et Seeber G.: "Design and control of linings of pressure tunnels and shafts based on measurements of the deformability of the rock", Congrès international des grands barrages de Rome, Q25 R91, CIGB, Paris, 1961.

Lane, K.S.: "Field test sections save cost in tunnel support", ASCE, 1975.

Londe, P.: "La mécanique des roches et les fondations des grands barrages", p. 104, CIGB, Paris, 1973.

Londe, P.: "Les mesures en galeries", Symposium international Field measurements in geomechanics, Kovari, Zurich, 1977.

Louis, C. et Maini, Y.N.: "Determination of in situ hydraulic parameters in jointed rock", Congrès international de mécanique des roches de Belgrade, Thème 1, Comm. 32, 1970.

Panet, M.: "La mécanique des roches appliquée aux ouvrages de génie civil", Anciens élèves de l'Ecole nationale des Ponts et Chaussées, Paris, 1976.

Plouviez, P., Fauchart, J. et Hueber, J.: "Essais au puits de Sangatte", plaquette Société Situmer, p. 44, 1974.

Pousse L. et Jacquin P.: "La galerie d'amenée en charge Roselend-La Bathie", Congrès international des grands barrages de Rome, Q25 R89, CIGB, Paris, 1961.

Rocha, M., Baptista, J. et Da Silva, J.: "A new technique for applying the method of the flat jack in the determination of stresses inside the rock masses", Congrès international de mécanique des roches de Lisbonne, Thème 4, Comm. 10, 1966.

Rossi, P.P.: "Les fondations du barrage de Ridracoli (Italie)", Revue française de Géotechnique, No 25, 1984.

Rukin, V.V., Kupferman, V.L. et Kardash, J.A.: "In situ tests of rock mass and tunnel lining at the Nurek hydroproject", Symposium international de Paris, Vol.2, p. 567, BRGM Orléans, 1983.

Sharp, J.: "Loading, unloading and shearing tests carried out in underground works including stress measurements", General report, Symposium international de Paris, Revue française de Géotechnique, No 23, 1983.

Sulem J.: "Comportement différé des galeries profondes", Thèse, Ecole nationale des Ponts et Chaussées, Paris, 1983.

Talobre, J.: "La mécanique des roches", 1ère édition (2ème édition nettement modifiée en 1967), Dunod Ed., Paris, 1957.

Teufel, L.W. et Warpinski, N.R.: "Effect of in situ stress on hydraulic fracture propagation and containment", Joint US-Japan seminar, Tokyo, 1982.

Tincelin, E.: "Mesure des pressions de terrain dans les mines de fer de l'est", Annales de l'ITBTP, No 58, Paris, 1952.

Tincelin, E.: "Ecrasement d'un pilier dans une mine de fer de Lorraine", Int. Büro für Gebirgsmechanik, Leipzig, 1968.

Toche: "Stockage souterrain de méthane liquide", Congrès Ass. technq. du gaz, Paris, 1961.

Ward, W.W., Coats, D.J. et Tedd, P.: "Performance of tunnel support in the Four Fathom Tunnel", Symposium international Tunnelling 76, Inst. of Civil Engineering, Londres, 1976.

Yang, Z.W. et Je, L.J.: "The radial flat jack test in China", Symposium international de Paris, vol. 2, p. 585, BRGM Orléans, 1983.

Deux expériences de laboratoires souterrains en conditions difficiles

P.V.DE LAGUÉRIE
Société GEOSTOCK, Paris-la-Défense, France

RESUME

Deux experiences de pilotes souterrains et leurs problèmes d'instrumentation sont décrits :

- la cavité pilote cryogénique de Schelle (Belgique) : réalisée dans l'argile de Boom, la galerie d'essai a été refroidie progressivement jusqu'à -196°C pendant 9 mois et est restée remplie d'azote liquide pendant 2 mois.

- la cavité pilote pour du fuel lourd dans le Vexin (France) : réalisée dans la craie, la galerie d'essai a été maintenue à 80°C pendant 1 mois.

On donne en conclusion, quelques principes tirés de ces expériences applicables pour les laboratoires souterrains liés à l'évacuation des déchets radioactifs.

ABSTRACT

Two tests for underground pilots and the relevant instrumentation problems are described :

- Schelle (Belgium) cryogenic pilot cavity : the test gallery built in Boom clay was progressively cooled down to - 196°C during 9 months, and remained filled with liquid nitrogen during 2 months.

- Vexin (France) heavy fuel pilot cavity : the gallery built in chalk was kept at 80°C during 1 month.

We are giving, in conclusion, some rules derived from these tests which could be applied to underground laboratories concerning radioactive wastes repository.

1. INTRODUCTION

Spécialisée dans le domaine du stockage souterrain d'hydrocarbures, la société GEOSTOCK a mis en oeuvre au cours des années précédentes la plupart des techniques existant actuellement pour le stockage de produits pétroliers.

Cavités lessivées dans le sel, mines reconverties, cavités minées ou aquifères, autant de techniques différentes qui ont permis le stockage de produits aussi variés que l'essence, le naphta, le butane et le propane liquéfié sous pression, le fuel lourd, le gaz oil, le gaz naturel etc...

Cette expérience a été mise à profit depuis plusieurs années dans le domaine du stockage de déchets nucléaires.

Dans cet article, on présente deux exemples de laboratoire souterrain réalisé pour éprouver deux nouveaux modes de stockage (cryogénique et fuel lourd) dont la caractéristique commune essentielle est la création de contraintes thermiques importantes pouvant remettre en cause la stabilité de l'ouvrage, comme c'est le cas dans les dépôts de déchets radioactifs.

Les deux expériences de laboratoire souterrain présentées ici et dont on décrit l'instrumentation concernent :

- La cavité pilote cryogénique de Schelle (Belgique) : cette cavité creusée dans l'argile de Boom a été refroidie progressivement pendant 9 mois de la température ambiante à -196°C et est restée plus de 2 mois remplie d'azote liquide.

- La cavité pilote de Géovexin : cette cavité réalisée dans la craie du Vexin (France) a été maintenue à 80°C pendant près d'un mois.

Dans le premier cas, le comportement élasto-viscoplastique de la roche, évoluant avec la température, conduit à un massif dans lequel on trouve à la fois une zone visco-plastique gonflante à caractéristiques variables, et une zone élastique rétractante à bonnes caractéristiques mécaniques.

Dans le second cas au contraire, on est en présence d'un comportement typiquement élasto-plastique.

2.   CAVITE PILOTE DE STOCKAGE CRYOGENIQUE DE SCHELLE (BELGIQUE)

L'objectif de cette cavité pilote était de simuler et de démontrer le fonctionnement d'un stockage industriel de gaz liquéfié à basse température (propane : -45°C, éthylène : -105°C, gaz naturel liquéfié : -161°C), les débouchés d'une telle technique sont nombreux si on considère en particulier l'économie réalisée et l'accroissement de sécurité par rapport aux stockages aériens actuels.

La cavité pilote de Schelle a été réalisée par la société GEOSTOCK en association avec la société belge DISTRIGAZ et a bénéficié du soutien de la Commission des Communautés Européennes et de l'ANVAR (Agence Nationale pour la Valorisation de la Recherche).

2.1. Description du pilote (figure 1 et 2)

La cavité pilote est cylindrique et réalisée dans l'argile de Boom à une profondeur de 23 mètres avec un diamètre intérieur de 3 m et une longueur de 30 mètres. Le site retenu est celui d'une carrière d'argile de 20 mètres de profondeur, ce qui permet l'accès à la cavité pilote par une galerie horizontale de 70 mètres creusée à partir de cette carrière.

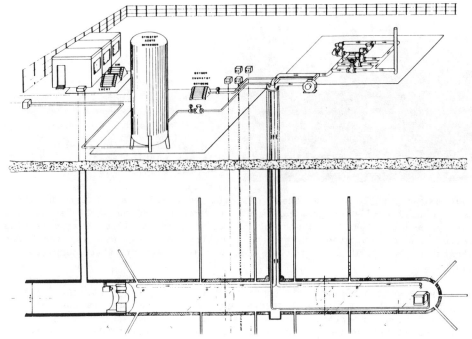

Fig. 1 : Schéma de la cavité pilote de Schelle

Les 100 mètres de la galerie d'accès (70 m) et de la cavité pilote (30 m) ont été creusés au moyen d'un bouclier équipé d'une machine à attaque ponctuelle. Le revêtement est constitué de voussoirs en béton résistant aux basses températures.

La galerie est reliée aux installations de surface par un puits d'un mètre de diamètre, par lequel transitent les canalisations permettant d'effectuer la mise en froid, à savoir une ligne d'injection d'azote liquide, une ligne d'aspiration de l'azote gazeux et une ligne de recirculation. La mise en froid est assurée par une injection d'azote liquide régulée sur une consigne de température à l'intérieur de la cavité.

La mise en froid a été progressive, de la température ambiante jusqu'à −196°C. L'ensemble des essais a duré 15 mois (octobre 81 − janvier 93)

- 8 mois de descente en température de 15°C à −196°C avec plusieurs paliers prolongés à −45°C (propane) et −105°C (éthylène)
- 2,5 mois pendant lesquels la cavité a été maintenue à moitié pleine d'azote liquide
- 4 à 5 mois pour le suivi du réchauffage de la cavité.

2.2. Description succincte du comportement de l'argile aux basses températures

La caractéristique essentielle de l'argile pour un tel type de stockage est le gonflement de celle-ci lors du gel (teneur en eau de l'argile de Boom : 25% ; augmentation de volume de l'eau libre lors du gel : 9%) et

23

Fig. 2 : Vue de la galerie
d'essai partiellement
remplie d'azote liquide

surtout sa propriété de gonflement différé. En effet, la totalité de l'eau
contenue dans l'argile ne gèle pas à 0°C ni même à -3°C ou -4°C, car du fait
de la présence d'eau adsorbée et d'eau fortement liée aux molécules
d'argile, les forces mises en jeu sont telles que ce n'est pas avant des
températures de -50°C à -60°C que les argiles peuvent être complètement
gelées.

Si cette caractéristique de gonflement lors du gel est particulièrement
favorable vis-à-vis du stockage cryogénique, en ce sens que la tendance est
à l'accumulation importante de contraintes de compression jusqu'à des
températures très basses (par opposition aux contraintes de traction
induites dans les roches dures), l'analyse globale du comportement de la
cavité reste particulièrement ardue pour les raisons suivantes :

- évolution brutale des caractéristiques mécaniques de l'argile avec
  l'abaissement de température (ex : Rc = 0,5 MPa à 15°C et 60 MPa à
  -100°C)
- modèle de comportement différent suivant les zones de température :
  visco-plastique gonflant de 0°C à -60°C, visco-plastique neutre de
  -60°C à -100°C, élastique rétractant de -100°C à - 196°C

En ce qui concerne le comportement thermique l'analyse est également
assez complexe, puisqu'il faut prendre en compte la chaleur latente de
gel aux différentes températures, ainsi que l'évolution particulièrement
sensible des caractéristiques de conductivité et de chaleur spécifique.

2.3. **Mesures réalisées**

2.3.1. **Objectifs des mesures**

Les deux critères essentiels permettant de conclure quant à la
faisabilité d'un stockage industriel à basse température étaient les
suivants :

- la stabilité mécanique à long terme du stockage et en particulier
  l'absence de toute fissure dans laquelle le gaz liquide viendrait se
  loger en risquant d'étendre celle-ci de manière incontrôlable,
- des pertes thermiques (boil off du gaz liquéfié) acceptables d'un point
  de vue technico-économique.

Les différentes mesures réalisées ont été :

- mesures thermiques : mesures des températures dans le terrain et dans l'atmosphère de la galerie, comptage des quantités d'azote injectées.

- mesures géophysiques : écoute de microbruits, campagne de sismique réfraction pour la détermination de la position du front de gel.

- mesures géotechniques : mesures des déplacements horizontaux et verticaux dans le terrain, déformation périmétrale et longitudinale sur la paroi, mesure de convergence diamétrale, essais au pénétromètre électrique.

- suivi vidéo de l'intérieur de la cavité.

Le critère essentiel de choix de l'ensemble du matériel a été bien entendu le comportement vis-à-vis des basses températures ; comportement qui n'est pas bien connu pour le matériel géotechnique et géophysique, rarement soumis à des températures aussi basses.

2.3.2. Description du matériel de mesure

a) Equipement thermique

Près de 200 thermocouples répartis dans 19 forages réalisés depuis la cavité permettent la détermination du front de gel (voir fig 3) et des répartitions de température dans le revêtement et dans le massif (dont l'analyse permet la détection éventuelle de digitations imputables à une fracture). Des thermocouples de type Cuivre-Constantan gainé inox et avec un câble teflon + tresse inox ont été utilisés.

Dans l'atmosphère de la galerie, 12 thermistances (Pt 100 Ω à 0°C) sont installées et permettent la régulation de l'injection d'azote et le contrôle de l'homogénéité des températures.

b) Mesures géophysiques

(i) Ecoute de microbruits : 5 géophones situés dans la cavité d'essai et 24 en surface ont permis une écoute en continu des micro-bruits permettant la détection d'éventuelles fissures dans le

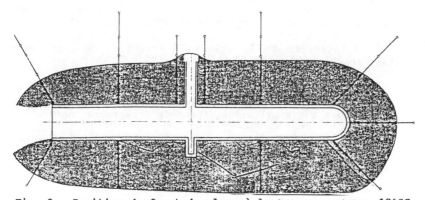

Fig. 3 : Position du front de glace à la fin de l'etape -196°C

25

massif. Les géophones, bien que non conçus spécialement pour résister aux basses températures, ont cependant fonctionné correctement même dans l'azote liquide (seule une baisse de qualité d'amortissement est à signaler). Les seuls micro-bruits sont apparus au moment des passages au dernier palier de mise en froid ; grâce a l'analyse des vitesses et des décroissances d'amplitude, et compte tenu de leur apparition limitée aux premières heures, leur origine a pu être formellement imputée à la remise en place des voussoirs, affectés en premier par la perturbation thermique.

(ii) Campagne de sismique réfraction : l'objet de ces mesures était la détermination du front de gel par une méthode indirecte souple et facile de mise en place. Cette méthode repose sur le contraste de vitesse existant entre l'argile non gelée (1500 m/s) et l'argile gelée (4500 m/s). Le pilote de Schelle a permis de mettre au point la méthode et de la caler sur les résultats donnés par les sondes de température.

c) Equipement géotechnique

Un premier choix a dû être fait entre les types de mesure en général : mesures de pression, mesures de déplacement, quel type de mesure de déplacement ? Il s'est avéré pratiquement impossible de réaliser des mesures de pression totale à des températures nettement inférieures à 0°C. Les caractéristiques des membranes et des fluides ne permettaient guère d'envisager des mesures quelque peu précises à des températures inférieures à -20°C.

Parmi les mesures de déplacement, seuls ont été retenus des capteurs du type inductif pour des températures peu négatives et surtout des capteurs à corde vibrante pour des températures inférieures à -100°C.

Les capteurs à corde vibrante peuvent être réalisés en inox résistant aux basses températures, le câblage électrique étant lui en téflon. Le problème de correction de température qui avec la résistance mécanique du capteur est souvent délicat à solutionner, est résolu assez facilement avec les cordes vibrantes puisque la connaissance de la température est donnée par la résistance de la bobine excitatrice et la correction se fait par application du terme $E \alpha \Delta\theta$ (E et α sont constants avec la température).

A l'intérieur de la galerie l'essentiel de l'instrumentation était constitué de cordes vibrantes :

- des extensomètres à corde vibrante placés entre les voussoirs pour mettre en évidence leurs déplacements relatifs. Deux sections ont ainsi été équipées de 9 capteurs chacune.

- des extensomètres à corde vibrante du même type placés directement sur l'argile à travers un voussoir spécial permettant l'accès au terrain, ceci pour ausculter l'argile confinée.

- deux blocs d'argile (30 x 30 x 30 cm) ont également été équipés d'extensomètres à corde vibrante et laissés dans la cavité afin de mesurer en parallèle le comportement de l'argile non confinée.

- des mesures de convergence diamétrales dans deux sections de galerie ont été réalisées : convergence de l'anneau de voussoir dans la première section, convergence de l'argile dans la deuxième section.

Ces mesures ont été réalisées à partir de plots ancrés dans le béton ou l'argile suivant la section auscultée et au moyen de fils Invar. Dans une première phase, ces mesures ont été faites manuellement à l'aide d'un appareil du type Distomatic de Télémac assurant une tension constante sur le fil Invar et la mesure grâce à un dynamomètre de précision. Ces mesures manuelles ont pu être exécutées jusqu'à la température de −80°C grâce à un équipement spécial de type scaphandre.

Lorsque la température est devenue trop basse pour que l'on puisse entrer dans la galerie, des convergencemètres automatiques, dont l'organe de mesure était également des cordes vibrantes, ont été installés. Les mesures ont alors pu être assurées en continu jusqu'à la fin des essais.

Dans le massif, à l'extérieur de la galerie et donc à des températures relativement peu froides (−30°C), ont été installés:

- des extensomètres verticaux à induction (Distofor de la société Télémac). Deux cannes de 30 mètres de longueur et comportant 6 capteurs inductifs chacune ont été installées dans des forages réalisés depuis la surface à 3 et 9 mètres de la cavité.

- des clinomètres fixes à corde vibrante de Télémac au nombre de 4, placés également dans deux forages situés à 3 et 9 mètres de la cavité.

L'ensemble des appareils était relié à une centrale de mesure située en salle de contrôle permettant une auscultation automatique.

Enfin une campagne d'essais au pénétromètre électrique depuis la surface a été réalisée à la fin de l'essai en azote liquide. Celle-ci a permis la détermination des caractéristiques mécaniques et de fluage de l'argile en place sur toute la gamme de températures depuis l'ambiante jusqu'aux températures inférieures à −100°C.

d) Bilan des mesures géotechniques

On peut dire que l'ensemble du matériel mis en place s'est comporté de manière satisfaisante vis-à-vis des basses températures et a permis de tirer des renseignements tout à fait précieux.

Les mesures données par les extensomètres situés sur la paroi d'argile et sur les blocs d'argile libres, ainsi que les mesures de convergence en paroi ont pu être corrélées au modèle de calcul. Par contre, en ce qui concerne les clinomètres, les résultats ont été plus délicats à analyser en raison de la difficulté de modéliser le couplage de l'appareil au terrain. La même difficulté a été rencontrée dans l'analyse des extensomètres posés sur les voussoirs (modélisation d'un anneau articulé de voussoirs).

Enfin nous avons été confrontés à un problème relativement fréquent en expérimentation géotechnique : l'absence de véritable mesure de contrainte in situ. En effet, alors que dans le cas du milieu homogène élastique de caractéristiques connues et fixes la démarche déplacement vers contrainte est relativement aisée, celle-ci devient beaucoup plus délicate dans le cas d'un milieu dont les lois et les paramètres de comportement (viscoplastique, élastique) évoluent de façon particulièrement rapide.

En conclusion, les mesures réalisées ont permis de trancher sans ambiguité sur le modèle de comportement et d'estimer, plus ou moins précisément, les valeurs des paramètres à introduire dans ce modèle : les paramètres sensibles aux déplacements sont bien estimés alors que ceux sensibles aux contraintes le sont moins bien, pour certains du moins.

Toutefois, les essais au pénétromètre électrique in situ et les essais de fluage réalisés sur carottes à basse et très basse température au Laboratoire d'Ingénierie Nordique de Montréal (Pr. Ladanyi) permettent de préciser ces dernières caractéristiques mécaniques.

Fig. 4 : Vue de la cavité expérimentale GI2 – Vexin

Le modèle de calcul permet alors d'une part un calage sur l'ensemble des valeurs de déplacement obtenus et apporte d'autre part, par la connaissance de l'état de contraintes à court terme et son extrapolation à long terme, la preuve de la stabilité du stockage.

Ces résultats confortés par l'évolution régulière des isothermes, du front de gel et de la consommation d'azote liquide, et par l'absence de micro-bruits, permettent d'affirmer qu'aucune fissure ne s'est propagée à travers l'anneau gelé et que cet état est stable à long terme. Le succès de l'expérience est donc suffisamment démontré.

3. CAVITE EXPERIMENTALE GI2 - PROJET DE STOCKAGE DE FUEL LOURD DANS LE VEXIN

Le projet de stockage de fuel lourd du Vexin concernait la réalisation de 2,5 millions de m3 à 150 m de profondeur dans la craie turonienne, à 50 km à l'ouest de Paris.

Avant de lancer la construction proprement dite de l'ouvrage, un puits et plusieurs galeries expérimentales furent réalisés. Un des problèmes majeurs du stockage de fuel lourd est le figeage de celui à température ambiante qui conduit à envisager des températures de fonctionnement de 60 à 80°C au moment du remplissage puis au moment de la vidange (circulation d'eau chaude ou de fuel chaud pour défiger le produit en place). Une des galeries expérimentales a donc été affectée pour les essais de comportement thermique et mécanique du rocher.

3.1. Description du pilote

La galerie expérimentale GI2 est implantée à la cote -150 m et présente un diamètre moyen de 2 m et une longueur totale de 30 m. Les essais ont été réalisés sur les 12 derniers mètres de la galerie.

La craie turonienne du Vexin est une roche tendre, homogène et peu fracturée (c = 0,8 à 1,8 mPa, φ = 24 à 31°, E = 6000 MPa, ν = 0,22). L'ensemble des mesures réalisées dans les autres galeries d'essai géotechnique a permis de définir précisément le comportement élasto-plastique de la craie et grâce à l'utilisation de modèle de calcul, des galeries de 8,60 m de haut et 6,40 m de large ont pu être réalisées pour le stockage proprement dit, ce qui, compte tenu de la roche, constitue une belle réussite.

La galerie d'essai a été remplie en eau à 80°C ; cette eau a été réchauffée en permanence par circulation à fort débit pendant 25 Jours.

3.2. Matériel de mesure

a) Température : 4 thermocouples ont été installés dans la galerie et 20 autres dans le massif à partir de 5 forages radiaux. L'enregistrement en continu des températures a permis l'analyse de la conduction thermique dans le terrain et la détermination des coefficients d'échange thermique.

b) Mesure des variations de contrainte en paroi
    Des essais classiques au vérin plat ont été effectués dans le
    radier et en parement de la galerie avant essai pour mesurer les
    contraintes initiales. Les bases de mesures ont ensuite été équipées
    d'extensomètres à corde vibrante et pendant toute la durée de l'essai
    la variation de contrainte a pu être suivie par maintien de la
    déformation en ajustant la pression dans les vérins. Après les
    corrections de températures sur le vérin lui-même et sur la dilatation
    de la craie, on a pu en déduire l'évolution des contraintes
    thermomécaniques en paroi pendant toute la phase de chauffe puis
    pendant le refroidissement.

c) Déformation diamétrale résiduelle après essai :
    Ces mesures de convergence résiduelle ont été faites sur trois
    sections à l'aide d'une canne de convergence escamotable.

d) Déformation diamétrale en continu :
    Une section a été auscultée en continu avec un élongamètre à lame
    fléchie équipé de 2 capteurs à corde vibrante.

e) Déformation périmétrale locale
    18 capteurs à corde vibrante ont été installés sur tout le périmètre
    d'une section.

f) Déformation radiale en sondage
    Un forage a été équipé d'une tête élongamétrique différentielle à corde
    vibrante permettant la mesure de déplacement à 0,5 m, 1 m et 1,5 m.

3.3. Résultats - conclusion

    A titre d'illustration, la figure 4 présente l'état de déformation
radiale du massif.

    L'ensemble des résultats a permis de confirmer le comportement
élasto-plastique de la craie, d'améliorer la connaissance des transferts
thermiques et de donner des valeurs précises aux paramètres thermomécaniques.

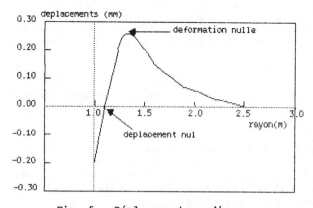

Fig. 5 : Déplacements radiaux

En raison de difficultés technico-économique, le stockage du Véxin a finalement été converti en un stockage de 130.000 m3 de propane liquéfié sous pression de 0,8 MPa.

4. AUTRES EXEMPLES

Un certain nombre d'autres expériences in situ, mettant en oeuvre des contraintes thermiques, ont été réalisées par GEOSTOCK parmi lesquelles on peut citer :

- comportement thermique et mécanique d'une galerie d'évacuation des fumées de cimenterie,
- comportement thermique des bouchons de béton obturant les galeries de stockage d'hydrocarbure,
- suivi thermique d'un stockage de fuel lourd en Corée.
- essai cryogénique dans le granite (forage Ø 1 m).

5. APPLICATION AU DOMAINE DES DECHETS NUCLEAIRES

Notre expérience dans le domaine du stockage souterrain est mise à profit depuis plusieurs années pour les études concernant les stockages de déchets nucléaires pour un certain nombre d'organismes (CEA, ANDRA, CCE, etc...). On citera en particulier l'étude de conception d'un stockage dans le granite pour le compte de la CCE (1), réalisée en collaboration avec la société SGN.

(1)   Etude de conception générale d'une installation permettant l'évacuation des déchets radioactifs dans une formation granitique. CEE - GEOSTOCK - Contrat n°107.79.1 WASF (1980).

Enfin, tout récemment GEOSTOCK a effectué pour le compte d'un client étranger, l'étude de conception d'un laboratoire souterrain dans le granite pour le stockage de déchets de haute activité. Les caractéristiques essentielles de ce pilote sont les suivantes :

Implanté à 500 mètres de profondeur, le pilote comporte d'une part les galeries d'essai proprement dites, dans lesquelles sont placées les sources électriques chauffantes et d'autre part, associées à chaque galerie d'essai, des galeries d'instrumentation.

- l'implantation des points de mesure, aussi bien géotechnique, hydrogéologique, que pour l'étude de la migration des radionucléïdes a été établie en fonction de l'orientation et de la densité des joints, discontinuités ou fractures.

- une répartition des mesures de contraintes au plus près des sources chauffantes a été prévue pour avoir les pics de variation.

- outre les mesures de déplacements proches du canister, des mesures au loin (50 à 100 m) ainsi qu'un point de référence ramené depuis la surface jusqu'en galerie ont été prévus.

- le délicat problème de mesure de la contrainte a été réglé par l'utilisation de rosettes de cordes vibrantes implantées en forage. L'étalonnage est assuré d'une part par des essais à température ambiante dans des sections de galerie en cours d'abbatage, et d'autre part, en continu, par la possibilité de créer un état de surcontrainte artificielle grâce à un dilatomètre implanté dans un forage voisin.

- Pour les mesures hydrogéologique, deux longueurs de galerie de 10 mètres chacune sont obturées. Les mesures de débit d'exhaure et les prises de pressions interstitielles depuis ces galeries permettront un suivi précis de l'évolution des perméabilités et des écoulements.

6. CONCLUSION

Des deux pilotes décrits ici on retiendra pour la conception de l'instrumentation d'un laboratoire souterrain de déchets nucléaires les principes suivants :

- choisir un matériel simple éprouvé mécaniquement aux températures de fonctionnement sans oublier les câbles électriques, connections, etc

- opter pour l'instrumentation dont l'étalonnage et la correction de température est sans ambiguité (cas de la corde vibrante).

- adapter les types d'appareil au terrain et à son comportement présumé pour faciliter l'interprétation (couplage appareil-terrain).

- adapter les chaines de mesures aux conditions de fonctionnement. (ex : impédance des cordes vibrantes = 90 Ω à 15°C et 15 Ω à -196°C).

An overview of strategies for in-situ investigations for radioactive waste disposal

RUDOLF ROMETSCH
Nagra, Nationale Genossenschaft für die Lagerung radioaktiver Abfälle, Baden, Switzerland

ABSTRACT

In-situ investigations aim at the demonstration of the validity of radioactive waste disposal concepts. The most important type of experiment produces scientific results for validation of predicted behaviour of system components. Pilot disposal operations, on the other hand, are primarily meant to support public acceptance, as is also true for the most extensive type of in-situ experiments, consisting of extended observations of a real repository.

RESUME

Les investigations faites in-situ ont pour but de confirmer la validité des concepts du stockage définitif. Les expériences les plus importantes sont celles qui produisent des résultats scientifiques pour la confirmations du comportement prévu des composantes du système. Les entreprises pilotes ont plutôt pour but de contribuer à l'acceptance du stockage définitif par le public. Ceci est encore plus prononcé pour les expériences globales consistant en observations prolongées d'un stockage définitif réel.

There is only one real problem in radioactive waste disposal, or in fact in disposal of any wastes with a long lasting or permanent toxicity: the proof of the long-term safety. That this proof should be established by experimental demonstration is a plea repeated again and again by certain participants in the nuclear debate. However, the scientists and engineers involved in the research and development work on waste disposal realise that proof by direct experiment, as often requested, is not possible. This is evident when one considers the necessary duration of directly conclusive experiments. Nevertheless, researchers are ready to make tremendous efforts in order to establish the proof indirectly by modelling the long-term behaviour of repository systems and feeding data into the models from short duration experiments or from the observation and interpretation of natural processes. In that way, they try to validate their models by the closest possible comparison with real situations. In this process, in-situ experiments play a most important rôle.

1. The place of in-situ experiments in repository research and development

Conceiving a waste repository is primarily an act of imagination; not ordinary uncontrolled imagination, but imagination checked by the general

33

TABLE 1

Waste Matrix

- composition of glasses
- devitrification temperature
- cement mixtures
- characteristics of bitumen or resins
- leaching coefficients
etc.

Buffer Materials

- chemical composition
- swelling properties
- interactions (with waste matrix and rock)
- transformation by heat
- sorption of nuclides
etc.

Geological Surrounding

- permeability
- water movements
- chemical interactions
- sorption of nuclides
- migration of nuclides
etc.

knowledge on the behaviour of materials in initally well-defined surroundings which then may vary with time. In course of this process, which I like to call "safety synthesis", it frequently becomes apparant that our knowledge on the materials involved is not sufficient to derive valid conclusions on component behaviour over long times, as a function of the varying surrounding conditions. Laboratory experiments may provide a first approximation to the missing data. These have been performed on barrier materials in the near field as well as on barrier functions of buffer and surrounding materials. Some examples of such investigations are listed in Table 1.

In order to develop valid models of the repository system, the results of the laboratory investigations have to be extrapolated over orders of magnitude with regard to time, space and often also with regard to temperature, pressure and other surrounding conditions which can not be simulated closely enough in the laboratory.

These extrapolations are bound to create uncertainties. The extrapolation in scale, i.e. over large masses of material can be reduced by repeating some of the laboratory experiments using large masses of material in situations which are as close as possible to those to be found in the envisaged repository system. This would be the first type of in-situ experiment. They are planned for the validation of repository components and subsystems. In-situ experiments in rock laboratories with such strategic aims are being performed or built up in several countries. Examples for laboratories in crystalline rock are the Stripa mine in Sweden, the underground lab in Canada or the Grimsel Rock Laboratory in Switzerland; a most impressive in-situ experimental station in clay has been established at Mol; an example for in-situ experimentation in rock salt is the heating experiments in the lower part of the Asse Salt Mine in Germany.

2. Validation of system components by in-situ experiments

I would like to describe briefly the scope and aims of the Grimsel Rock Laboratory as an example of in-situ experimentation for the validation of repository system components. Grimsel is not a repository site. It is an accessible mass of granite - accessible from the inside because of an existing tunnel some 3 km length under the "Juchlistock" and the Grimsel lake. The granite belongs to the same crystalline basement which lies in the northern part of Switzerland at a depth suitable for repository construction. The Grimsel granite is similar but not the same and not under the same conditions as the crystalline rock at a repository site still to be localized in an area some 100 km distant. Besides petrographic variations it is expected that the different stresses in the alpine region would influence the pattern of fissures and the different heads the water flow pattern. This means that the in-situ experiments would have to be repeated at the repository site once it has been identified and accessed. However, the results of the Grimsel in-situ experiments are expected to yield useful results. In addition, they provide an excellent opportunity to develop and adapt the methods to be used later under the conditions at the actual repository location.

Thus, the strategic aim of the Grimsel in-situ experiments is two-fold:

- development of methodology, and

- validation of system components under conditions approximating as closely as possible to these at the repository site.

Accordingly, the initial programme of experimentation covers the topics mentioned in Table 2.

To perform these experiments a system of cavities has been excavated (Figure 1) on the west side of the access tunnel to the Grimsel II plant which is a 300 MW pumping/turbine station. There is no intention of making experiments with radioactive wastes in this rock laboratory, i.e. the in-situ experiments performed therein have not the character of a pilot or demonstration operation. However, migration studies with tracer quantities are not excluded, although they would require a special permit.

3. Pilot operation

A second completely different type of in-situ experiment is meant to provide opportunity for pilot operation with real radioactive waste conditioned for disposal in a repository. Such experiments are barely distinguishable from real disposal operations.

35

FLG Grimsel Test Site

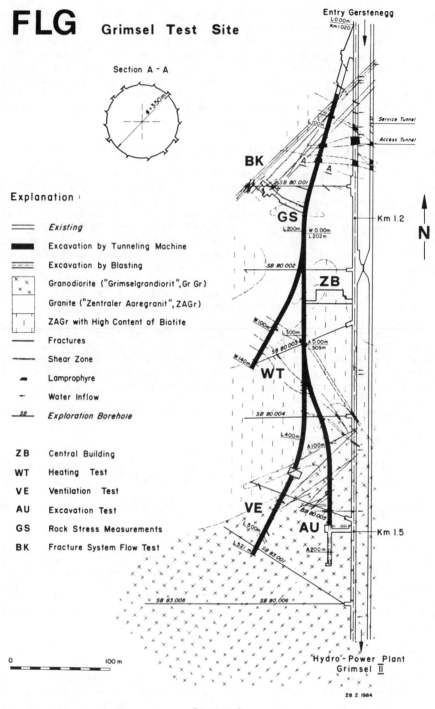

Section A - A

Explanation:

Existing	
Excavation by Tunneling Machine	
Excavation by Blasting	
Granodiorite ("Grimselgrandiorit", Gr Gr)	
Granite ("Zentraler Aaregranit", ZAGr)	
ZAGr with High Content of Biotite	
Fractures	
Shear Zone	
Lamprophyre	
Water Inflow	
SB Exploration Borehole	

ZB	Central Building
WT	Heating Test
VE	Ventilation Test
AU	Excavation Test
GS	Rock Stress Measurements
BK	Fracture System Flow Test

0   100 m

Figure 1.

TABLE 2	
Geophyiscs	Testing and adapting non-destructive methods for the identification of discontinuities influencing hydrodynamics and rock mechanics;
Neotectonics	Testing and adapting methods for identification of active disturbed zones;
Hydrogeology	Determination of basic values for hydrodynamic modelling in low permeable rock;
Migration	Determination of additional parameters influencing nuclide migration in fissures in crystalline rock;
Disturbed Zones	Investigation of modified rock properties around artificial cavities;
Rock mechanics	Adaptation of methods to determine rock stresses in deep boreholes;
Thermal Influences	Investigation of temperature induced processes in fissured crystalline rock masses.

Well known examples are the experimental disposal of low and intermediate level wastes in some of the cavities of the Asse Salt Mine in Germany and the "Climax" facility for pilot disposal of irradiated fuel. The Asse experiment was planned to build up experience with regard to all phases of the real disposal operation. Drums with solidified low level waste have been used to fill completely some of the cavities. The space between the drums and the walls of the cavity have been filled with crushed rock salt in order to achieve compact incorporation of the waste into the rock salt mass as envisaged for the final disposal operation. For intermediate level waste a special type of waste drum using recoverable shielding has been installed. The Asse in-situ experiment has successfully shown how the disposal operation can be mastered by simple straightforward technology. It is most convincing and appears to provide an opportunity for extrapolating the behaviour of the disposal system into the future. It has only one drawback: there is no measurable nuclide migration, and therefore no basis for extrapolation to support safety analysis. This is an inherent disadavantage of properly performed pilot operations as a form of in-situ experiment.

There is, of course, a temptation to adapt the strategy of in-situ experimentation by pilot disposal of wastes in such a way as to obtain measurable migration of radionuclides within some years of observation. However, this would mean simulating events in a repository which one would carefully exclude in an ordinary disposal operation, or at least making these events occur with an artificially high speed. The conclusions of the experiment would thus be weakened by this, as well as by the other inherent difficulty of pilot operations, namely assuring sufficient similarity of the experimental site to the disposal site.

37

When analogy conclusions are considered acceptable, then the best in-situ experiments are those provided by nature. The study of confinement of e.g. uranium in certain types of rock, or rather the limits of that confinement and the resulting migration do allow certain conclusions to be drawn. The interpretation is not easy, as the differences in the physical-chemical characteristics beween uranium and important radionuclides e.g. neptunium which have a relatively high probability of escape from a repository, have to be carefully weighed in the comparison. In spite of this difficulty, I suggest including observation of uranium confinement and migration in the second type of in-situ experiments. The "pilot" operation with uranium has at least the advantage of covering a long enough period of time, perhaps even longer than needed to compare with waste repositories. Uranium ore veins in the Grimsel granite might serve as a basis for migration studies.

4. Incomplete disposal operations as in-situ experiment

I would like to discuss briefly a third type of in-situ experiment. It is a spin-off from public discussions on the philosophy of disposal in final repositories. To my mind, it has no real scientific or technical value but rather tries to respond to the public concern over renunciation of control over final repositories. It appears that quite a percentage of the population is ready to support this idea, which boils down to requesting extended experimental operation of a future repository.

The idea developed in a public information meeting within a Swiss community. A large known marl formation on the land of this community has been proposed for further exploration with a view to compare it with other potential sites for the construction of a final repository for low and inter-mediate level waste. For geolocial investigations connected with repository planning federal licences are required. The procedure for granting such licences envisages consultations on three levels of the political adminis-tration. As a basis for the consultation a public information meeting was held.

In spite of the fact, that the actual question concerned only geological site investigations, the meeting soon turned into a discussion of disposal problems. Since 1979 the Swiss law requires disposal of radioactive waste in final repositories. The guidelines issued by the federal authorities define final repositories. They must be constructed in such a way that it is possible at any time to close and seal them within a few years. After closing and sealing it must be possible to forego all maintenance and surveillance measures. Nevertheless, the safety of the repository must be assured for all times. In fact, it must be proven that released radionuclides can at no time cause an individual radiation dose of more than 10 mrem/year (and should normally be much lower).

Since this defintion has become a matter of public discussion, the question has again and again been asked: Why renounce surveillance measures and give up any possibility of maintenance? - quite contrary to the previous requirement for a permanent solution of the waste problem, which led to legislation on final repositories.

Of course, compromise proposals are made from different sides. They all lead to the same suggestion: leave the closing and sealing of the final repository sufficiently incomplete to allow observation of the repository over a non-defined period of time. In a way this is a proposal for a large scale in-situ experiment which is barely distinguishable from the real disposal operation. The effect of such an experiment, if any, would be mainly psychological.

5. Summary

Generally, in-situ experiments aim at the demonstration of the validity of concepts for final repositories for radioactive wastes. Distinguishing three major types of such experiments, namely

- validation of system components
- pilot operations to demonstrate feasibility, and
- keeping a repository as long as possible under observation

one detects in this sequence an increasing proportion of non-rational elements in the justification.

It appears necessary today to recognise these psychological problems and treat them seriously. They differ in value from the scientific and technical arguments but not in importance. There is a need to build up credibility and to increase confidence in order to overcome the rational and irrational anxieties based on radioactive wastes.

United States Department of Energy strategy for in-situ testing at candidate repository sites

MARK W.FREI
Office of Geologic Repository Deployment, US Department of Energy, Washington, DC
DAVID L.SIEFKEN
Geosciences and Technology Department, OCRWM Technical Support Team, Roy F.Weston, Inc.

ABSTRACT

The United States Department of Energy is preparing to conduct in-situ testing as part of site characterization studies at three candidate sites for the first repository in accordance with the Nuclear Waste Policy Act of 1982 (PL 97-425). The Act established a siting process and an attendant schedule for the development of geologic repositories for the disposal of civilian high-level radioactive waste, transuranic waste, and spent fuel.

The Department of Energy is required in that siting process to recommend three sites to the President for approval before beginning site characterization. The site characterization activities at each site will include additional testing from the surface, construction of an exploratory shaft or shafts, and large-scale in-situ testing of the host rock from underground workings at the proposed repository depth and/or other depths, as appropriate. Applicable regulations promulgated by the Nuclear Regulatory Commission (10 CFR Part 60) require a performance confirmation program, including in-situ monitoring, laboratory and field testing, and in-situ experiments, beginning with site characterization and continuing through construction and operation of the repository until closure.

The Department of Energy is in the process of developing an overall strategy for in-situ testing in the geologic repository program that is consistent with the requirements of the Nuclear Waste Policy Act of 1982 and 10 CFR Part 60. This overall strategy calls for a five-phase program for in-situ testing. The first phase consists of pilot research projects in rock types similar to those under consideration for a geologic repository. Subsequent phases consist of site-specific in-situ testing, beginning with site characterization activities at three sites and continuing at the site selected for development as the first geologic repository through construction, operation, and closure of the geologic repository - a time period on the order of up to 100 years.

The strategy is to initiate tests at each phase, where needed, to satisfy the information needs of the various stages of site selection, development and licensing; to continue, where appropriate, the same tests from one phase to the next in the five-phase program; and, to increase the scope, dimensional scale, time duration, and interactive processes considered in the testing over those tests performed at each previous phase. The testing at each phase will build upon the tests performed and the data collected in the preceding phases. The major finding at each phase will be the determination or confirmation, with increasing confidence and certainty at each successive phase, of the suitability of the site for a geologic repository which will meet applicable standards for public health and safety.

1. INTRODUCTION

The United States Department of Energy (DOE) is preparing to conduct in-situ testing as part of site characterization studies at three candidate sites for the first repository, in accordance with the Nuclear Waste Policy Act of 1982 (the Act).[1] The Act established a siting process and an associated schedule for the development of geologic repositories for the disposal of civilian high-level radioactive waste, spent fuel, and transuranic waste. This paper presents the Department's overall strategy for the conduct of in-situ testing in support of the geologic repository program. Before discussing this strategy, a brief description of the siting process and its status will be given to put the in-situ testing approach in the proper perspective.

DOE currently has nine potentially acceptable sites under consideration for the first geologic repository for civilian high-level radioactive waste, transuranic waste, and spent fuel (Figure 1). The nine potentially acceptable sites include two bedded salt sites in the Palo Duro Basin in Texas, two bedded salt sites in the Paradox Basin in Utah, three domal salt sites (two in Mississippi and one in Louisiana), one site in flood basalts on the Hanford Site in Washington, and one site in welded tuff adjacent to the Nevada Test Site.

As mandated in the Act, DOE will nominate, with accompanying environmental assessments, at least five of these potentially acceptable sites as suitable for site characterization and then recommend to the President for his approval three of the nominated sites for site characterization. The sites recommended for site characterization must be in at least two different rock types in order to meet the requirements

Figure 1.   Potentially Acceptable Sites for the First Repository

41

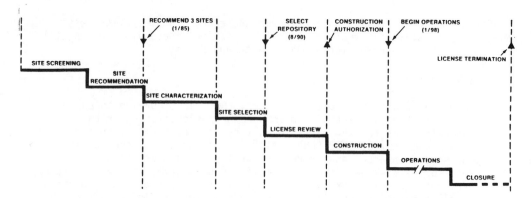

Figure 2.   Draft Mission Plan Repository Schedule

of the Nuclear Regulatory Commission (NRC) promulgated in 10 CFR Part
60.[2]

DOE has developed, pursuant to the Act, general guidelines to be used in
this site nomination and recommendation process (10 CFR Part 960).[3]
The guidelines are currently under review by the NRC. The site
nomination and recommendation process will not proceed until the NRC has
concurred on the guidelines. As shown in Figure 2, DOE anticipates
concurrence on the guidelines from the NRC by this summer, nomination of
sites as suitable for site characterization (with accompanying
environmental assessments) by this fall, and recommendation of sites to
the President for site characterization by January 1, 1985 as mandated in
the Act.

After DOE has Presidential approval of the three sites recommended for
site characterization, DOE will submit a Site Characterization Plan (SCP)
for each site to the NRC for their review, pursuant to the requirements
of the Act and 10 CFR Part 60. The SCP's will consist of three major
parts in accordance with the NRC's Regulatory Guide 4.17[4] - the first
part being a description of the criteria and decision process used to
select the sites, the second part being descriptions of the site, waste
package and form, and conceptual design of the repository, and the third
part being an identification of issues related to the suitability of the
site for a geologic repository and detailed plans for testing during the
site characterization phase to resolve these issues. The site
characterization activities at each site will include additional testing
from the surface, construction of an exploratory shaft or shafts, and
large-scale in-situ testing of the host rock from underground workings at
the proposed repository depth and/or other depths, as appropriate.

As required in the Act and 10 CFR Part 60, DOE will submit semiannual
updates of the SCP's to the NRC, for their review, on the progress of
site characterization and on waste form and packaging research and
development. The SCP's and updates will also be submitted to the
Governors and legislatures of the States in which the candidate sites are
located, or to the governing body of an affected Indian tribe where such
candidate site is located, as the case may be, for their review. The
semiannual updates will (1) discuss the results of site characterization

activities to date, (2) identify new issues not previously mentioned in the SCP, plans to resolve these new issues, those studies originally planned that are no longer considered necessary and therefore eliminated from the site characterization program, decision points reached during site characterization, and modifications to schedules, and (3) report progress in developing the design of a geologic repository operations area appropriate to the site.

The major objective of the site characterization program, including in-situ testing, is to determine the suitability of the three sites for a geologic repository. At the end of the site characterization phase, DOE will recommend to the President, in a Site Selection Report (SSR) and an accompanying Environmental Impact Statement (EIS), one site from among the three characterized sites for development as a geologic repository. The initial results from the in-situ testing conducted during the site characterization phase will support the SSR, the EIS, and the construction authorization application (CAA) to the NRC for the site recommended for development as a geologic repository. DOE anticipates recommendation of one site for development as a geologic repository in the 1990 time frame, consistent with the need to have the first repository operational by January 1998 as mandated in the Act. DOE will miss the March 31, 1987 date in the Act for recommendation of one site for development as a geologic repository in order to not sacrifice or compromise the technical work leading to the selection of that site.

2. DOE STRATEGY FOR IN-SITU TESTING

DOE is in the process of developing an overall strategy for in-situ testing in the geologic repository program. This process includes project-specific development of exploratory shaft test plans and site characterization plans, interactions with the NRC on information needed to support the applications for construction authorization and a license to receive and emplace wastes, and promulgation of a program-wide position on testing at the various stages of licensing.

This paper presents the strategy in its current stage of development, specifically as it relates to the first repository program. As the strategy becomes more firmly established, the evolving project-specific test plans will become more aligned and consistent with the overall program strategy.

The strategy, in general terms, calls for a five-phase program for in-situ testing (Figure 3) as follows:

1. the pilot research phase of in-situ testing in rock types similar to those sites currently under consideration
2. the site characterization phase for each of the three sites recommended for site characterization
3. the construction authorization application (CAA) review phase for the one site recommended for development as a geologic repository
4. the construction phase, beginning with authorization for construction, and
5. the operational phase, beginning with emplacement of wastes in the repository.

43

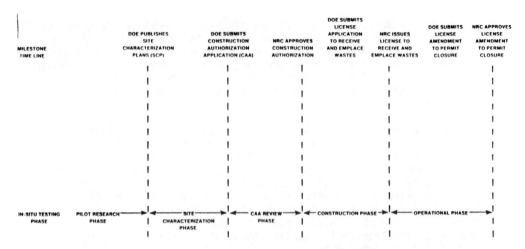

Figure 3. In-Situ Testing Phases

Definitions of these phases and the purposes of testing associated with each phase are described more fully in subsequent sections of this paper.

The DOE strategy for in-situ testing has been to use the pilot research phase as a means of developing testing methods, instrumentation, and experience in similar rock types in order to support the in-situ testing in each of the four subsequent phases at each of the candidate sites. The DOE strategy for the four subsequent phases is to continue to expand the data base needed for site selection, development and licensing. Where appropriate, the same tests will be continued from one phase to the next and the scope, dimensional scale, time duration, and interactive processes (coupled effects) considered in the testing will be increased, as appropriate, over those tests performed at each previous phase. The testing at each phase will build upon the tests performed and the data collected in the preceding phases. The testing at each phase will be structured to provide the information needed to proceed to the next phase of site selection, development, or licensing. For example, testing during the construction phase will focus on demonstration of the technology associated with operation of the repository. This strategy takes advantage of the increasing time period and increasing size and/or extent of underground workings available at each successive phase. The longer time period, increased dimensional scale, expanded scope, and examination of interactive processes (such as coupled hydrologic and thermal effects) during successive testing phases should significantly increase the confidence level and reduce the uncertainty associated with the results from the preceding phases.

The five-phase program is consistent with 10 CFR Part 60, which states that the performance confirmation program, including in-situ monitoring, laboratory and field testing, and in-situ experiments, shall have been started during site characterization and will continue through construction and operation until closure of the repository.

The major finding at each phase (starting with the second phase) will be the determination or confirmation of the suitability of the site, with increasing confidence and certainty at each successive phase, for a geologic repository. The emphasis on demonstration at each phase of the suitability of the site as a geologic repository is not intended to diminish the significance of other objectives to be achieved by the separate in-situ testing phases. Rather, it reaffirms the DOE commitment to eliminate a site from consideration for a geologic repository at any time when the results of testing indicate the site is unsuitable for a geologic repository.

3. PILOT RESEARCH PHASE

DOE believes that the results of pilot research projects[5] conducted in similar rock types in the United States and other countries can be used to supplement the in-situ testing program during the site characterization and CAA review phases. The major results of in-situ testing in these pilot research projects have been the development of testing methods, instrumentation, and experience in rock types similar to those under consideration in the geologic repository program. In addition, data bases have been started on the physical, thermal, mechanical, and thermomechanical properties of the separate rock types, including changes in these properties under thermal and mechanical loading, e.g., the jointed block test at the Near-Surface Test Facility in basalt at the Hanford site. These data bases, along with the results

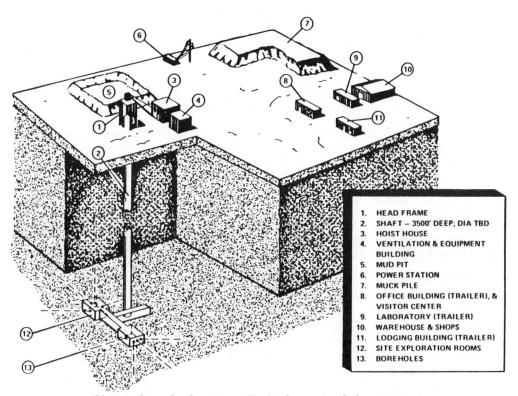

1. HEAD FRAME
2. SHAFT – 3500' DEEP; DIA TBD
3. HOIST HOUSE
4. VENTILATION & EQUIPMENT BUILDING
5. MUD PIT
6. POWER STATION
7. MUCK PILE
8. OFFICE BUILDING (TRAILER), & VISITOR CENTER
9. LABORATORY (TRAILER)
10. WAREHOUSE & SHOPS
11. LODGING BUILDING (TRAILER)
12. SITE EXPLORATION ROOMS
13. BOREHOLES

Figure 4.   Exploratory Shaft Conceptual Arrangement

of testing from the surface and in the laboratory, provide technical support for the nomination of sites as suitable for site characterization and the accompanying environmental assessments.

In the United States, simulated underground repository experiments have been performed in basalt at the Hanford Site in Washington, tuff and granite at the Nevada Test Site, domal salt in Louisiana, and bedded salt in Kansas. In-situ testing has also been performed at the Colorado School of Mines facility in Idaho Springs, Colorado. Internationally, in-situ testing is being performed at the Stripa facility in Sweden, the Asse mine in the Federal Republic of Germany, and the Underground Research Laboratory in Canada. In addition, although not associated with the civilian radioactive waste management program, the construction of and testing in the Waste Isolation Pilot Project (WIPP), a demonstration facility for the evaluation of long-term storage of defense-generated transuranic waste and experiments with defense high-level wastes, in bedded salt in New Mexico may offer significant insight and data to supplement the in-situ testing at salt sites under consideration for the first geologic repository.

4. SITE CHARACTERIZATION PHASE

As stated previously, the major objective of the in-situ testing in the site characterization phase is to determine the suitability of the site for a geologic repository. This phase of in-situ testing will focus on answering the following two questions that are key to a determination of site suitability:

1. Is there anything that would prevent the safe, economic construction and operation of a geologic repository?

2. Is there anything that would preclude the long-term containment and isolation of waste in that geologic repository?

This will require evaluations of the pre-closure occupational health and safety of workers in the repository during construction, operation and closure, and both the pre-closure and long-term post-closure radiological health and safety of the public. Applicable regulations include the Environmental Protection Agency (EPA) standards for high-level waste disposal (40 CFR Part 191)[6], the NRC regulation for geologic repositories for disposal of high-level wastes (10 CFR Part 60), the NRC standards for protection against radiation (10 CFR Part 20)[7], the DOE general guidelines (10 CFR Part 960), mine safety and health regulations (30 CFR Part 57)[8], and State and local standards as appropriate.

The schedule shown previously in Figure 2 allots 67 months from the recommendation of three sites to the President for approval for site characterization until the recommendation of one site, from among the three characterized sites, to the President for development as a geologic repository. The in-situ testing program in this time period will include sinking of the exploratory shaft or shafts, driving of drifts and other underground openings, drilling of boreholes, establishing a representative baseline of pre-test conditions, and performing in-situ testing. Figure 4 illustrates the exploratory shaft test facility concept which will provide the access to conduct this testing program.

```
IN-SITU TESTING
PHASE:          SITE CHARACTERIZATION

PURPOSES:       DETERMINATION OF SITE SUITABILITY

                RECOMMENDATION OF ONE SITE FOR DEVELOPMENT

                SUPPORT FOR SITE SELECTION REPORT (SSR)

                SUPPORT FOR ENVIRONMENTAL IMPACT STATEMENT (EIS)

                SUPPORT FOR CONSTRUCTION AUTHORIZATION
                APPLICATION (CAA)

                ADVANCEMENT OF REPOSITORY DESIGN - TITLE I

TYPICAL TYPES
OF TESTING:     GEOLOGIC MAPPING

                LARGE-SCALE SAMPLING

                ROCK EXCAVATION TECHNIQUES

                ROCK SUPPORT

                FLUID CONTROL

                ROOM CLOSURE

                GEOMECHANICAL PROPERTIES

                THERMAL, THERMOMECHANICAL RESPONSE

                IN-SITU STRESS

                HYDRAULIC CONDUCTIVITY

                BOREHOLE AND SHAFT SEALING
```

Figure 5.   Site Characterization Phase Testing

The purpose of in-situ testing during the site characterization phase, shown on Figure 5, is to provide the data needed to support the recommendation of one site for development as a geologic repository in the Site Selection Report and the Environmental Impact Statement, support the construction authorization application for the selected site, and advance the conceptual design of the repository to the level of design needed in the CAA. In addition, some tests of a long-term nature will begin in the site characterization phase and continue into subsequent phases.

The types of tests will be derived through a process of 1) identifying geoscience and design issues to be resolved, 2) determining the analyses appropriate for evaluation and resolution of the issues, 3) identifying the data needed for these analyses, 4) determining the appropriate testing methods to develop the needed data, and 5) specifying the appropriate conditions under which to perform the tests. Accordingly, the types of in-situ tests performed in the exploratory shaft and associated underground workings will be largely site- and geologic media-specific. The types of in-situ tests being planned for each of the first repository projects are described in detail in subsequent papers at this conference.

There will be, however, certain types of testing initiated during the

47

site characterization phase which may be common to all projects. These could include examination, geologic mapping, and sampling of the host rock exposed in the exploratory shaft and/or the underground workings (e.g., to determine with reasonable assurance the lateral continuity and uniformity of the host rock), and evaluations such as rock excavation techniques, rock support requirements and techniques, fluid control measures, room closure, geomechanical properties of the rock mass, thermal and thermomechanical response of the rock mass, in-situ state of stress, large-scale hydraulic conductivity, and borehole and shaft sealing.

5.   CAA REVIEW PHASE

In-situ testing will be continued, at the site selected for development as a repository, during the NRC review of the construction authorization application. This time period is herein designated the CAA review phase. The Act mandates that the NRC review of the construction authorization application not exceed three years (with a possible one year extension at the request of the NRC).

The purposes of testing in the CAA review phase are summarized in Figure 6. The results of this phase of in-situ testing will provide data for confirmation of the suitability of the site for development as a repository, verification and optimization of the repository design, and confirmation of the data and analyses submitted in the construction authorization application. Attendant benefits of CAA review phase testing include developing data for activities such as validation of computer codes.

The CAA review phase testing will be primarily a continuation of long-term tests initiated during the site characterization phase to confirm the results reported in the CAA by extending the time duration of the data. Additional testing may be performed in selected subject areas, such as verification of waste package design, development of

IN-SITU TESTING
PHASE: **CAA REVIEW**

PURPOSES: **CONFIRMATION OF SITE SUITABILITY**

                      VERIFICATION AND OPTIMIZATION OF REPOSITORY DESIGN

                      CONFIRMATION OF CAA DATA AND ANALYSES

                      VALIDATION OF COMPUTER CODES

TYPICAL TYPES
OF TESTING: **CONTINUATION OF SITE CHARACTERIZATION PHASE**

                      WASTE PACKAGE

                      INSTRUMENTATION DEVELOPMENT

                      GROUND-WATER CHEMISTRY

                      GEOCHEMICAL INTERACTIONS

                      EXPLORATORY DRIFTING

Figure 6.   CAA Review Phase Testing

IN-SITU TESTING
PHASE: CONSTRUCTION

PURPOSES: CONFIRMATION OF SITE SUITABILITY

SUPPORT FOR LICENSE APPLICATION TO RECEIVE AND
EMPLACE WASTES

DEMONSTRATION OF TECHNOLOGY ASSOCIATED WITH
REPOSITORY OPERATION

EVALUATION OF EQUIPMENT AND INSTRUMENTATION
PERFORMANCE

DEVELOPMENT OF SCALING FACTORS TO FULL
REPOSITORY SIZE

VALIDATION OF MODELING RESULTS

EVALUATION OF COUPLED THMC EFFECTS

REDUCTION OF UNCERTAINTY IN DESIGN BASES

DEVELOPMENT OF OPERATIONAL AND EMERGENCY
PROCEDURES

DEMONSTRATION OF SAFE OCCUPATIONAL CONDITIONS

TRAINING AND CERTIFICATION OF PERSONNEL

Figure 7.   Construction Phase Testing

instrumentation for inclusion in the performance confirmation program, determination of ground-water chemistry, and evaluation of near-field geochemical interactions. In addition, the CAA review phase may include extensive horizontal drilling and/or exploratory drifting to confirm the lateral continuity of the host rock and to determine the presence and significance of discontinuities or heterogeneities in the host rock.

DOE is currently considering the construction of a separate, colocated Test and Evaluation Facility (TEF), at the site selected for development as a repository. As provided for in the Act, subsurface construction and testing of a colocated TEF could begin with the designation of the repository site prior to submittal of the construction authorization application for the site selected for development as a repository. In-situ testing in the colocated TEF at that time would simulate operating conditions for the repository in order to demonstrate the technology associated with the operation of a geologic repository. While a TEF is authorized by the Act, DOE has not yet reached a decision as to whether to proceed with a TEF. This decision is currently planned to be made in the 1987 time frame.

6.   CONSTRUCTION PHASE

Construction-phase testing will be performed after authorization of construction by the NRC. The construction-phase testing will be, in part, a continuation of the CAA review phase testing using an expansion of the same underground workings, an expansion into new underground workings opened for the repository, and the TEF, if built.

The construction-phase testing will continue throughout both the construction period and the period of NRC review of the license application to receive and emplace wastes. Based on the recently issued draft Mission Plan schedule[9] shown in Figure 2, the construction-phase testing is anticipated to extend from 1993 to 1998.

The major objective of the construction-phase testing, as shown in Figure 7, is to provide confirmation of the suitability of the site for a geologic repository. In addition, the construction-phase testing will provide the data needed to support the application to the NRC for a license to receive and emplace wastes and to demonstrate the technology associated with the operation of a geologic repository, e.g. waste handling, waste emplacement, backfilling, sealing, and retrievability.

The construction-phase testing could also provide the data needed for evaluation of equipment and instrumentation performance, development of scaling factors to full repository size, validation of modeling results, evaluation of coupled thermal-hydrologic-mechanical-chemical effects, as appropriate, and reduction of uncertainty in the design bases. The construction-phase testing could also allow development of operational and emergency procedures for routine and abnormal operations, demonstration of safe occupational conditions, and training and certification of personnel.

7.   OPERATIONAL PHASE

DOE will continue in-situ testing, as appropriate, as part of the performance confirmation program throughout the approximately 30-year period of operation and the following period of up to 50 years after the initiation of waste emplacement when the capability to begin retrieval operations must exist. As stipulated in Subpart F of 10 CFR Part 60, the general requirement for this performance confirmation program is to provide data which indicate, where practicable, whether 1) actual subsurface conditions encountered and changes in those conditions during construction and waste emplacement are within the limits assumed in the licensing review, and 2) natural and engineered systems and components required for repository operation, or which are designed or assumed to operate after closure of the repository as barriers against radionuclide migration, are functioning as intended and anticipated.

The operational-phase testing therefore will provide not only continued confirmation of the suitability of the site for a geologic repository but also confirmation of the performance of the repository subsystems (Figure 8). These findings will be the major basis of support for the application to the NRC for a license amendment to permit closure of the repository.

8.   SUMMARY

In summary, the United States Department of Energy is in the process of developing an overall strategy for in-situ testing in the geologic repository program that is consistent with the requirements of the Nuclear Waste Policy Act of 1982 and 10 CFR Part 60. This overall strategy calls for a five-phase program for in-situ testing. Site-specific in-situ testing will begin with site characterization activities at three sites and will continue at the site selected for development as the first geologic repository through construction,

```
IN-SITU TESTING
PHASE:              OPERATIONAL

PURPOSES:           CONFIRMATION OF SITE SUITABILITY

                    CONFIRMATION OF PERFORMANCE OF REPOSITORY
                    SUBSYSTEMS

                    SUPPORT FOR LICENSE AMENDMENT TO PERMIT
                    CLOSURE
```

Figure 8.   Operational Phase Testing

operation and closure of the geologic repository - a time period on the order of up to 100 years.

The strategy is to initiate tests at each phase, where needed, to satisfy the information needs of the various stages of site selection, development and licensing; to continue, where appropriate, the same tests from one phase to the next in the five-phase program; and, to increase the scope, dimensional scale, time duration, and interactive processes considered in the testing over those tests performed at each previous phase. The testing at each phase will build upon the tests performed and the data collected in the preceding phases. The major finding at each phase will be the determination or confirmation, with increasing confidence and certainty at each successive phase, of the suitability of the site for a geologic repository which will meet applicable standards for public health and safety.

REFERENCES

[1] Nuclear Waste Policy Act of 1982, Public Law 97-425, United States of America (1983).

[2] U.S. Nuclear Regulatory Commission, Disposal of High-Level Radioactive Wastes in Geologic Repositories, 10 CFR Part 60.

[3] U.S. Department of Energy, General Guidelines for Recommendation of Sites for Nuclear Waste Repositories, 10 CFR Part 960.

[4] U.S. Nuclear Regulatory Commission, Standard Format and Content of Site Characterization Reports for High-Level Waste Geologic Repositories, Regulatory Guide 4.17 (1982).

[5] Stein, R. and Collyer, P. L., Pilot Research Projects in the United States for Underground Disposal of Radioactive Wastes, IAEA International Conference on Radioactive Waste Management, Seattle, Washington, IAEA-CN-43/459 (1983).

[6] U.S. Environmental Protection Agency, Environmental Radiation Protection Standards for Management and Disposal of Spent Nuclear Fuel, High-Level and Transuranic Radioactive Wastes, 40 CFR Part 191 (draft).

[7] U.S. Nuclear Regulatory Commission, Standard for Protection Against Radiation, 10 CFR Part 20.

[8] U.S. Department of Labor, Mine Safety and Health Administration, Safety and Health Standards - Metal and Non-Metal Underground Mines, 30 CFR Part 57.

[9] U.S. Department of Energy, Draft Mission Plan (May, 1984).

Summary of discussion

This discussion, chaired by Mr. Barthoux (ANDRA), was concentrated on two main topics.

1. Questions related to instrumentation procedures

Following Mr. Duffaut's presentation, the performances of existing flat-jacks for in situ stress measurements were discussed (Messrs. Nataraja, Durand, Dietz, Hunsche). Presently, rock pressures up to 10-15 MPa can be measured by flat-jacks whose area may reach $2m^2$. It was shown (Mr. Durand) that other stress measurement methods (doorstopper, hydrofrac) lead to comparable results.

Plate-bearing tests may become too heavy in very stiff rocks such as basalt (Mr. Dietz). In this case, pressurized chamber tests can be performed in steel-lined tunnel sections (Mr. Duffaut).

As regards measurements at low or high temperatures, adequate concepts and calculation tools are now available for any temperature range (Mr. Boulanger). Instrumentation in these cases raises the same problems of correction of the measurements, and of reliability of data aquisition systems.

2. Objectives of underground laboratories

The bulk of the comments and questions concerned the paper by Mr. Rometsch, and mainly the concept of leaving a repository open for some time after waste disposal (i.e. 50 years), in order to allow some monitoring and thus increase the public opinion's confidence in the soundness of the method. It must be emphasised that this concept has not yet been studied in detail; it should, however, help in overcoming the contradiction between the need for eventual sealing of a repository, and the need for some monitoring.

Mr. Kühn pointed out that this concept is in conflict with German position about disposal in salt. Leaving a repository open for decades could enhance the possibility of water intrusion thus leading to an undesirable situation.

Mr. Baetslé suggested the possibility of performing in situ migration tests from disposed, non-containerised waste, which would increase the possibility of detecting actual, in situ migration. Mr. Lake and Mr. Gera pointed out that it should be necessary to remove not only the container, but also the buffer and backfill materials as well; this arrangement would so largely differ from the "complete" repository system that any comparison could hardly be valid.

Mr. Tyler recalled that purposely ill-conceived canisters will be emplaced in the WIPP and monitored for 25 years. It is hoped that the isolation by salt will be complete and that public confidence will be obtained.

This comment was also supported by M. Frei, for whom the isolation capabilities of the geological barrier will be the dominant factor for selection of a repository site according to NRC's regulations.

For migration experiments, higher confidence however could be gained from natural systems such as:

    (a) migration of natural chromium in sediments (Mr. Rometsch);
    (b) transfer of anthropogenic isotopes (tririum, strontium) or even cosmogenic isotopes in the geosphere (Mr. Müller).

Session 2/Séance 2
Reviews of experiments in the major rock types
Revue des expériences dans les principaux types de formations

Chairman/Président:
F.FEATES
Department of the Environment, London, UK

In situ experiments in granite in underground laboratories
A review

G.R.SIMMONS
Atomic Energy of Canada Ltd, Whiteshell Nuclear Research Establishment, Pinawa, Manitoba

ABSTRACT

In situ experimental facilities in support of nuclear fuel waste disposal are operating, or are under construction, within granite rock bodies in Canada, Sweden, Switzerland and the United States. The objectives, the experimental plans, and the degree of success are briefly discussed for experiments in each facility.

1. INTRODUCTION

Underground laboratories in granite are of two types: those associated with existing mines and those constructed specifically for testing purposes. The experiments being undertaken are very largely near-field, that is, focussed on the conditions in, and the response of, the rock mass immediately adjacent to the excavations. A few experiments consider the response of a very large volume of rock. The experiments are designed to measure rock mass properties, to study physical and chemical phenomena of interest, and to assess and validate mathematical models of physical situations expected to be encountered in nuclear fuel waste disposal.

This paper summarizes experiments in, or planned for, underground laboratories in Canada, Sweden, Switzerland and the United States of America.

2. THE UNDERGROUND RESEARCH LABORATORY, CANADA

2.1 General

The Underground Research Laboratory (URL) is an integral part of the geosciences research within the Canadian Nuclear Fuel Waste Management Program [1], and the focus of a two-phase experimental program [2]. The URL program objectives that relate to this Workshop are

- to perform experiments relevant to the Nuclear Fuel Waste Management Program that require underground access to a plutonic rock environment,

- to assess airborne, surface and borehole survey techniques for characterization of the subsurface geological and hydrogeological environment, and

- to assess the changes to the physical and chemical conditions in the
rock mass and groundwater caused by excavation of the URL.

The URL is located on 3.8 km^2 of land leased for twenty-one years from
the Province of Manitoba. The lease area is a previously undisturbed portion of
the Lac du Bonnet batholith, a large granitic intrusive. The layout of surface
facilities and the conceptual arrangement of the underground facilities are
shown in Figure 1. The URL shaft will be 255 m deep with a shaft station at the
130-m depth and the main testing area at the 240-m depth. This arrangement will
provide access to the two main phases of granite in the area, a grey phase
containing few fractures and a pink, more fractured phase, and to two
hydraulically conductive fracture zones. This will provide a wide variety of
testing conditions.

2.2   Experimental Program

The URL experimental program comprises site evaluation and underground
experimental activities [3]. Two of the site evaluation activities are directly
related to developing the URL underground facilities. A model of the rock mass
through which the URL will be excavated, based on airborne, surface and borehole
geological and geophysical survey data, will be assessed against surveys of the
excavation surfaces. The groundwater flow systems determined from surface and
borehole information are being used to develop and calibrate flow system models.
These models are being used to predict the perturbation that excavation of the
URL will introduce, and their accuracy will be assessed against field measurements of the perturbation [4]. The collection of results began in 1984 May, and
the geological and hydrogeological system response predictions will be
documented by 1984 June.

The underground experimental activities have focussed on procedure and
methodology development and equipment testing in the URL shaft collar. This is
the top 15 m of shaft, excavated as part of head frame construction. This
program has included testing of CSIRO[1] Hollow Inclusion cells, IRAD
vibrating-wire stress meters, IRAD rigid probe sonic extensometers and thermistors under excavation conditions. The methods to be used for assessing excavation damage, rock mass response to excavation, and rock mass stress by overcoring (USBM[2] and CSIRO or CSIR[3] borehole deformation gauges), for
removing bulk rock samples by coring (with and without fractures) and by diamond
sawing, and for mapping and taking stereophotographs of the excavation walls
were applied in the shaft collar [5].

Underground experiments will be done during the construction and operating
phases of the URL. The construction phase activities (from 1983 to 1986) focus
on collecting geological, hydrogeological and geomechanical information on the
rock mass surrounding the excavation. The techniques used in the shaft collar
program, discussed above, are being applied during this phase.

The operating phase experiments (from 1986 to 2000) are being developed to
measure the in situ properties of the rock mass. These experiments are now in

---

(1)   Commonwealth Scientific and Industrial Research Organization, Australia
(2)   United States Bureau of Mines
(3)   Council of Scientific and Industrial Research (South Africa)

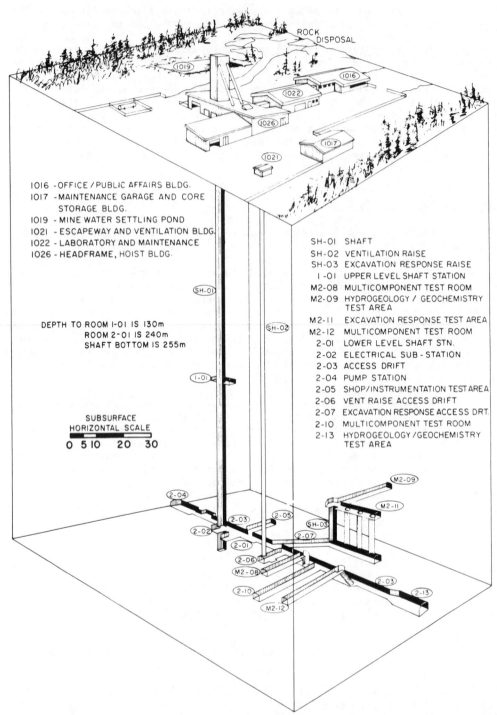

ROCK DISPOSAL

1016 - OFFICE / PUBLIC AFFAIRS BLDG.
1017 - MAINTENANCE GARAGE AND CORE
        STORAGE BLDG.
1019 - MINE WATER SETTLING POND
1021 - ESCAPEWAY AND VENTILATION BLDG.
1022 - LABORATORY AND MAINTENANCE
1026 - HEADFRAME, HOIST BLDG.

SH-01   SHAFT
SH-02   VENTILATION RAISE
SH-03   EXCAVATION RESPONSE RAISE
 1-01   UPPER LEVEL SHAFT STATION
M2-08   MULTICOMPONENT TEST ROOM
M2-09   HYDROGEOLOGY / GEOCHEMISTRY
        TEST AREA
M2-11   EXCAVATION RESPONSE TEST AREA
M2-12   MULTICOMPONENT TEST ROOM
 2-01   LOWER LEVEL SHAFT STN.
 2-02   ELECTRICAL SUB-STATION
 2-03   ACCESS DRIFT
 2-04   PUMP STATION
 2-05   SHOP/INSTRUMENTATION TEST AREA
 2-06   VENT RAISE ACCESS DRIFT
 2-07   EXCAVATION RESPONSE ACCESS DRT.
 2-10   MULTICOMPONENT TEST ROOM
 2-13   HYDROGEOLOGY /GEOCHEMISTRY
        TEST AREA

DEPTH TO ROOM 1-01 IS 130m
        ROOM 2-01 IS 240m
        SHAFT BOTTOM IS 255m

SUBSURFACE
HORIZONTAL SCALE
0 5 10   20   30

FIGURE 1:   LAYOUT OF THE UNDERGROUND RESEARCH LABORATORY, CANADA

the conceptual planning stage and experiments being considered include

- a pressure chamber experiment,
- a mine-by (excavation response) experiment,
- an in situ migration experiment,
- a moisture balance experiment,
- a macropermeability experiment (may be integrated with other experiments),
- container - buffer - rock interaction experiments at ambient and elevated temperatures,
- borehole plugging experiments,
- shaft sealing experiments, and
- a heated block test.

The URL will also be used for long-term reliability and performance testing of instruments.

Detailed plans for these experiments have not been completed. The final experimental plan will make use of experience gained internationally from in situ experiments.

3.   THE STRIPA MINE, SWEDEN

3.1   Introduction

Since 1976, an in situ experimental program has been underway in Sweden. The Stripa Iron Ore Mine, which operated from 1485 to 1976, was taken over by the Swedish Nuclear Fuel Supply Company (SKBF) as a laboratory. The mine is located about 150 km west of Stockholm. The in situ tests are being conducted in newly excavated rooms at a depth of 340 to 420 m in a granite body adjacent to the iron ore body. The layout of the experimental rooms is shown in Figure 2. The testing rooms are below the water table.

Three major testing programs have been initiated in the Stripa Mine: the Swedish-American Cooperative Program from 1977 to 1980, the Multinational Stripa Project - Phase 1 from 1980 to 1984, and the Multinational Stripa Project - Phase 2 beginning in 1983. These programs are summarized briefly below.

3.2   The Swedish-American Cooperative Program

The Lawrence Berkeley Laboratory (LBL) and the Swedish Nuclear Fuel Safety Program (KBS) jointly operated this program [6] from 1977 to 1980. The major tests in the program dealt with the thermal, mechanical and hydrogeological properties of fractured, water-saturated granite.

3.2.1 Thermomechanical Tests

A time-scaled heater test was carried out to investigate the long-term thermomechanical response of the rock to thermal loading. The laws of heat conduction were used to design a test to compress the time scale by a factor of 10 [7, 8]. The test array consisted of eight 1-kW heaters, each 1 m in length, emplaced 10.5 m below the floor of the drift. The heaters were in two rows spaced 8 m apart along the axis of the drift and 3 m apart laterally. Instrumentation was installed to monitor temperature and rock deformation.

59

Two full-scale heater experiments were installed to investigate the short-term near-field effects of an applied thermal load on the rock mass. Two heaters, each 3 m long and 0.3 m in diameter, were emplaced in vertical boreholes drilled to a depth of 5.5 m into the floor of the Full-Scale Heater drift. Power outputs of the two heaters were 5 kW and 3.6 kW. The two emplacement holes were 22 m apart and were, thus, thermally isolated from each other. Instrumentation installed in the floor of the drift included rod extensometers, USBM gauges and IRAD vibrating-wire gauges to monitor borehole displacements and stresses, and numerous thermocouples to measure temperature. Several horizontal rod extensometers were also installed from a parallel drift.

As an addition to the 5-kW heater test, eight 1.0-kW heaters, which had been spaced around a 0.9-m radius from the main 5-kW heater, were switched on after 204 days of main heater operation. This increased the rock temperature by approximately 100°C with a corresponding increase in compressive stress. Temperatures on the wall of the central borehole were calculated to be 300 to 350°C during this period. Spalling of the wall of the 5.0-kW heater borehole occurred within a few days after the auxiliary heaters were turned on, and it increased both in extent along the length of the borehole and in size of the rock chips. This time-dependent spalling behaviour was not predicted by thermoelastic theory.

By analyzing the data from these tests, experimenters have drawn these conclusions [9]:

- Conduction is the dominant heat transfer mode and calculations of temperature distribution based on conduction theory and intact rock thermal properties agree reasonably well with experimental data.
- Displacement calculations using linear elastic theory do not agree well with experimental data. The differences indicate that the thermal expansion coefficient in situ is lower than the values measured in the laboratory and the magnitude varies with temperature.

The stress data from these tests have not been published; hence, no comparison of calculated versus actual changes in stress is available.

The thermocouples and rod extensometers apparently functioned satisfactorily for the duration of the experiments. The effectiveness of the IRAD vibrating-wire borehole stress meters and USBM borehole deformation gauges has not been determined because the data have not yet been analyzed.

3.2.2 Hydrogeological Studies

Hydrogeological characterization studies investigated the factors controlling the movement of groundwater in the fractured rock mass. In this program, the fracture systems in the Stripa granite were identified, the groundwater systems surrounding the testing area were characterized, and a macropermeability experiment was run.

Three inclined boreholes were drilled from the surface to the vicinity of the heated test rooms. The holes were oriented to intersect the major fracture sets, and the cores were carefully reconstructed to determine variations in fracture geometry. Injection tests in packed-off sections of the surface boreholes were performed to develop fracture aperture distribution data. Pressure tests in the surface boreholes have shown that drainage into the underground workings has decreased the water pressure to below hydrostatic at depths below about 100 m.

The large-scale macropermeability experiment was performed to improve techniques for characterizing the permeability of large volumes of low-permeability rock. Fifteen radial boreholes were drilled from the test room to measure pressure gradients. This technique created a large-scale pressure sink to perturb a relatively large volume of the fracture flow system. The ventilation drift was sealed off and equipped with a constant-temperature, circulating air flow system that was controlled to evaporate all water seeping into the room. The water inflow rate was then determined from measurement of the mass flow rate and the difference in the humidity of the entering and exiting air stream.

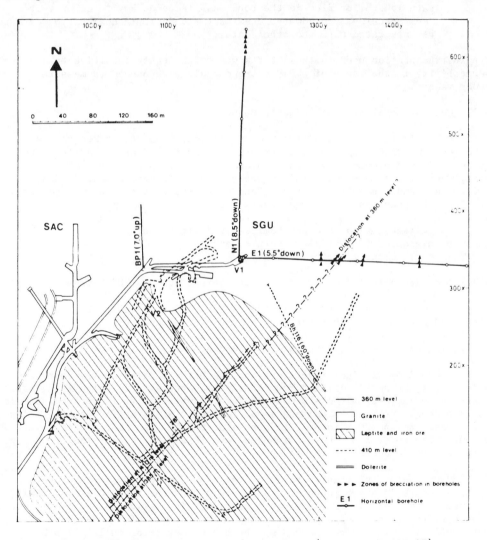

FIGURE 2: LAYOUT OF STRIPA TEST AREAS (FROM REFERENCE 27)

The following general conclusions have been drawn from these testing activities [10]:

- Although the fractures within a rock body can be reasonably defined from surface, subsurface and borehole data, much work is required to measure hydrogeological parameters suitable for system modelling.
- Borehole injection testing and the room-scale macropermeability experiment were used independently to assess rock mass permeability, and the results agree within an order of magnitude.
- A well-planned and coordinated surface and underground hydrogeological assessment program is essential to obtain adequate site characterization.
- Additional study of the anisotropy of permeability and the effective path length for flow in the rock mass is necessary to allow rock mass porosity, groundwater chemistry and isotopic composition data to be integrated with the other data.

The instrumentation used in these tests was sufficiently sensitive to measure permeability in the range of 10^{-10} m/s and could be enhanced to measure lower values.

3.3   Multinational Stripa Project - Phase 1

Following completion of the Sweden-American Cooperative Program, an autonomous OECD/NEA multinational program was initiated at the Stripa Mine. Phase 1 of this project began in 1980 and will be completed in 1984. The participating countries with full membership are Finland, Japan, Sweden, Switzerland and the United States. Canada and France are associate members. The Phase 1 experimental program consists of

- hydrogeological and geochemical investigations,
- migration in a single fracture, and
- a large-scale buffer mass test.

3.3.1 Hydrogeological and Geochemical Investigations to 1200 Metres

The objectives of this activity are [11, 12] to develop the methodology and equipment for hydrogeological and hydrogeochemical investigations in an underground facility and to continue hydraulic, chemical and isotopic characterization of the Stripa granite and groundwaters.

The hydrogeological investigations are being carried out in four boreholes: V1, E1 and N1 drilled from the 360 m depth and V2, which is a deepened existing borehole from the 410 m depth (see Figure 2). V1 is vertical and is 505.9 m long. V2 is vertical and is 822 m long. N1 is angled at 8.5° below horizontal and is 300 m long. E1 is angled at 5.5° below horizontal and is 300 m long. The cores from these boreholes were logged and geophysical well-log surveys were conducted in the boreholes.

Hydrogeological studies have confirmed that most flow through the Stripa granite occurs in a few zones of fracturing within the rock mass, particularly at increasing depths. Flow through intact rock is negligible and flow through discrete fractures is of minor importance. In the very few conductive zones, permeabilities higher than 4×10^{-8} m/s have been measured, while all other zones have permeabilities less than 10^{-9} m/s.

Cross-hole geophysical testing (mise à la masse technique) between bore-holes V1 and V2 provided unique data necessary for interpreting hydrogeological and geological information on interconnecting fractures.

Preliminary modelling of the system did not reproduce the field data particularly well. Shortcomings include the use of two-dimensional analyses and limited in situ property data.

A major conclusion of this activity is that nondestructive techniques are required to identify the locations and sizes of fracture zones in a rock mass between widely spaced boreholes.

The isotopic and chemical composition of Stripa area groundwaters has been studied continuously [13, 14] since the beginning of the Swedish-American Cooperative Program. In Phase 1 this has been extended to depths of 1200 m in borehole V1. This program has identified that there are chemically distinct groundwater systems at various depths in the Stripa granite. All data indicate that salinity and pH increase, and alkalinity decreases, with depth. The groundwaters sampled in the upper 300 m at Stripa respond to changing con-ditions on the surface and are difficult to characterize chemically. The chemistry of deep groundwaters is a result of the mixing of meteoric waters with saline fluid inclusions leached from rock pores and/or deep fossil seawaters. The hydrogeochemistry system is dynamic and therefore changes with time. Various chemical analyses of the Stripa groundwaters have provided data essential to understanding the evolution of these groundwaters. These data and all available information on the groundwater flow systems must be considered in hypothesizing the origins of the various waters.

The equipment developed in this phase of the program is prototypical. Obtaining representative deep groundwater samples is still difficult.

3.3.2 Migration in a Single Fracture

The main objectives of this experiment are [15]

- to observe the movement of nonsorbing and sorbing tracers under con-trolled and well-defined conditions in a real environment,
- to interpret the movement of tracers in such a way that the results become useful for the prediction of radionuclide migration,
- to obtain a basis for comparing laboratory data on sorption with observations in a real environment,
- to develop techniques for small volume sampling of water and fracture surfaces with sorbed tracers, and
- to gather experience with stable tracers before using radioactive tracers.

The field program implemented to meet these objectives involves character-izing and conducting tracer tests in a discrete fracture. In preparation for the main test, injection and collection equipment was assembled and tested in a fracture near the wall of an existing drift. The injection system consists of controlled pressure and volume pumping equipment and straddle-packer assemblies to seal the fracture under test within the injection borehole. The volume of the injection compartment is made small by closely spacing the packers and keeping the mandrel large. The collecting devices are single-packer assemblies with a funnel on the collector side to direct all water in the collection

63

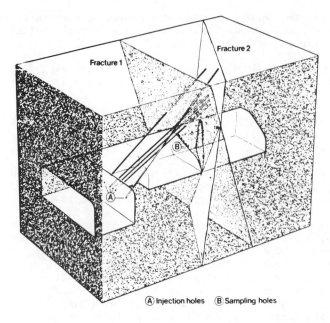

(A) Injection holes (B) Sampling holes

FIGURE 3:
LAYOUT OF STRIPA
PHASE 1 DISCRETE
FRACTURE EXPERIMENT
(AFTER REFERENCE 15)

borehole into the withdrawal or sampling point of the packer. These devices are
installed in short holes drilled into the exposed fracture surface where it
intersects a drift. The sampling equipment for the collectors is designed to
maintain an oxygen-free environment.

From the preparatory investigations, it was determined that the equipment
performed satisfactorily. This test also showed that the permeability of the
fracture, even over distances of 1 to 2 m, was extremely variable. Flow rates
at the six sample points during an injection test were 83, 28, 5, 0.2, 0 and 0
mL/h.

For the main tests, a suitable fracture was identified, and a new room was
excavated to provide access for testing. In this test, five injection holes
were utilized; four intersect the test fracture about 5 m from the drift face
and the fifth intersects it about 10 m from the drift face (see Figure 3). A
second discrete fracture intersects the test fracture and the rock face in the
test room. Thirty collection devices were installed in the drift face at the
two fractures. Only eleven of these are hydraulically connected to the injection holes. Injection tests were done using both sorbing and nonsorbing
tracers, and the arrival times and concentrations were recorded.

The fracture was excavated by drilling numerous large-diameter cores along
its length. The surface concentrations of sorbing tracers are being analyzed.
The results indicate that the paths followed by the flowing water are very
tortuous and that they occupy a small part of the fracture area. The following
are some conclusions that have been drawn from this test:

- Geological and geophysical methods were inconclusive in characterizing
  the structure of the discrete fractures.
- Discrete fractures are very anisotropic with very large variations in
  hydraulic connectivity and permeability over short distances.

64

- The primary tracer sorption surface is only a small portion of the total fracture surface.

In operating the discrete fracture migration test, there were some mechanical difficulties with the injection equipment.

3.3.3 Buffer Mass Test

The engineered barriers in the Swedish disposal concept include the use of bentonite as a packing material around waste packages. The objective of the buffer mass test is to verify the suitability and predict the performance of bentonite-based buffers and soil-based backfills under geological and elevated temperature conditions [16]. The major issues being studied are emplacement performance, temperature distributions, water uptake, swelling pressure, water pressures, and the influence of geological conditions on the wetting of buffer materials.

The field test has been installed in the test room previously used for the macropermeability experiment. This room has been extensively fracture-mapped and the average permeability has been estimated at 10^{-10} m/s. The extensive survey information from prior work provided a good data base for the design of and predictions for the Buffer Mass Test. The experimental layout is shown in Figure 4. Six large-diameter vertical holes were drilled, equally spaced, along the test room. Each hole contains a simulated waste container with an electric heater and is surrounded by precompacted bentonite blocks. A bulkhead was constructed to isolate the two holes, #1 and #2, farthest into the room, and that portion of the room was backfilled with a pneumatically compacted sand-bentonite mixture.

The emplacement holes were instrumented for moisture content, swelling pressure and temperature. The geology and hydrogeology conditions in the test room have a significant effect on this test. The natural fractures in the rock mass have provided three "wet" boreholes (#1, #2 and #5) and three "dry" boreholes. In placing the compacted bentonite in the "wet" holes, a 10-mm annular slot was left between the buffer and the rock to allow for distribution of the inflowing water over a large amount of buffer. In the dry holes, a 30-mm annular slot was filled with lightly compacted bentonite. Each installation was designed to provide a bulk buffer density of 2100 kg/m^3.

The buffer mass test was initially run with all heaters drawing 600 W of power. At this power level, the temperatures measured in the test compared well with predictions. The temperature in the wet holes was lower and the bentonite swelling pressure was higher than in the dry holes. The water uptake measurements in buffer and backfill have given mixed results and have been difficult to interpret. As water uptake continues, small drops in the system temperatures are being recorded.

Two dry boreholes (#3 and #4) were re-excavated and moisture migration into the buffer was studied. Very little water penetrated into the highly compacted buffer. Borehole #6 has been in contact with water longer and is scheduled to be excavated to observe the conditions in a hole after a longer period of water uptake.

A dye injection test is scheduled for borehole #5, in which methyl blue will be injected into the buffer-rock interface to assess migration into the buffer and H_2S will be injected at the buffer-heater interface to assess the vaporization-condensation cycles that may exist in the buffer near the heater.

3.4   Multinational Stripa Project - Phase 2

As the Phase 1 experimental program approached completion, a Phase 2 program was planned, with scheduled start up in 1983. The participating countries are those of Phase 1 and the United Kingdom. In the following sections, the experiments in Phase 2 are discussed, based on information from references 11 and 12.

3.4.1 Cross-Hole Techniques To Detect And Characterize Fracture Zones Near a Repository

The purpose of the program is to develop techniques and instrumentation capable of detecting fracture zones in crystalline bedrock. The methods should be able to determine the location, extent, and thickness of fracture zones and also give a quantitiative measure of the bedrock quality and water-bearing capacity. The large-scale studies are directed at surveying through 500 m of rock and identifying features 10 m thick. These features will be of structural and hydrogeological interest. The small-scale studies are directed at identifying structural features 0.1 m thick when surveying through 50 m of rock. These features are significant in locating test rooms.

The large-scale testing program is being conducted at the Gidea field testing site away from the Stripa Mine. This location provides a borehole array consisting of 10 boreholes 700 m deep and 25 boreholes 100 m deep. At this site, cross-hole seismic surveys have been conducted and interpreted. The results have agreed with previous surveys of the area.

A small-scale testing area is being developed within the Stripa mine in the area used for drilling the holes for hydrogeological assessment in Phase 1. To provide a suitable test array, six F-series boreholes are being drilled around borehole E1. As of 1983 December, boreholes F1, F2 and F3 were completed. The core from each hole is being logged and geophysical surveys are being done to characterize the borehole wall conditions. The testing program in these boreholes includes development and assessment of borehole radar, and cross-hole hydraulics.

3.4.2 Three-Dimensional Tracer Experiment

The purpose of this study is

  - to develop techniques for large-scale tracer experiments in low-permeability fissured rock,
  - to determine flow porosity,
  - to study longitudinal and transverse dispersion in fissured rock,
  - to study channelling, and
  - to obtain data for model verification and/or modification.

To address these objectives, a test array, as shown in Figure 5, is being developed and characterized. A set of nonsorbing tracers will be injected simultaneously in the far end and at intermediate points of the otherwise sealed injection holes. Different tracers will be used at different injection points. As the tunnel is well below the water table, the tracers will flow towards the

66

tunnel. The arrival of the tracers at the tunnel will be monitored in the collection holes near the tunnel. The collection holes are arranged in such a way that they cover a large volume of the rock. They are placed where they intersect larger "wet" fissures. The tracer from each injection point can then be monitored in a large number of holes arranged in such a way that the transverse spreading of the tracer can be measured. By injecting different tracers in the different injection holes, the intersection of flow paths can also be studied.

It is expected that at least three injection holes will be drilled and that at least 40 collection holes will be necessary. The lengths of the injection holes will be about 50 m and the lengths of the collection holes will be 5 to 10 m. Non-active tracers will be used as much as possible. The nonsorbing tracers used may include uranine, bromine, and iodine. If the flow paths are such that sorbing tracers can be expected to arrive in a reasonable time, cesium and strontium are two candidate tracers that have been tested in natural fissures in the laboratory and that will have been used in the fissure in Stripa.

Tracer injection and collection can be done by techniques similar to those that have been developed and are used in the ongoing single-fracture experiment.

3.4.3 Borehole and Shaft Sealing Tests

Nuclear waste disposal at great depths requires plug systems to prevent water flow and radionuclide transport through boreholes, shafts and tunnels. Effective sealing requires that the plug be at least as impermeable as the rock that is replaced by the plug. This implies than no passages are formed along the rock-plug interface.

In a pilot borehole sealing test, two vertical 76-mm diameter boreholes are being used to demonstrate plug placement and performance assessment methodology. In addition, an existing 100-m long, 56-mm diameter, horizontal borehole has been plugged and is being tested. In each borehole, a bentonite plug has been installed and low head tests for moisture migration into the seals are in progress.

In the shaft sealing test program, a 15-m long by 1.5-m diameter vertical shaft made partly by a slot-drilling technique and partly by blasting will be used. It extends between two drifts in the mine. In the shaft, a two-stage test for shaft seal effectiveness is being undertaken. In the first stage of the program, now in progress, concrete seals separated by sand are being in-stalled in the shaft. Water under low head will be injected into the sand layer and will be collected by pipes around the periphery of the shaft. Hydraulic pressures will be measured at several locations within the seal. Later this seal will be removed and a steel bulk head, sand, and a bentonite seal will be installed and tested.

The final part of this project will study the isolation of a significant water-bearing zone that crosses a drift, tunnel or shaft. For the test, such a zone has been created by placing a series of perforated pipes around the periphery of a tunnel, which has been excavated in fractured rock. The pipes are embedded in sand backfill between two concrete bulkheads that support the seal of highly compacted bentonite. Water will be injected at a pressure of up to 300 m of head into the central sand zone, which simulates crushed rock, and the pressure and leakage will be measured. The swelling pressure of the seal will also be measured. Seal installation began in 1984 March.

3.4.4 High-Temperature Buffer Mass Tests

The buffer mass test being conducted in phase 1 has been extended in phase 2 to include high-temperature tests. In the phase 1 testing program, borehole #3 was disassembled and studied. It was reassembled with new buffer in early 1983 and the power output of the heater in borehole #3 was raised to 1200 W. The predicted and calculated temperature distributions agree quite well. The swelling pressures became measurable at the higher power level. To observe the effects of even higher power levels, the heater power was increased to 1800 W in 1984 February. Again the temperature distributions were as expected and the swelling pressures were low.

Because the high power tests in borehole #3 were successful, the power level of the heater in borehole #1 beneath the backfill was increased to 1800 W in 1984 March. Tests on the high-temperature performance of the buffer and backfill systems are continuing.

4.   GRIMSEL ROCK LABORATORY, SWITZERLAND

4.1   General

In Switzerland, NAGRA, the National Cooperative for the Storage of Radio-active Waste, is developing an underground research laboratory in the Alps about 50 km south of Lucerne [18, 19]. In 1983, construction began on the Grimsel Rock Laboratory under 450 m of overlying rock, using existing tunnels associated with a large pumped storage facility for access. The conceptual layout of the laboratory is shown in Figure 6. A full-face tunnel-boring machine is being used to excavate the 3.7-m diameter tunnels, except in areas for radionuclide migration experiments, for excavation damage assessment and for services. These latter areas are being excavated using the drill and blast technique.

The Grimsel Rock Laboratory will provide NAGRA with experience in geo-physics, hydrogeology, migration of radionuclides, rock mechanics, sealing, and thermally induced processes in an Alpine granite. The objectives are to study aspects specific to NAGRA repository concepts, such as container design, heat output, and emplacement configuration, and to build up appropriate expertise and practical experience with the techniques, instruments and equipment necessary for later characterization of an actual repository site.

In 1980, six horizontal cored boreholes were drilled from the existing tunnels to characterize the zone into which the rock laboratory would be excavated. Studies included core logging, downhole geophysical techniques, hydrogeological testing, head measurements, water sampling, and rock properties testing on the core. Based on information from these boreholes and the access available to the test area, experiments were planned for the layout shown in Figure 6.

4.2 Experimental Program Plan

The experimental program planned for the Grimsel Rock Laboratory is a bilateral Swiss and Federal Republic of Germany program. The preliminary

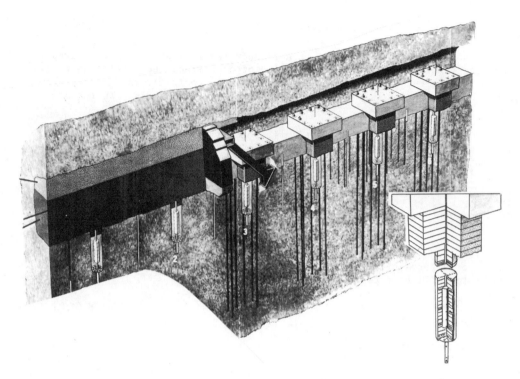

FIGURE 4: LAYOUT OF BUFFER MASS TEST, STRIPA (AFTER REFERENCE 27)

program for the Rock Laboratory, as outlined by Pfister [18] and Pfister and
Nold [19], is described below. In situ experiments are now underway.

4.2.1 Geophysical, Geological and Tectonic Investigations

Two promising geophysical techniques, high-frequency electromagnetics for
small-scale investigations and low-frequency seismic for large-scale
investigations, are being tested as nondestructive tools capable of surveying
large volumes of rock. These techniques are being applied in a cross-hole
configuration in areas of the Rock Laboratory that can be excavated later. The
effectiveness of these techniques will be assessed against actual geological and
hydrogeological conditions found in the excavations.

Detailed geological maps will be prepared for all exposed rock surfaces
and will be used to develop a complete picture of conditions near the excavations. These will be compared with mapping projections prior to excavation.

The crustal tilt and the continuing motion of rock blocks are being
measured in the rock laboratory. Point measurements will be made with high
resolution tilt meters. These meters have a usable range of ± 25 rad and a
resolution of 0.15 nrad. The tilt meters will be calibrated frequently to
account for drift and the field readings will be corrected accordingly. Line
measurements will be made using special extensometers.

4.2.2 Hydrogeological Investigations

An adequate understanding of a groundwater flow system in fractured rock requires information on the distribution of open fractures within the rock mass, the permeability, effective porosity and storage coefficients of the fractures and the distribution of hydraulic heads within the rock mass. In the Rock Laboratory, two hydrogeological tests are being planned.

One test is a combined ventilation and tracer experiment to extend the macropermeability test run in the Swedish-American Cooperative Program at Stripa. The water inflow rate to an isolated section of a drift will be determined using a controlled ventilation system with extensive instrumentation for determining total water flow. The water flow in the rock mass surrounding the room will be monitored by measuring the head distribution and the movement of injected tracer in a series of boreholes.

The motion of water through rock will be studied in an array consisting of a central injection hole surrounded by a series of monitoring boreholes. The anisotropic permeability will be studied and then the effect of temperature on the system will be determined.

The hydraulic head distributions in the rock mass surrounding the Rock Laboratory will be measured.

4.2.3 Geochemistry

Laboratory studies of rock mass sorption properties and computer simulations of mass transport in a fractured rock mass must be assessed by conducting in situ tests with sorbing tracers. Both laboratory and in situ tests are planned. In the first phase, blocks of the rock mass will be removed for controlled testing in a laboratory. In the second phase, an in situ migration experiment will be conducted in a single fracture to provide a two-dimensional test. This will be compared with computer simulations based on property information from small-scale laboratory tests.

Another facet of the geochemistry program is the sampling of groundwaters to study stable isotope hydrology and chemistry and groundwater age. The rock will be sampled to study the fracture fillings and rock chemistry.

4.2.4 Rock Mass Response to Excavation

The effect of excavation technique on the properties of the surrounding rock mass will be assessed for bored-tunnel and blasted excavation in the same test area. The arrangement of excavations allows the test area to be characterized before and after tunnel excavation. The rock mass fracturing, displacements, stresses and hydrogeological conditions will be assessed.

An important consideration in the planning and assessment of all experiments is an understanding of the state of stress and deformation properties of the rock mass. In the Rock Laboratory, stress and deformation of the tunnels and rooms will be measured using a variety of techniques including triaxial and biaxial overcoring, hydraulic fracturing, flat jacks, extensometers, deflectometers, convergence meters and dilatometers.

70

4.2.5 Rock Mass Response to Heating

A single-heater test will be run in the Rock Laboratory to measure the effect of heat on the rock mass and specifically on the aperture of fractures.

5.   GRANITE FIELD TESTS - UNITED STATES OF AMERICA

5.1   Introduction

In situ experiments in granite are in progress at two locations in the United States, under the direction of the Department of Energy (DOE):   at the Colorado School of Mines (CSM) Edgar Mine in Colorado and the Climax Mine on the Nevada Test Site. The testing programs at CSM are being administered for the DOE by the Office of Crystalline Repository Development (OCRD). The programs in both facilities are in support of high-level waste disposal.

5.2   OCRD/CSM Test Site

5.2.1 General

The program at the OCRD/CSM test site has been described in references 20 and 21. The test room is approximately 100 m below ground surface, and is 20 m long, 3 m high and 5 m wide. The facilities for experimental activities associated with this room are three 33-m long, NX (76-mm) diameter boreholes drilled parallel to the room at three distances away from the room wall, 45 NX (76-mm) diameter boreholes drilled radially in groups of seven in each of the first seven blast round zones, and a 2 m x 2 m x 2 m block in the floor at the far end of the room. The general layout is shown in Figure 7.

The objectives of the testing program in this facility are

- to demonstrate and evaluate careful excavation techniques in hard rock,
- to characterize the disturbance zone around the excavation,
- to assess the suitability of a heated block test for characterizing hard rock, and
- to create an underground research laboratory in which a wide variety of geomechanics experiments can be performed.

5.2.2 Careful Excavation and Damage Assessment

The excavation of the test room was done using ten carefully designed, drilled and loaded blast rounds to minimize the damage to the surrounding rock. By minimizing the zone affected by excavation damage, the stability of the opening is improved and requirements of long-term sealing are reduced. The disturbed zone in the test room has been assessed in radial NX (76-mm) holes covering seven excavation blast rounds by monitoring for variations in modulus (E), permeability (K), cross-hole ultrasonic velocities (vp), rock quality designation (RQD) for the core, and measured stress field around the opening. These parameters were measured using the CSM borehole pressure cell, double-packer gas permeability equipment, cross-hole ultrasonic equipment, and core logging/TV and boroscope examination of the hole. Overcore stresses were determined using the USBM gage, CSIRO Hollow Inclusion Cell, and Modified Leeman (Lulea) Cell.

The extent of the disturbance zone was determined by assessing the RQD, E, vp and K data, and the results from each assessment were similar. In one particular zone, the depth of disturbance was less than 1 m. The depth of disturbance was of the same order of magnitude as predicted using formulas developed by Sve De Fo [22] for blast round design and applied to room excavation. A great deal of analysis remains to be done to determine the meaning of the individual results as well as the relationships between them.

The experience gained during this assessment program has shown that the NX (76-mm) version of a CSM cell is a simple and reliable borehole tool, that the cross-hole ultrasonic technique is highly dependent on frequency and on the discontinuities in the intervening rock, and that double-packer permeability equipment performed adequately.

Three overcoring stress measurement techniques were used in four parallel, horizontal holes drilled into a wall of the test room. The results were quite variable and are not easily interpreted, partly due to the jointed nature, the foliation, the anisotropy and the inhomogeneity of the rock mass. Some difficulties were also attributed to performance of the instruments. No conclusion was drawn on the relative suitability of instrumentation.

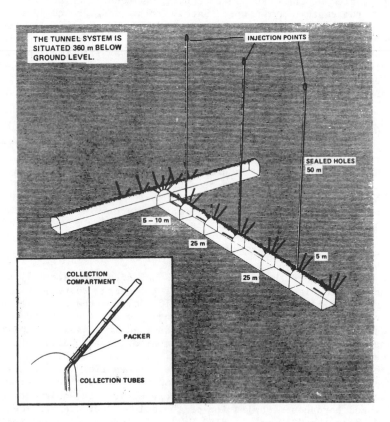

FIGURE 5:   LAYOUT OF STRIPA 3-D MIGRATION EXPERIMENT
(FROM REFERENCE 12)

5.2.3 The Heated Block Test

This test was designed and operated by Terra Tek Inc. [23] in the OCRD/CSM test site. The objective of this test was to develop and evaluate a suitable test method for quantifying the mechanical and thermal behaviour of a jointed rock mass. A block, 2 m on a side, was line drilled to a depth of 2 m and remained connected to the rock mass at the bottom. The block had three sub-vertical natural fractures, one of which was instrumented. The slots around the block were filled with hydraulic flat jacks to load the block in two direc-tions. The orientation of the instrumented fracture was diagonal across the block so that normal and biaxial loadings were possible. Nine electric heaters were installed in a line across the center of the block in one direction. Two heaters were outside the block on each side and five were within the block.

The test plan was developed to measure the stress-strain behaviour of the entire block and the fracture at ambient and elevated temperatures, and the thermal heating effects on the block and the fracture under various confining pressures and temperatures. The block was instrumented on the surface and in the interior, as listed below, to measure deformation, strain, temperature, and fracture permeability.

Surface Instrumentation
- 8 horizontal strain indicators (HSI)
- 3 bonded strain gauge rosettes
- 4 vibrating-wire strain meters (IRAD)
- 52 pairs of Whittemore gauge pins

Interior Instrumentation
- 80 type K thermocouples
- 2 Hollow Inclusion triaxial cells (CSIRO)
- 1 USBM borehole deformation gauge
- 2 vibrating-wire borehole stress meters (IRAD)
- 4 four-anchor rod extensometers
- 2 monitoring holes in the test fracture for permeability measurements

These instruments were monitored through a multiphase test and the results were analyzed. Data from the following instruments were then used to determine the parameters of interest:

Property	Surface Instruments	Interior Instruments
Thermal Conductivity	not used	thermocouples
Vertical Thermal Expansion	not used	extensometers
Vertical Deformation Modulus	not used	extensometers
Joint Normal Stiffness	Whittemore Pins	not used
Horizontal Poisson's Ratio	Whittemore Pins	not used
Vertical Poisson's Ratio	Whittemore Pins	extensometers
Stress Changes	not used	not used

Significant difficulties with instrumentation were identified during the block tests:

- lack of a stable zero reference for the Whittmore pin displacement measurements,
- very small surface strain readings from the bonded strain gauges,

73

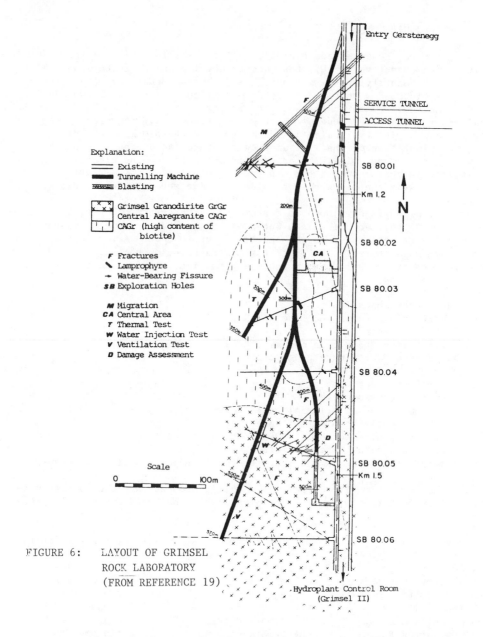

Explanation:

≡≡≡ Existing
▬▬▬ Tunnelling Machine
▨▨▨ Blasting

Grimsel Granodirite GrGr
Central Aaregranite CAGr
CAGr (high content of biotite)

F Fractures
＼ Lamprophyre
→ Water-Bearing Fissure
SB Exploration Holes

M Migration
CA Central Area
T Thermal Test
W Water Injection Test
V Ventilation Test
D Damage Assessment

Scale

0   100m

FIGURE 6:   LAYOUT OF GRIMSEL
ROCK LABORATORY
(FROM REFERENCE 19)

Entry Gerstenegg
SERVICE TUNNEL
ACCESS TUNNEL
SB 80.01
Km 1.2
N
SB 80.02
SB 80.03
SB 80.04
SB 80.05
Km 1.5
SB 80.06
Hydroplant Control Room
(Grimsel II)

- widely varying results from the horizontal strain indicators,
- debonding of the CSIRO cells from the rock during the test, and
- inconsistent data from the IRAD stress meters.

A program is now underway at CSM with a solid frame mounted over the block to provide a zero reference for displacement readings. A series of ambient-temperature load tests are being run with the Whittemore pins, and proximity detectors that are monitoring rods anchored within the block are providing rotation and displacement readings. The results are being used to identify the motion of various block segments.

A continuing program of activities is being planned for this test area.

5.3   Spent Fuel Test-Climax

The Spent Fuel Test-Climax (SFT-C) is being conducted by the Lawrence Livermore National Laboratory (LLNL) for the United States Department of Energy (DOE) at the DOE Nevada Test Site [24, 25].

The overall objective of the SFT-C is to evaluate the feasibility of emplacement and retrieval of used reactor fuel assemblies at a plausible repository depth in a typical granitic rock. The test has two main technical objectives:

- to simulate the effects of thousands of canisters of nuclear waste emplaced in geological media, using only a small number of used fuel assemblies and electrical heaters, and
- to evaluate the difference, if any, between the effect of an actual radioactive waste source and an electrical simulator on the test environment.

There are also some secondary technical objectives:

- to compare the magnitude of displacement and stress effects from mining alone with that of thermally induced displacements and stresses that occur after the used fuel is introduced,
- to document quantitatively the amount of heat removed by mine ventilation,
- to compare the thermal load response of relatively fractured and less fractured rock, and
- to evaluate the performance of geotechnical instrumentation in a simulated repository environment.

In 1977 and 1978, simple heater tests and permeability tests were carried out to assess the in situ properties of the Climax granite. The two heater tests, H-1 and H-2, were designed to measure the thermal properties of the rock in two orthogonal directions. These tests were operated at several times the expected thermal flux and gradient and contained thermal instrumentation only.

The test results indicated that the thermal conductivity of the rock is isotropic and only slightly dependent on temperature. The in situ thermal conductivity was measured as 3.0 W/(m.K), about 15 percent higher than measured in unconfined lab samples. The in situ thermal diffusivity of 1.2 mm^2/s was in good agreement with measurements on intact core samples.

The test site was then reconfigured to perform gas permeability tests at elevated temperatures. All deduced permeabilities were less than 1 ndarcy and consistently decreased with increasing temperature. These in situ data are consistent with laboratory measurements made on other granitic rocks.

Development of the SFT-C took advantage of existing facilities, which included a personnel and materials shaft, hoist and headframe, and associated

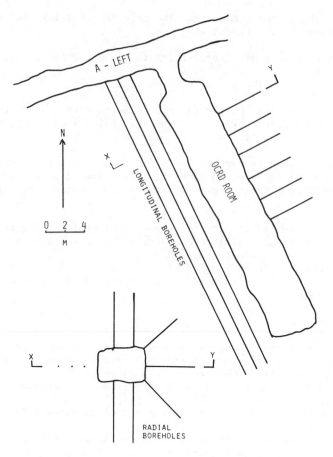

FIGURE 7:   LAYOUT OF OCRD/CMS TEST ROOM (FROM REFERENCE 20)

surface plant. A shaft, 0.76 m in diameter by 420 m deep, was bored to lower
the encapsulated spent fuel from the surface to the underground workings.
Underground construction included driving two 3.4 x 3.4 m heater drifts and a
4.6 x 6.1 m high canister storage drift (see Figure 8). Seventeen canister
emplacement holes, 0.61 m in diameter by 5.2 m deep, were drilled vertically
downward on 3-m centers to accommodate 11 used fuel canisters and 6 electrical
simulators. An eighteenth hole was provided for practice fuel-handling
operations. These holes are lined with 0.46-m diameter carbon steel liners.

Nearly 1000 instruments have been installed to measure temperatures,
displacements, stresses, air quality, radiation doses to granite and personnel,
and acoustic emissions. Data from most of these instruments are collected by a
dual-disc-based HP-1000 minicomputer system.

The program objective of handling and retrieving used fuel was
demonstrated by loading and removing 11 encapsulated, used, commercial power
reactor fuel bundles from the facility.

The thermal and thermomechanical responses of the rock mass to test room excavation, experiment heating and post-experiment cool down have been modelled and have been, or are being, measured. The agreement between models (elastic continuum or fractured rock) of the test room excavation and field measurements during test room excavation was poor. The actual response of the rock to heating agreed well with the finite-element calculations used to estimate the response. The results for the cool-down phase are not yet available. A preliminary conclusion is that the response of the rock mass to heating is less affected by geological structure in the rock mass than is the response of the rock mass to excavation.

Temperature calculations have been in good agreement with field measurements, except for calculations of the heat removed by ventilation.

The temperature and displacements have been calculated for the cool-down period and are being monitored. The results have not yet been published.

The SFT-C has demonstrated the need for a thorough study of the in situ material properties and state of stress to properly analyze experiments. Particular emphasis is being given to measurement of the state of stress, of thermal properties including variation with temperature, and of the rock mass deformation modulus at ambient and elevated temperatures. The actual heat output of the used fuel assemblies will be confirmed by post-test calorimetry to confirm the power generation curves used in the mathematical analysis of the experiments.

An important activity in the post-test program is an examination of the petrographical and chemical composition of the rock around electrically heated boreholes and of similar rock around radiogenically heated boreholes. The program is now in progress.

A major contribution of the SFT-C to in situ experimentation is the extensive testing and evaluation of instruments that have taken place. In general, instrumentation used at SFT-C has been high-quality, off-the-shelf equipment, and performance has been good on all but the vibrating-wire borehole stress meters and one group of potentiometers on multirod extensometers [26]. The mechanical design of the vibrating-wire stress meter was modified to hermetically seal the units against water ingress. Following this modification, the units functioned adequately, but difficulties exist with interpretation of the results. The linear potentiometers on one group of extensometers became nonlinear for an unexplained reason. The failed potentiometers were replaced with four other types of transducers to assess their relative performance in sealed, vacuum purged and dry-nitrogen-flushed head assemblies. Three of the replacement types performed well.

The instruments that performed well include the rod extensometers, the convergence wire extensometers (LLNL design), and the instruments that measured temperature (thermocouples and resistance temperature detectors (RTD)), humidity and air flow. The Hewlett-Packard data acquisition system also performed well.

The following conclusions and recommendations [26] resulted from the SFT-C tests:

- Further research is required to fully understand the observed failures of linear potentiometers used in sealed, or partially ventilated,

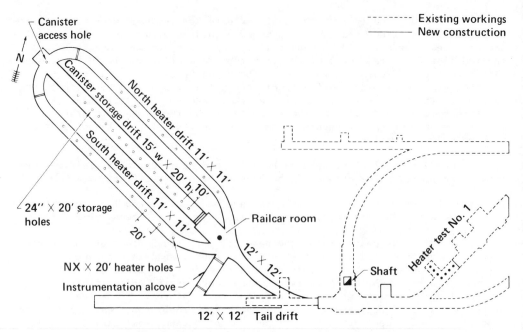

FIGURE 8:   LAYOUT OF SPENT FUEL TEST-CLIMAX
             (FROM REFERENCE 24)

  extensometer head assemblies.
- Based on field observations, LVDTs[4] or proximeters are recommended
  as extensometer transducers where a sealed head assembly is required.
  The user must be cognizant of the potential drift and thermal
  instabilities of the units.
- The first-generation wire extensometers and fracture monitors developed
  for this test are an accurate, reliable means for measuring convergence
  and discrete joint motion, respectively.
- The improved hermetically sealed vibrating-wire stress meters function
  reliably. However, calibration of the gauge remains difficult and
  further work is warranted in this area.
- Utilization of a single lot of sheathed thermocouples in a zone box
  configuration is a cost-effective, accurate, and reliable means of
  measuring temperatures in the repository environment. Care must be
  taken to tailor the sheath composition to the thermal and chemical
  environment.

6.   CONCLUSION

    Each of the projects discussed has made a contribution to understanding
those aspects of waste disposal related to engineered barriers and the geo-
sphere. The objective of in situ testing is to develop sufficient understanding of these systems to make accurate mathematical and physical predictions

---

(4)   Linear Variable Differential Transformer

for use during the licensing process for a disposal vault. This understanding has not been fully developed yet, and further in situ testing is required.

This paper has attempted to review all active subsurface experimental programs in granite environments that support nuclear fuel waste management programs. The material has been summarized from published documents. Errors and omissions are the responsibility of the author.

REFERENCES

[1]   Rummery, T.E., Lisle, D., Howieson, J., and Charlesworth, D.H.: "Radio-active Waste Management Policy and Its Implementation in Canada", Proc. International Conference on Radioactive Waste Management, IAEA, in preparation; also Atomic Energy of Canada Limited Report, AECL-8033, 1983.

[2]   Simmons, G.R. and Soonawala, N.M.; editors: "Underground Research Laboratory Experimental Program", unrestricted, unpublished Atomic Energy of Canada Limited Technical Record, TR-153, 1982. availabe from SDDO, Atomic Energy of Canada Limited Research Company, Chalk River, Ontario, K0J 1J0.

[3]   Simmons, G.R., Brown, A., Davison, C.C., and Rigby, G.L.: "The Canadian Underground Research Laboratory", Proc. International Conference on Radioactive Waste Management, IAEA, in preparation; also Atomic Energy of Canada Limited Report, AECL-7961, 1983.

[4]   Davison, C.C.: "Monitoring Large-Scale Hydrogeological Conditions at the Site of Canada's URL", Proc. Design and Instrumentation of Experiments in Underground Laboratories, CEC and OECD/NEA, Brussels, in preparation.

[5]   Thompson, P.N., Baumgartner, P., and Lang, P.A: "Planned Construction-Phase Geomechanics Experiments at the Underground Research Laboratory" Proc. Design and Instrumentation of Experiments in Underground Laborat-ories, CEC OECD/NEA, Brussels, in preparation.

[6]   Whitherspoon, P.A., Cook, N.G.W., and Gale, J.E.: "Geologic Storage of Radioactive Waste: Field Studies in Sweden", Science, 211, 894-900, (1981).

[7]   Witherspoon, P.A., Cook, N.G.W., and Gale, J.E.: "Progress with Field Investigation at Stripa", Lawrence Berkeley Laboratory Report, LBL-10559, 1980.

[8]   Cook, N.G.W. and Hood, M.: "Full-Scale and Time-Scale Heating Experiments at Stripa: Preliminary Results", Lawrence Berkeley Laboratory Report, LBL-7072/SAC11, 1978.

[9]   Cook, N.G.W., Witherspoon, P.A., Wilson, E.L., and Myer, L.R.: "Progress with Thermomechanical Investigations of the Stripa Site", Proc. Geological Disposal of Radioactive Waste - In Situ Experiments in Granite, pp. 19-31, OECD, Paris, 1983.

[10]   Gale, J.E., Witherspoon, P.A., Wilson, C.R., and Rouleau, A.: "Hydro-
       geological Characterization of the Stripa Site", Proc. Geological Dis-
       posal of Radioactive Waste - In Situ Experiments in Granite,
       pp. 79-98, OECD, Paris, 1983.

[11]   Carlsson, L., Norlander, H., and Olsen, T.: "Hydrogeological
       Investigation in Boreholes", Proc. Geological Disposal of Radioactive
       Waste - In Situ Experiments in Granite, pp. 109-120, OECD, Paris,
       1983.

[12]   Carlsson, Hans S.: "The Stripa Project", Proc. International Confer-
       ence on Radioactive Waste Management, IAEA, Vienna, in preparation,
       1984.

[13]   Fritz, P., Barker, J.F., and Gale, J.E.: "Isotope Hydrology at the Stripa
       Test Site", Proc. Geological Disposal of Radioactive Waste - In Situ
       Experiments in Granite, pp. 133-142, OECD, Paris, 1983.

[14]   Nordstrom, D.K.: "Preliminary Data on the Geochemical Characteristics of
       Groundwater at Stripa", Proc. Geological Disposal of Radioactive Waste
       In Situ Experiments in Granite, pp. 143-153, OECD, Paris, 1983.

[15]   Abelin, H., Gidlund, J., and Neretnieks, I.: "Migration Experiments in a
       Single Fracture in the Stripa Granite. Preliminary Results.", Proc.
       Geological Disposal of Radioactive Waste - In Situ Experiments in
       Granite, pp. 154-163, OECD, Paris, 1983.

[16]   Pusch, R. and Borgeson, L.: "Preliminary Results from the Buffer Mass
       Test of Phase 1, Stripa Project", Proc. Geological Disposal of Radio-
       active Waste - In Situ Experiments in Granite, pp. 173-183, OECD,
       Paris, 1983.

(17)   Carlsson, Hans S.: "The Proposed Phase II of the International Stripa
       Project", Proc. Geological Disposal of Radioactive Waste - In Situ
       Experiments in Granite, pp. 187-196, OECD, Paris, 1983.

[18]   Pfister, E.: "Proposals for In-Situ Research in the Proposed Laboratory
       at Grimsel in Switzerland", Proc. Geological Disposal of Radioactive
       Waste - In Situ Experiments in Granite, pp. 220-229, OECD, Paris,
       1983.

[19]   Pfister, E, and Nold, A.: "The Grimsel Rock Laboratory For In-Situ
       Experiments in Crystalline Rock", Proc. International Conference
       on Radioactive Waste Management, IAEA, Vienna, in preparation, 1984.

[20]   Hustrulid, W.: "CSM/OCRD Hard Rock Test Facility at the CSM Experi-
       mental Mine, Idaho Springs, Colorado", Battelle Memorial Institute/Of-
       fice of Crystalline Repository Development Report, BMI/OCRD - 4(1),
       1983.

[21]   Hustrulid, W. and Ubbes, W.: "Results and Conclusions from Rock
       Mechanics/ Hydrogeology Investigations: CSM/ONWI Test Site", Proc.
       Geological Disposal of Radioactive Waste - In Situ Experiments in
       Granite, pp. 57-75, OECD, Paris, 1983.

[22] Holmberg, R.: "Hard Rock Excavation at the CSM/ONWI Test Site Using the Swedish Blast Design Techniques", Office of Nuclear Waste Isolation Report ONWI 140-3, 1981.

[23] Hardin, E., Barton, N., Lingle, D., Board, M., and Voegele, M.: "A Heated Flatjack Test Series to Measure the Thermomechanical and Transport Properties of In Situ Rock Masses ("Heated Block Test")", Office of Nuclear Waste Isolation Report, ONWI-260, 1981.

[24] Ballou, L.E., Patrick, W.C., Montan, D.N., and Butkovich, T.R.: "Test Completion Plan for Spent Fuel Test - Climax, Nevada Test Site", Lawrance Livermore National Laboratory Report, UCRL-53367, 1982.

[25] Patrick, W.C., Ramspott, L.D., and Ballou, L.B.: "Experimental and Calculational Results from the Spent Fuel Test-Climax", Proc. Geological Disposal of Radioactive Waste - In Situ Experiments in Granite, pp. 45-56, OECD, Paris, 1983.

[26] Patrick, Wesley C., and Rector, Norman L.: "Reliability of Instrumentation in a Simulated Nuclear-Waste Repository Environment" Proc. Field Measurements in Geomechanics International Symposium, Zurich, in preparation, also Lawrence Livermore National Laboratory Report, UCRL-88806, 1983.

[27] "Stripa Project Annual Report, 1982", Swedish Nuclear Fuel Supply Co/ Division Nuclear Fuel Safety, Technical Report 83-02, 1983.

The in situ test program for site characterization of basalt

H.B.DIETZ
Basalt Waste Isolation Project, Rockwell Hanford Operations, Richland, WA, USA

ABSTRACT

The Exploratory Shaft test program has been formulated for in situ characterization of basalt as part of the Basalt Waste Isolation Project. This
paper identifies the tests currently anticipated as required to meet the needs
of the repository license application. The first phase of the Exploratory
Shaft Program is planned to provide access to the preferred candidate repository horizon for limited testing to support Repository Design. The second
phase consists of underground construction and conduct of geologic, hydrologic,
and geomechanics tests which can only be performed in the candidate repository
horizon.

INTRODUCTION

In 1976, the U. S. Energy Research and Development Administration (ERDA),
the predecessor to the U. S. Department of Energy (DOE), established the
National Waste Terminal Storage (NWTS) Program to investigate a number of
geologic rock types to determine their suitability for the disposal of radioactive waste. Its mission was to provide multiple repository facilities in
various deep geologic formations within the United States for the terminal
storage (disposal) of nuclear waste. The Columbia River basalts that underlie
the Hanford Site were among those selected for study. Rock types under investigation in addition to basalt are granite, tuff, and bedded and domal salt.

The DOE has continued the study of sites for mined geologic-disposal
systems suitable for terminal and retrievable storage of commercially
generated high-level and transuranic radioactive wastes (in addition to
examining subseabed disposal as an alternative disposal method) along with
the research and development of technology necessary to ensure the safe long-
term containment and isolation of these wastes. The Nuclear Waste Policy Act
of 1982 requires that the DOE recommend three sites for characterization for
the first geologic repository by January 1985, from among five sites nominated
by DOE prior to that time. The President of the United States will then select
one of the three sites as a candidate site to begin the licensing process.
Rockwell Hanford Operations (Rockwell) is the prime contractor responsible for
the Basalt Waste Isolation Project (BWIP), which is conducting the Hanford
Site investigations.

The goal of the project is to determine if potential geologic repository
sites exist in basalt under the Hanford Site and to identify and develop the
associated facilities and technology required for the permanent isolation of

radioactive waste in one of these potential sites. If feasibility is shown, the DOE may proceed, consistent with the Nuclear Waste Policy Act of 1982, with the detailed design, construction, and operation of a geologic repository (i.e., a Mined Geologic Disposal System) licensed to store nuclear waste within the Columbia River basalts and to conduct the engineering studies required to design such a facility. Studies have been conducted to reduce the area being considered for siting of a Nuclear Waste Repository in Basalt (NWRB) to a reference repository location in the west-central part of the Hanford Site in the State of Washington (see Fig. 1). The Cohasset flow is the preferred candidate repository horizon. The NWTS program guidance calls for the development of test facilities in phased increments that support progressive assessment of the suitability of a reference location for a repository. One such facility planned for the Hanford Site will be an Exploratory Shaft (ES) of a size that provides personnel access to the preferred repository horizon for the purpose of in situ characterization testing to provide data for use in determining site suitability.

This paper will discuss the in situ test program for site characterization of basalt through an ES.

The ES will be developed in two phases. Phase I is to provide access to the preferred candidate repository horizon and conduct preliminary characterization tests. Phase II of the test program includes extensive in situ tests for site characterization. Both Phase I and Phase II of the ES test program must be completed in time to provide input to the repository license application. The in situ tests, conducted as part of Phase I and Phase II of the ES, will be supplemented by laboratory and other field tests.

SITE CHARACTERIZATION

The BWIP goal is to determine, through site characterization, if suitable geologic repository sites exist in the basalt flows beneath the Hanford Site. This site characterization is controlled by regulatory criteria and the Nuclear Waste Policy Act of 1982. Using this guidance, the BWIP is preparing the Site Characterization Plan (SCP) to establish the series of work elements and data needs required for the preparation of repository license application and a site recommendation report (SRR). The ES testing is only one of several data sources identified in the SCP. Figure 2 shows a simplified logic for this site characterization.

The data required to resolve the work elements, identified in the SCP, will be obtained from the following data sources:

Exploratory Shaft

This includes the geologic, hydrologic, geomechanic, and constructibility testing that will be carried out from the ES and underground facility.

Near Surface Testing

The Near Surface Test Facility (NSTF) located in Gable Mountain on the Hanford Site provides underground access to the interior of the Pomona basalt flow. Full-scale heater tests, a large block test, equipment and instrumentation development, and test procedure development are all part of the work at the NSTF.

83

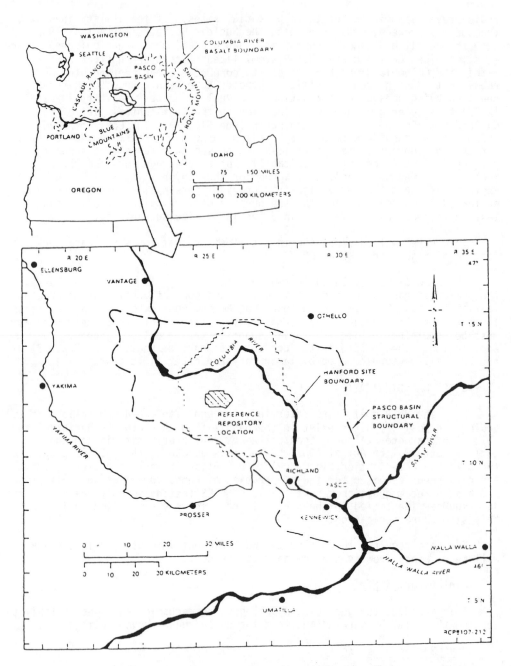

FIGURE 1. Reference Repository Location Map.

84

Surface Boreholes

There are 24 boreholes, of which 15 were continuously cored, that penetrate the candidate repository horizons. Hydrologic testing, geologic logging, geophysical logging, and geomechanics testing are conducted in conjunction with the boreholes.

Surface Exploration

Geologic mapping and sampling, and several types of geophysical surveys are used for characterization from the surface.

Laboratory Testing

This includes geomechanic, water chemistry, geologic, waste package, backfill, and numerous other testing that will be carried out in a laboratory setting.

A series of test plans have been or are being prepared to define the testing and data acquisition in each of these areas.

As data is acquired it is analyzed and integrated with data from other sources. This data analysis can vary from the calculation of a mean value to numerical modeling. The data then goes to design and/or performance assessment and will eventually be incorporated into the SRR.

The final step in site characterization is the preparation of the SRR and repository license application. The data for these documents will come from the design, and performance assessment and directly from the analyzed and integrated data.

EXPLORATORY SHAFT TEST PROGRAM

The Nuclear Waste Policy Act of 1982 establishes the sequence of events leading to the start of repository construction. Section 113 of the Nuclear Waste Policy Act requires characterization of each candidate site and provides for the sinking of exploratory shaft(s) for access to at-depth horizons. The ES Test Program will provide in situ characterization data necessary for the SRR, repository design, and the license application.

The NRC has established criteria (10 CFR 60) governing the disposal of high-level radioactive wastes in geologic repositories. Section 60.10 is pertinent to the scope of the ES:

"(d) The program of site characterization shall be conducted in accordance with the following:

1. Investigations to obtain the required information shall be conducted in such a manner as to limit adverse effects on the long-term performance of the geologic repository to the extent practical.

2. The number of exploratory boreholes and shafts shall be limited to the extent practical consistent with obtaining the information needed for site characterization.

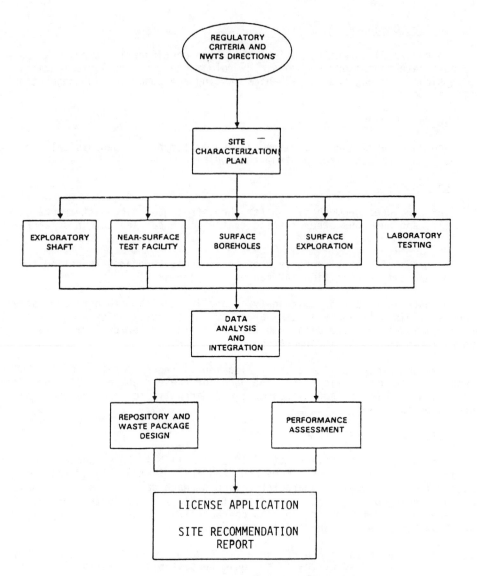

FIGURE 2.   Logic for Site Characterization

3.   To the extent practical, exploratory boreholes and shafts in the geologic repository operations area shall be located where shafts are planned for underground facility construction and operation or where large unexcavated pillars are planned.

4.   Subsurface exploratory drilling, excavation, and in situ testing before and during construction shall be planned and coordinated with geologic repository operations area design and construction."

Section 60.21 of the 10 CFR 60 identifies the site characterization data required to support a license application.

The BWIP has established the data needed to show compliance to criteria identified in 10 CFR 60. The data needed has been allocated to sub-parts of the BWIP.

EXPLORATORY SHAFT - PHASE I

Phase I provides access to the candidate repository horizon(s) to conduct preliminary characterization tests. To accomplish this, a principal borehole has been drilled and tested; surface facilities will be constructed; the ES will be blind bored, lined, and grouted; additional confirmatory data will be obtained from porthole tests; and a shaft station developed in the preferred candidate repository horizon. The Phase I design arrangement is shown in Figure 3.

EXPLORATORY SHAFT - PHASE II

Phase II accomplishes at-depth site characterization testing and provides in situ data needed to support the repository site recommendation, design, and license application. A conceptual arrangement is shown in Figure 4.

The logic governing the management of test data is shown in Figure 5. The logic shows the planned flow of information and activities leading to data reports.

OBJECTIVE

The ES Test Program will be conducted in two phases. The overall objective and the specific technical objectives for each phase are presented in Tables 1 and 2. These objectives summarize the aims or goals of the in situ test program and are a statement of what is to be achieved.

TABLE 1. PHASE I OBJECTIVES

Overall Objective

Provide access to the preferred candidate repository horizon to perform in situ testing.

Technical Objectives

Objective I-1, Principal Borehole. Provide the engineering and geologic data required for the design and selection of porthole locations and ascertain the overall suitability of the proposed location of an exploratory shaft at the Reference Repository Location.

Objective I-2, Shaft Constructibility. Assess the method of blind boring and liner installation for construction of repository shafts at the Reference Repository Location.

Objective I-3, Shaft Liner Sealing. Evaluate the shaft liner seal to assure the safety of personnel from potential water inflow during breakout.

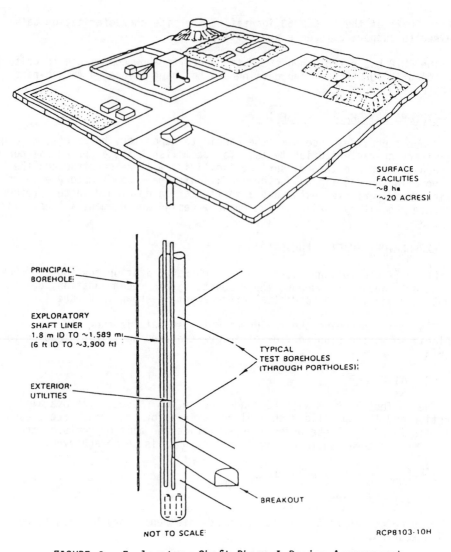

FIGURE 3.   Exploratory Shaft-Phase I Design Arrangement.

Objective I-4, Shaft Station Geohydrology. Assess the geohydrologic properties of the preferred candidate repository horizon to assure the safety of personnel from potential water inflow during breakout.

Objective I-5, Shaft Station Geomechanics. Obtain a preliminary evaluation of opening stability through observation, monitoring, and geomechanics characterization in the shaft station.

Objective I-6, Shaft Station Constructibility. Conduct a preliminary assessment of the method of constructibility of the shaft station at the preferred candidate repository horizon for application to the design of repository rooms.

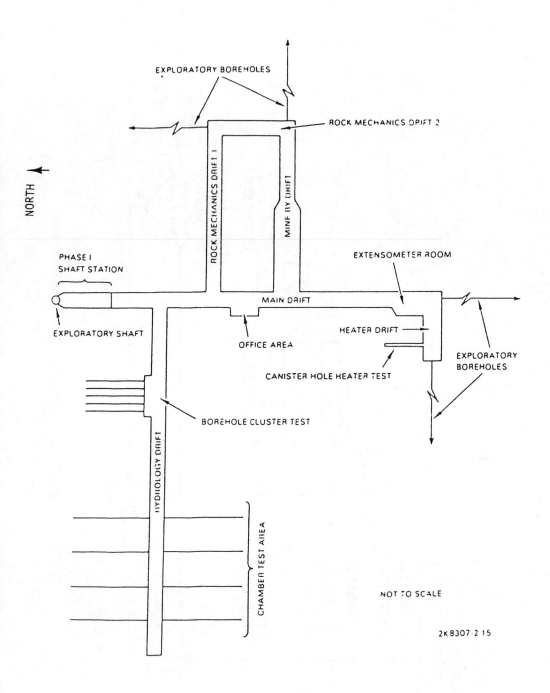

FIGURE 4. Exploratory Shaft-Phase II Conceptual Arrangement.

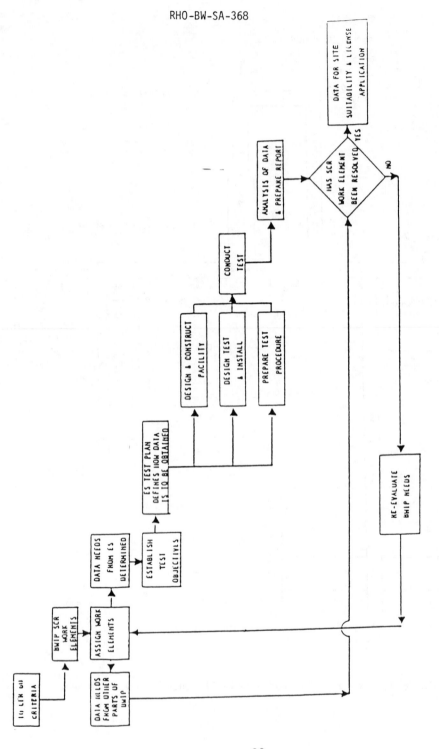

FIGURE 5. Logic: Exploratory Shaft Test Data Management

PLANNED TESTS	PHASE I						PHASE II			
TECHNICAL OBJECTIVES	I-1	I-2	I-3	I-4	I-5	I-6	II-1	II-2	II-3	II-4
	PRINCIPAL BOREHOLE	SHAFT CONSTRUCTIBILITY	SHAFT LINER SEALING	SHAFT STATION GEOHYDROLOGY	SHAFT STATION GEOMECHANICS	SHAFT STATION CONSTRUCTIBILITY	GEOLOGIC CHARACTERIZATION	HYDRAULIC CHARACTERIZATION	GEOMECHANICS CHARACTERIZATION	UNDERGROUND CONSTRUCTIBILITY
GEOLOGIC CHARACTERIZATION										
PRINCIPAL BOREHOLES TESTS	●									
BOREHOLE TESTS				●	○		●	○	○	
FACILITY TESTS					○	●	●	○	○	●
HYDROLOGIC CHARACTERIZATION										
PRINCIPAL BOREHOLE TESTS	●									
BOREHOLE HYDROLOGY TESTS				●				●	○	
CHAMBER TEST								●		
TRACER TESTS								●		
GEOMECHANICS CHARACTERIZATION										
PRINCIPAL BOREHOLE TESTS	●									
OPENING DEFORMATION MONITORING				●	●				●	
OPENING SUPPORT MONITORING				●	●				●	
ACOUSTIC EMISSION MONITORING				●					●	
BOREHOLE JACKING TEST									●	
CROSSHOLE SEISMIC TEST									●	
PLATE BEARING TEST									●	
LARGE FLATJACK TEST									●	
ROOM-SCALE ENLARGEMENT									●	
CANISTER HOLE DRILLING TEST									●	
HEATER TEST									●	
SMALL FLATJACK TEST									●	
OVERCORING TEST					●				●	
HYDRAULIC FRACTURING TEST									●	
CONSTRUCTIBILITY REPORTING		●	●			●				●

LEGEND
● PRIMARY DATA   ○ SUPPORTING DATA

TABLE 1.   Exploratory Shaft Objectives and Planned Tests.

TABLE 2. PHASE II OBJECTIVES

Overall Objective

Conduct in situ testing to obtain information necessary for repository
site suitability recommendation; development and design; and license application.

Technical Objectives

Objective II-1, Geologic Characterization. Assess the geologic characteristics and their variability at the preferred candidate repository horizon in
the vicinity of the ES to determine site suitability.

Objective II-2, Hydrologic Characterization. Assess the hydrologic
properties of the ES site to help evaluate, through performance assessment
using all data sources, the isolation capability of basalt to meet the proposed
regulatory criteria.

Objective II-3, Geomechanics Characterization. Determine through testing
and performance evaluation that stability of openings at the preferred candidate
repository horizon can be maintained through the operational life of the
repository without adversel affecting isolation capabilty and retrievability.

Objective II-4, Underground Constructibility. Assess the method of constructibility of the Phase II excavations at the Reference Repository Location
for application to final design of repository rooms.

TEST CHOSEN

The tests selected to meet the technical objectives are shown in
Table 1.

Expériences en place en terrains argileux

R.H.HEREMANS
ONDRAF/NIRAS, Brussels, Belgium

RESUME

Plusieurs pays ont pris en considération les formations pélitiques comme milieu récepteur pour l'évacuation de certains types de déchets radioactifs. Toutefois pour les hautes activités, elles n'ont connu jusqu'à présent qu'un intérêt limité. Si quelques laboratoires en Amérique du Nord et en Europe ont lancé des programmes de R&D, la plupart cependant se sont limités à la caractérisation d'échantillons et à quelques expériences de transfert de chaleur à faible profondeur. Deux exceptions toutefois, l'Italie qui poursuit ses recherches depuis de nombreuses années présente maintenant un projet de construction d'un laboratoire souterrain dans une couche d'argile marneuse en Sicile et la Belgique qui vient de terminer un tel ouvrage à 240 m de profondeur dans une couche d'argile plastique.
Quelques détails sur les travaux de construction ainsi que sur les programmes de mesures et d'expériences en place, prévus ou en cours sont donnés dans le présent document.

ABSTRACT

Several countries have considered pelitic formations as host medium for some radioactive waste diposal. However, as far as high activities are concerned, they have only encountered limited interest. Some laboratories in North America and in Europe have launched R&D programmes, but most of them were limited to sample characterisation and some heat transfer experiments at shallow depth. Two exceptions however, Italy which has pursued, its investigations since several years now develops a construction project for an underground laboratory in a marly clay in Sicily and Belgium which has just completed such a facility at 240 m depth in a plastic clay layer. Some details concerning the construction and the measurement and in place experiments forseen or underway are given in the present paper.

GENERALITES

Les formations pélitiques en général, qu'elles se caractérisent comme schistes indurés ou argiles plastiques, ont été prises en considération dès les années 50 en vue de l'évacuation de certains déchets radioactifs.
Ces formations sont intéressantes du fait de leur faible perméabilité qui leur confère le pouvoir d'une "barrière hydrodynamique"et de leur bonne capacité de rétention leur permettant d'assurer le rôle de "barrière géo-

93

chimique" à la migration de bon nombre de radio-éléments.

C'est surtout dans le cadre de l'évacuation à faible profondeur de déchets de basse radioactivité que de nombreux travaux de recherche en laboratoire ont été entrepris sur des matériaux argileux et divers site d'évacuation actuellement en service tirent avantage des propriétés "retard" de ces formations.

Un exemple typique, et unique à notre connaissance, dans le monde occidental est celui du système d'évacuation de liquides faiblement radioactifs par fracturation hydraulique développé à la fin des années 50 par les "Laboratoires Nationaux d'Oak-Ridge" aux E.U. et appliqué à l'échelle industrielle depuis de très nombreuses années dans des conditions de sûreté tout-à-fait acceptables [1] [2]. La formation pélitique mise à contribution dans ce cas est un schiste paléozoïque ayant de l'ordre de 400.10^6 d'années d'âge. Lorsque l'on passe en revue les nombreux programmes de R&D sur l'évacuation à grande profondeur des déchets de haute radioactivité on constate que les matériaux argileux ne furent pas jusqu'à présent l'objet d'investigations aussi avancées que celles que connurent les formations salines et les roches granitiques. Il y a, sans nul doute, plusieurs raisons à cela. Celle la plus souvent évoquée est leur comportement mal connu vis-à-vis de la température mais la raison principale n'est-elle pas que les matériaux argileux présentent une composition et une structure qui peuvent différer considérablement suivant leur origine, leurs conditions de sédimentation, leur âge, l'épaisseur de leur couverture etc.

Dans une même formation les caractéristiques de la roche peuvent varier d'un endroit géographique à un autre et dans ce cas plus que dans tout autre une extrapolation de résultats obtenus sur un site expérimental vers un autre est plus que hasardeux.

Etablir un bilan des recherches entreprises en vue d'évaluer les possibilités d'évacuation de déchets hautement radioactifs donc émetteurs de chaleur, dans des matériaux argileux, n'est pas le but de cet exposé. Il est toutefois intéressant de rappeler ici les résultats d'une enquête faite par l'AEN en 1979 dont il est ressorti que la Belgique, le Canada, l'Italie, la Suisse, le Royaume-Uni ainsi que les Etats-Unis avaient tous dans ce domaine, lancé soit des programmes de caractérisation d'échantillons et des recherches diverses en laboratoire ou "in situ", soit des études d'évaluation de réalisations techniques et ou de sûreté.

Les formations considérées étaient très différentes les unes des autres, dans l'échelle des temps géologiques elles se situaient entre l'Ordovicien (500.10^6 d'années) et le Néogène (30.10^6 d'années). En Belgique, en Italie et aux Etats-Unis les expériences "in situ" à faible profondeur (jusqu'à $\sim$ 30 m) se rapportaient toutes au transfert de chaleur et phénomènes associés. Dans ce but, des sources thermiques simulant le conteneur de déchets vitrifiés de haute radioactivité avaient été enterrés et entourés de divers appareils de mesure placés suivant différentes configurations géométriques. Quelques informations particulières à chacune de ces expériences sont données dans le Tableau I, les numéros dans la dernière colonne sont repris dans la liste bibliographique se trouvant en dernière page de ce document.

Une étape importante dans la recherche sur l'évacuation en formation géologique profonde est celle qui permet la réalisation d'expériences "en place", celle qui permet d'explorer le massif "en direct", de suivre son comportement "de visu". Plusieurs pays ont eu la possiblité de s'engager dans cette voie en créant des laboratoires souterrains. Il est incontestable que ceux-ci contribueront largement à une solution sûre et à long terme du problème de l'élimination des déchets radioactifs en particulier mais aussi au développement de la science et de la technique en général. Divers laboratoires souterrains dans le sel, le granit, le basalt sont en

construction ou en exploitation dans des pays membres de l'Agence de l'
Energie Nucléaire.   La construction d'un laboratoire souterrain dans une
formation argileuse reste toutefois et à ce jour l'apanage de la Belgique
et du Centre d'Etude de l'Energie Nucléaire en particulier.   On peut espérer néanmoins, que dans le courant de cette même année 1984, le Comité
National pour la Recherche et le Développement de l'Energie Nucléaire et
des Energies Alternatives (ENEA) pourra en Italie mettre son projet de laboratoire souterrain dans une formation d'argile marneuse en exécution.
La suite de ce document traitera donc exclusivement de ces deux installations.

PROJET DE CONSTRUCTION D'UN LABORATOIRE SOUTERRAIN DANS LA MINE DE PASQUA-
SIA. (Contrat entre ENEA et ISMES)

Pour diverses raisons les formations argileuses ont été considérées comme
pouvant apporter une solution acceptable pour l'évacuation de déchets conditionnés en Italie.   Les premiers travaux de recherche furent concentrés sur
le site de la Trisaïa dès la fin des années 60. [8]   Par après la caractérisation d'échantillons en provenance d'autres formations argileuses a été
entreprise mais un consensus s'est rapidement créé autour de la nécessité
de disposer aussi d'une installation souterraine permettant la mise au
point de techniques d'investigation "in situ" et la réalisation d'expériences et de mesures susceptibles d'apporter des informations précieuses pour
le développement ultérieur éventuel d'un système d'évacuation de déchets
conditionnés de haute radioactivité dans une formation argileuse comapra-
ble.
Compte tenu des possibilités pratiques du moment le choix s'est porté sur
le site de la mine de PASQUASIA en Sicile.

Géologie du site et construction du laboratoire.

La formation argileuse connue sous le nom "argile bleue" est d'âge Pliocène ;
au droit du site l'épaisseur de la couche est comprise entre 90 et 100 m.
Les sables supérieurs sont argileux et d'âge Pliocène supérieur, la couche
sous-jacente est un calcaire Pliocène marneux.   Une coupe stratigraphique
simplifiée est donnée dans la figure 1.
L'argile bleue est une argile marneuse contenant un faible pourcentage de
silt, 50 % de sa composition est constituée de minéraux argileux du type
illite et kaolinite, la teneur en carbonate varie entre 25 et 30 %. Le matériau est rigide, sa résistance au cisaillement non drainé est de 10
kg/cm^2, la cohésion varie entre 1 et 1,5 kg/cm^2 et l'angle de friction entre
30 et 34°.   Au niveau de la future galerie, la pression totale verticale du
terrain (pression lithostatique) est de 36 kg/cm^2 tandis que la pression
effective est de 24 kg/cm^2 soit une différence entre les deux de 12 kg/cm^2
correspondant à la pression de l'eau interstitielle.   Des déterminations
plus précises et plus complètes des propriétés géomécaniques d' échantillons
prélevés sur place sont bien entendu prévues.
La préconsolidation du terrain n'est pas connue mais elle a du être importante
et la formation est intersectée par divers systèmes de joints et montre
des signes de glissement de plans.
La construction de la galerie se fera à partir du plan incliné d'accès à la
mine au niveau -160 m.   La zone expérimentale de 25 m sera separée du plan
d'accès par une galerie tampon de 36 m.   Une chambre de 6 m de longueur et de
section légèrement plus grande destinée ultérieurement à abriter l'instrumentation scientifique de controle et de mesure assurera la liaison entre galerie
tampon et galerie laboratoire.   L'orientation de la galerie est vers le Nord

Pays	Opérateur	Formation	Site	Source	Températ.	Ref.
E.U.	SANDIA	Canasauga	TENNESSEE	φ 0,3 m h 3 m KW 3,5	∿ 385 ℃	3
	SANDIA	Eleana Argillite	NEVADA	φ 0,3 m h 3 m KW 3,8	∿ 350 ℃	4
Italie	CNEN	Pliocène-Cala-brian	TRISAIA	φ 0,2 m h 2 m KW 1,2	∿ 200 ℃	5
	ENEA	Plio-Pléisto-cène	MONTE RO-TONDO	φ 0,06 m h 6 m KW 1,0	∿ 200 ℃	6
Belgique	CEN/SCK	Argile de Boom	TERHAEGEN	φ 0,3 m h 1,5 m KW 2,5	∿ 300 ℃	7

Tableau I

Expériences thermiques "in situ" dans des matériaux argileux

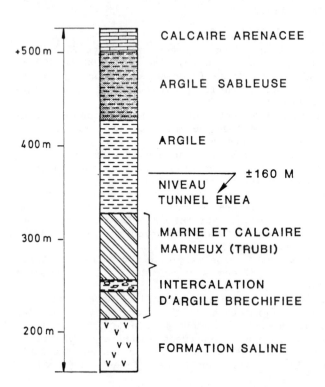

FIG.1 COUPE STRATIGRAPHIQUE SIMPLIFIEE DU SITE DE LA PASQUASIA

et perpendiculairement à la direction des couches. La localisation du laboratoire souterrain dans la formation géologique, sa voie d'accès ainsi que ses dimensions sont données dans la figure 2.

Le revêtement prévu pour la galerie est composé de cintres métalliques type NP16 espacés de 60 cm, d'un treuilli de 50 x 50 x 4 mm le tout recouvert d'une couche de béton projeté de 10 cm d'épaisseur et probablement d'une couche de finition en ciment.

Programme expérimental. (contrat entre ENEA et ISMES).

Un premier programme expérimental consistera en la mesure d'une série de paramètres mécaniques propres au matériau argileux "en place", en cours et après creusement. L'instrumentation prévue est composée de cellules de mesure de pression totale, de mesure de pression d'eau dans les pores, d'inclinomètres et d'extensomètres .

Plus de détails concernant ce programme sont donnés dans un autre document présenté au cours de cette réunion [9]. Il est donc superflu de s'étendre d'avantage sur ce sujet de même que sur les dispositifs d'acquisition de données.

Dans une seconde série d'expériences l'ENEA à prévu l'étude des effets thermiques sur le massif argileux, le programme est actuellement en élaboration. Ci-après quelques premières précisions :

- la source de chaleur sera un élément électrique de 1,5 KW installé dans un cylindre de 2 m de hauteur et de 0,10 m de diamètre ;

- une soixantaine de thermomètres à résistance permettront la mesure de température à différents endroits dans le massif (jusqu'à 2 m de distance de la source chaude) la précision de la mesure étant le 1/10° de degré ;

- quatre piézomètres électriques mesureront la pression d'eau interstitielle à quatre endroits différents, ils permettent des mesures jusqu'à 200°C ;

- les variations de pression totale seront captées par six cellules hydrauliques précises à 0,5 % entre 0 et 50 bar ;

- plusieurs autres dispositifs permettront la mesure en divers points des déformations verticales, des déformations en surface et de la convergence de trous de forage.

D'autres expériences succèderont à celles décrites ci-avant et tout comme pour les autres programmes elles se poursuivront durant plusieurs années.

REALISATION D'UN LABORATOIRE SOUTERRAIN DANS L'ARGILE DE BOOM (CEN/SCK - Mol).

En Belgique la décision de construire un laboratoire souterrain date de la fin des années 70, c'est-à-dire de la fin du premier plan quinquennal de R&D sur les possibilités d'évacuation de déchets radioactifs dans une couche d'argile plastique d'âge Tertiaire située dans le N.E. du pays. Ce qui distinguat ce projet de la plupart des autres c'est que le laboratoire souterrain serait construit sur un site qui ultérieurement pourrait être pris en considération pour la construction d'une installation finale d'évacuation. Le site étant celui du Centre d'Etude de l'Energie Nucléaire, il n'y eut pas de problèmes pour l'obtention des autorisations administratives de construction et d'exploitation.

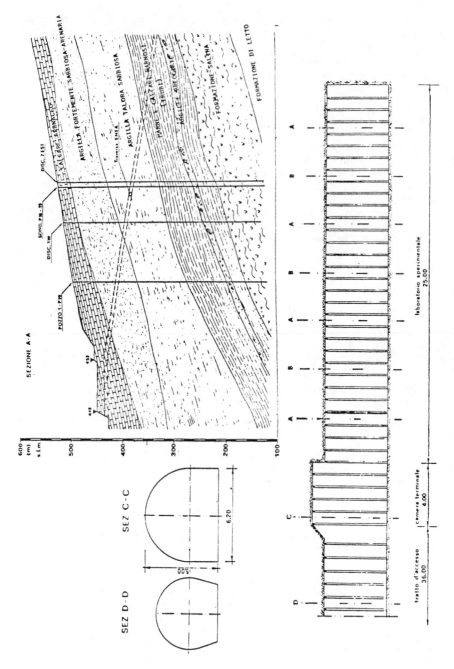

FIG. 2. — IMPLANTATION, VOIE D'ACCES ET DIMENSIONS DU LABORATOIRE SOUTERRAIN DE LA PASQUASIA (ITALIE)

Géologie du site et construction du laboratoire.

La formation argileuse de Boom, retenue dans l'inventaire national des formations géologiques susceptibles de convenir pour l'évacuation des déchets conditionnés de haute radioactivité et d'activité α, est d'âge Tertiaire et située dans l'Oligocène, elle s'étend en profondeur dans le N.E. du pays sur une épaisseur de l'ordre de 100 m. Elle est recouverte par des sables d'âge Miocène, Pliocène et Pléistocène tandis que les sables inférieurs sont également d'âge Oligocène. Ces sables sont aquifers et certaines des nappes sont même exploitables.

Une coupe stratigraphique simplifiée est donnée dans la figure 3. L'argile de Boom est compacte et homogène, 50 % environ de sa masse a une granulométrie inférieure à 2 μm et seulement 0,5 % une granulométrie supérieure à 200 μm. Les principaux minéraux argileux sont l'illite, la smectite, un minéral du type vermiculite et divers interstratifiés entre autres de chlorite. La teneur en carbonate est de l'ordre de 1,5 % et celle en matières organiques d'environ 3,5 %. Par comparaison avec l'argile marneuse de la Pasquasia l'argile de Boom à une résistance au cisaillement non drainé comprise entre 3,3 et 8,6 kg/cm^2, une cohésion moyenne de 1,1 kg/cm^2 et un angle de friction moyen de 19°. C'est donc un matériau plus plastique et moins résistant que celui de la Pasquasia. La pression verticale attendue était de l'ordre de 45 kg/cm^2 et la pression effective de 22 kg/cm^2. Des déterminations précises d'un grand nombre de propriétés géomécaniques d'échantillons prélevés lors de forages de reconnaissance ou en cours de construction ont été effectuées et les résultats comparés à ceux obtenus sur la même argile mais dans la zone d'affleurement [10]. La zone retenue pour l'emplacement du laboratoire avait été précédemment prospectée par une "sismique réflexion" et les formations concernées, trouvées exemptes de failles ou flexures. Les travaux sur terrain débutèrent en février 1980. Compte tenu des nappes aquifères à traverser pour atteindre la couche d'argile la technique de congélation préalable du terrain pour la construction du puits vertical d'accès fut retenue. De même, dès l'étude du projet il fut envisagé de congeler l'argile avant le creusement de la galerie horizontale, ceci parce qu'aucune réalisation de ce genre n'avait été entreprise antérieurement. Le bouchon de fond de galerie fut coulé fin juillet 1983. Schématiquement l'ensemble se présente comme le montre la figure 4. Des détails sur les phases de construction de l'ouvrage sont donnés dans quelques publications récentes dont deux sont reprises en référence [11] [12].

Programme expérimental.

Le manque d'information et d'expérience sur le comportement rhéologique d'une argile plastique à grande profondeur a amené le CEN/SCK à élaborer, en collaboration avec l'entrepreneur, l'architecte industriel et certains laboratoires de mécanique des sols, un vaste programme de mesures géotechniques "en place" et à attribuer la priorité à l'exécution de ce programme. Il convenait en effet de tenter de lever en tout premier lieu l'incertitude règnant sur les possibilités de creusement à l'horizontale dans l'argile avec des moyens technologiques économiquement acceptables. Le double objectif suivant fut fixé. Dès atteinte de la zone de transition entre sables aquifers et argile une instrumentation de mesure de pressions et de températures serait mise en place dans la zone gelée en vue de suivre la décongelation ultérieure du terrain et les déformations du revêtement qui en résulteraient éventuellement. Diverses techniques de mise en place des instruments et différents instruments pour la mesure des mêmes paramètres seraient utilisés en vue de sélectionner les conditions optimales pour de bonnes mesures autour

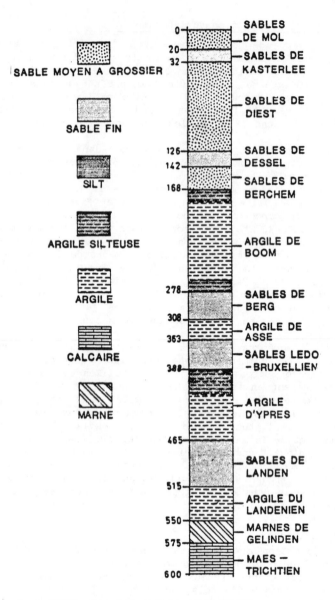

SABLE MOYEN A GROSSIER

SABLE FIN

SILT

ARGILE SILTEUSE

ARGILE

CALCAIRE

MARNE

0
20
32
126
142
168
278
308
363
388
465
515
550
575
600

SABLES DE MOL
SABLES DE KASTERLEE
SABLES DE DIEST
SABLES DE DESSEL
SABLES DE BERCHEM
ARGILE DE BOOM
SABLES DE BERG
ARGILE DE ASSE
SABLES LEDO - BRUXELLIEN
ARGILE D'YPRES
SABLES DE LANDEN
ARGILE DU LANDENIEN
MARNES DE GELINDEN
MAES - TRICHTIEN

FIG. 3. COUPE STRATIGRAPHIQUE SIMPLIFIEE DU SITE DE MOL

100

de la galerie horizontale. Afin de pouvoir suivre l'influence de la profondeur sur le comportement de l'argile ces instruments furent implantés à 5 niveaux différents, $\simeq$ 190, $\simeq$ 200, $\simeq$ 206, $\simeq$ 214 et $\simeq$ 227 m. La figure 5 représente schématiquement l'équipement instrumental d'une auréole de mesure. Les problèmes rencontrés lors de la mise en place et des campagnes de mesure font l'objet d'un autre document présenté au cours de cette même réunion de travail [13]. Dans et autour de la galerie horizontale de nombreux appareils ont également été mis en place, pendant ou immédiatement après sa construction. Il s'agit d'une instrumentation similaire à celle présentée dans la figure 5 : cellules de pression totale et de pression interstitielle, sondes thermiques et pour la mesure de déformation des voussoirs en fonte, des jauges de contrainte. Cet ensemble instrumental est maintenant en opération depuis plusieurs mois, il sera automatisé progressivement et dans la mesure des possibilités techniques. L'interprétation des résultats est en cours ainsi que la mise en corrélation avec les résultats obtenus en laboratoire en ce qui concerne les propriétés géotechniques d'échantillons. Comme il a déjà été mentionné, un des objectifs principaux d'un tel laboratoire est de pouvoir suivre le comportement du massif en place. Une très belle expérience et d'intérêt capital a ainsi pu être réalisée fin 1983. En vue d'étudier les possibilités de creusement dans l'argile non gelée à 240 m de profondeur il fut décidé de construire un puits vertical d'environ 20 m de longueur et de 2 m de diamètre au fond de la galerie horizontale. En vue du creusement, un instrumentation de mesure de déformations a été mise en place, elle se composait de 4 inclinomètres et de 3 extensomètres. En cours de construction, des cellules de pression totale furent également mises en place pour la mesure des pressions radiales ainsi que des vérins plats destinés aux mesures de pression dans le cuvelage constitué de blocs ou claveaux en béton à haute résistance. Cette construction est représentée sur la figure 4.

Dans le domaine de la géotechnique, l'expérience suivante sera la construction, toujours dans l'argile non gelée, à partir du fond de ce puits vertical de 1,60 m de diamètre utile, d'une petite galerie horizontale de même diamètre et de quelques mètres de longueur. On espère que sur la base de ces expériences pratiques et des résultats fournis par les nombreux appareils de mesure en place, il sera possible de définir dans une première phase les caractéristiques principales d'une machine qui permettra elle des essais de creusement à l'horizontale et à grand diamètre dans une argile telle que l'argile de Boom à grande profondeur.

Il convient encore d'ajouter ici que le creusement successif du puits d'accès (environ 30 m dans l'argile), de la grande galerie horizontale (36 m de longueur), du petit puits vertical (20 m de longueur) et de la petite galerie horizontale (quelques m) aura permis une reconnaissance poussée du massif grâce à l'examen visuel mais aussi grâce aux multiples analyses effectuées en laboratoire sur échantillons prélevés en cours de creusement.

A côté de ces investigations relatives à la mécanique des sols, d'autres recherches sont concernées par ce laboratoire souterrain, pour cette raison le revêtement de la galerie principale a été pourvu d'un certain nombre d'orifices obturés qui donnent accès au massif d'argile. Trois diamètres différents ont été prévus 76, 12,5 et 5 cm. Les expériences en développement concernent actuellement la corrosion de matériaux métalliques et de matrices de déchets. Un autre document présenté au cours de cette réunion donne plus de détails sur l'instrumentation construite dans ce but [14]. Les expériences de corrosion seront mises à profit pour une première évaluation de la valeur de certaines propriétés thermiques du massif argileux.

Dès à présent on travaille déjà à la préparation d'une expérience thermique

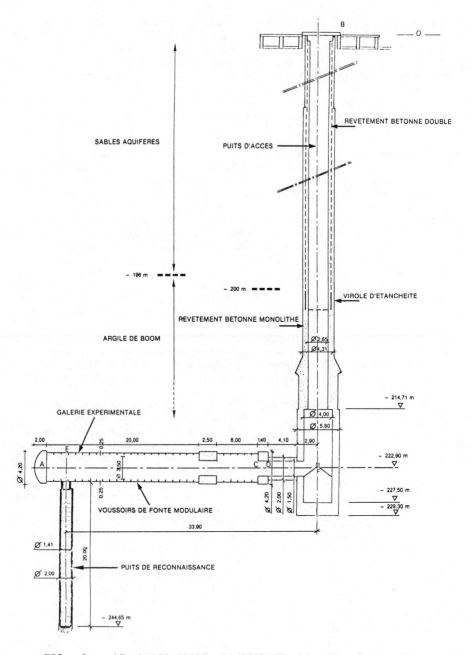

FIG. 4. LE LABORATOIRE SOUTERRAIN DE MOL (BELGIQUE)

102

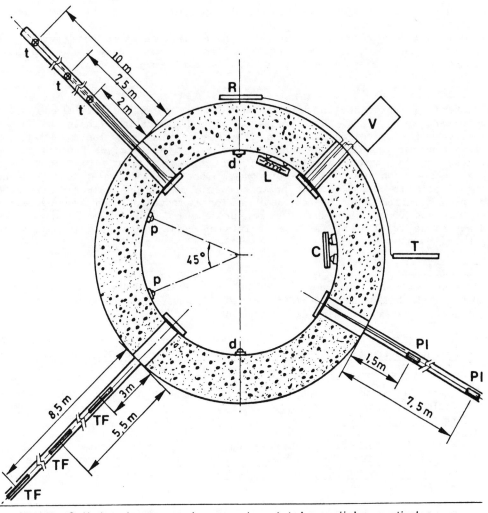

R,V,T = Cellules de mesure des pressions totales radiales, verticales, tangentielles

t = Thermistances

P I = Cellules de mesure des pressions interstitielles

TF = Tubes fendus pour essais dilato-pressiométriques

p,d = Plots pour mesures perimétrales et diamétrales

L,C= Extensomètre à variation d'inductance à corde vibrante

FIG. 5 EXEMPLE D'APPAREILS DE MESURE   DISPOSES AU
DROIT DES AUREOLES DE MESURE

de plus grande envergure, de mesures de mouvements éventuels d'eau dans l'argile, d'expériences de migration etc.

A plus long terme une extension de la galerie souterraine est envisagée en vue de la réalisation d'expériences à caractère technologique : manipulation de fûts, mise en place de matériaux de remplissage et ainsi de suite.

CONCLUSIONS

L'intérêt que présente la disponiblité d'une zone expérimentale en profondeur dans le type de roche envisagée pour l'évacuation finale de déchets radioactifs est indiscutable et a déjà été demontré par les résultats obtenus dans des installations comme celles de Stripa, Asse, Hanford ou autres. Compte tenu du manque d'information sur le comportement plus spécialement des argiles non indurées en profondeur et vis-à-vis d'actions externes que ce soit le creusement, la présence d'une source de chaleur ou de matériaux altérables cet intérêt se transforme indubitablement en nécessité. On ne peut donc qu'espérer que les projets en cours dans ce domaine et d'autres se réaliseront dans les délais et suivant les programmes élaborés par les chercheurs.

REMERCIEMENTS

L'auteur remercie Messieurs A. BRONDI de l'ENEA (Italie), P. MANFROY, A. BONNE et B. NEERDAEL du CEN/SCK (Belgique) pour les informations fournies qui ont permis la rédaction de ce document.

[1] Weeren H.O.: "Waste disposal by Shale fracturing at ORNL", Proceedings of the International Symposium on "The Underground Disposal of Radioactive Wastes". (Otaniemi/Finland 2-6 july 1979) IAEA - Vienna 1980.

[2] Final Environmental Impact Statement, Management of Intermediate Level Radioactive Waste
ERDA - 1553 (1977)

[3] Krumhansl, Sundberg : "The Conasauga near Surface Heater Experiment" Proceedings of the NEA workshop on the "Use of argillaceous materials for the isolation of radioactive waste - Paris 10-12 sept. 1979 (p 129) - ISBN 92664-02040.0

[4] McVey, Lappin and Thomas. : "Test Results and Supporting Analysis of a near Surface Heater Experiment in the Eleana Argillite". Proceedings of the NEA workshop on the "Use of argillaceous materials for the isolation of radioactive waste - Paris 10-12 sept. 1979 (p 93) - ISBN 92-64602050-0

[5] Tassoni E. : "An Experiment on the heat transmission in a clay rock" Proceedings of the NEA workshop on the "Use of argillaceous materials for the isolation of radioactive waste" - Paris 10-12 Sept. 1979 (p. 23) - ISBN 92664-02050-0

[6] Tassoni E. : "In situ and laboratory heating experiments in clay Oral communication : Technical Committee Meeting on "Effects of Heat from High-Level Waste on Performance of deep Geological Repository components" - IAEA - Stockholm, 28/08-2/09-1983.

[7] Heremans, R., Buyens, M., Manfroy, P. : "Le comportement de l'argile vis-à-vis de la chaleur". Proceedings of the NEA workshop on "In situ heating experiments in geological formations" - Ludvika/Stripa, 13-15 Sept. 1978

[8] "Relazione sugli Studi effecttuati sulla zona di Pantanello (Golfo di Taranto)". CNEN/EURATOM contratto 004665-3 WASI - Parte IV vol.1.

[9] Bruzzi, D., Gera, F., (ISMES - Italy) : "Geotechnical Instrumentation for Field Measurements in Deep Clays" - Proceedings of the CEC/NEA workschop on "Design and Instrumentation of in situ Experiments in Underground Laboratories for Radioactive Waste Disposal" Brussels 15-17 May 1984.

[10] Neerdael, Manfroy, Heremans R., : "Caractérisation géomécanique de l'argile de Boom." Proceedings of a technical session on rock mechanics "Advance in Laboratory sample testing" CCE - Brussels 27.04.83. A paraître

[11] Manfroy P., : "Rejet des déchets radioactifs en formation argileuse profonde - Construction d'un laboratoire souterrain pour un programme expérimental approfondi". Proceedings of the International Conference "Radioactive Waste Management" paper IAEA-CN 43/54 vol. 3 (Seattle/USA 16-20 May 1983) - IAEA - Vienna 1984.

[12] Funken, R., Gonze, P., Vranken, P., Manfroy, P., Neerdael, B., : "Construction of an experimental laboratory in a deep clay formation" Eurotunnel 83 Conference, paper 9, Basle (Switzerland). June 22-24, 1983

[13] Manfroy P., Neerdael B.,"Expérience acquise à l'occasion de la réalisation d'une campagne géotechnique dans un argile profonde". Proceedings of the CEC/NEA workshop on "Design and Instrumentation of in-situ Experiments in Underground Laboratories for Radioactive Waste Disposal" Brussels 15-17 May 1984.

[14] Casteels F., De Batist, R., Kelchtermans, J., Dresselaers, J., Timmermans, W., "In situ Testing and Corrosion Monitoring in Geological Clay Formation. Proceedings of the CEC/NEA workshop on "Design and Instrumentation of in-situ Experiments in Underground Laboratories for Radioactive Waste Disposal". Brussels 15-17 May 1984.

A review of in situ investigations in salt

KLAUS KÜHN
GSF, Institut für Tieflagerung, Braunschweig, FR Germany

ABSTRACT

In situ investigations for the disposal of radioactive wastes in rock salt formations have the longest history in the field. Well known names are Project Salt Vault (PSV) which was performed in the Lyons Mine, Kansas/USA, and the Asse salt mine in Germany.
The overall objective for in situ investigations is twofold :

1. To produce all necessary data for the construction and operation of repositories and

2. to produce all necessary data for a performance assessment for repositories.

1. Major events

Major events of in situ investigations in salt occurred in the following years :

1957   NAS/NRC Study in US on Radioactive Waste Disposal:
Identifies the Deep Geological Disposal in Salt Formations.
This was the first study in the world proposing disposal of radioactive wastes in geological formations.

1963   Tests by ORNL for USAEC in the Hutchinson Salt Mine, Kansas, for direct disposal of liquid high-level waste.
This was only a very short test, the idea of which was no longer pursued later on.

1964/65   Federal Republic of Germany decided for radioactive waste disposal in salt : Purchase of the Asse salt mine for research and development.
This mine was and is extensively used for in situ investigations.

1965-69   Project Salt Vault in the Lyons Mine, Kansas, performed by ORNL for USAEC.
This was one of the most successful in situ-tests which used spent MTR-fuel elements instead of solidified high-level wastes. In spite of this fact, its objective, however, was not the demonstration of the disposal of spent fuel.

1967-78 Disposal of low-level radioactive waste in the Asse salt mine.
About 120,000 containers -mainly 200 1 steel drums- were
successfully disposed of during this period.

1968 First experiment with electrical heaters in the Asse salt mine.
This was the first test in a series which has now come to
number six.

1972-77 Disposal of intermediate-level radioactive waste in the Asse
salt mine.
About 1,300 drums with solidified ILW were successfully
disposed of with shielded handling during this period.

1979-82 Brine Migration Tests in the Avery Island Salt Mine, Louisiana,
performed by RE/SPEC for ONWI.
These tests looked into the mechanisms of liberation of water
which is included in salt in minor amounts.

1979 Dry drilling of a 300 m borehole by ECN in the Asse salt mine
with subsequent convergence measurements and heating phases.
This Dutch experiment in the Asse salt mine looked into engineering aspects of drilling such holes and into creep behaviour of
rock salt at normal and elevated temperatures.

1982 Drilling of two shafts and subsequent mining at the WIPP site,
Carlsbad/New Mexico, by Sandia Laboratories for DOE with
forthcoming in situ-experiments.
This is the first construction in the world of a specially
designed underground laboratory and later repository.

1983 Start of US-German cooperative Brine Migration Test in the Asse
salt mine with major radioactivity.
This experiment is using for the first time large amounts of
radioactivity (about 36,000 Ci of Co-60) to investigate the
simultaneous effects of heat and radiation on rock salt.

1986-87 Exploratory shaft; Test and Evaluation Facility in the US.
At least one site in salt will be explored during this time
according to the US Waste Policy Act of December 1982.

1987 Start of test disposal of high-level radioactive glass sources
in the Asse salt mine.
30 glass blocks spiked with a total amount of some 13 MCi of
Cs-137 and Sr-90 will be used in this test in order to check-
out a complete handling and emplacement system for high-level
waste and in order to investigate a broad spectrum of detailed
scientific questions.

2. Major objectives of in situ experiments

The major objectives of in situ experiments in salt are :

. Development and test of disposal techniques for different types
  or categories of radioactive wastes

107

. Procurement of data and parameters for design and construction of a repository

. procurement of techniques and procedures for shut down and sealing of a repository

. Procurement of input data and parameters for computer models in order to predict the long-term behaviour of a repository

. Procurement of data and parameters for safety analysis or performance assessment

. (Site investigation, characterization and confirmation)

In general, these objectives are the same for all types of geological formations. There are, however, some special aspects for salt.

2.1 Disposal techniques

. Low-level radioactive wastes (contact handled)

. Intermediate-level radioactive wastes (remotely handled)

A number on if situ tests have been and will be performed for the development of disposal techniques. In the Asse salt mine about 120,000 drums of contact-handled low-level wastes were emplaced. This technique has been taken over by PTB for the engineering design of the Gorleben repository. The same is true for remotely handled intermediate-level waste, 1,300 drums of which were disposed of in the same mine. One main objective of the WIPP-facility in the US State of New Mexico is also to prove the handling and emplacement techniques for contact- and remotely handled TRU-wastes.

. High-level radioactive wastes (remotely handled)

A complete set of equipment for handling HLW (here simulated by MTR spent fuel) was already used and proved during Project Salt Vault. Also, at WIPP a limited amount of HLW will be emplaced for test and investigation purposes. At Asse, a complete handling and emplacement system is presently under design, so that the already mentioned 30 high-level glass blocks can be disposed of testwise in early 1987.

. In situ-solidification of liquid low- and intermediate-level radioactive wastes

An alternative disposal technique for LLW and ILW into a cavity is presently being developed and tested in the Asse salt mine in the Federal Republic of Germany.

2.2 Design and construction

A great number of in situ investigations were performed in order to produce data for the design and construction of repositories.
The main investigated items are :

. Mechanical behaviour of rock salt

. Dry drilling of deep boreholes with large diameters

. Heat dissipation in rock salt

. Thermo-mechanical behaviour of rock salt

. Convergence of boreholes at ambient and elevated temperatures

Special attention was paid to the thermo-mechanical behaviour of rock salt because of the creep phenomenon.

2.3 Shut down and sealing

The general issue of shut down and sealing is common to all types of geologic formations. Salt has, however, two special aspects :
(1). It is soluble in water and (2). it shows visco-plastic behaviour (creep), especially at elevated temperatures. The areas of shut down and sealing, namely :

. Backfilling of disposal rooms

. Closing of disposal rooms

. Sealing of disposal boreholes

. Closing of drifts

. Dams

. Shaft Seals

have to take care of these special aspects. In situ-investigations for shut down and sealing are still at an early stage.

2.4 Long-term behaviour

Many of the following items have been mentioned in chapter 2.2 :

. Heat dissipation in rock salt

. Thermo-mechanical behaviour of rock salt, other salt rocks and overburden rocks

. Release of water or gases from salt rocks

. Radiolysis

. Permeability of rock salt

. Heat induced fracturing of rock salt

. Stability against seismic events

It is especially important to predict all these near-field and far-field phenomena over a long term period up to several hundreds and thousands of years.

2.5 Performance assessment

Besides the HLW canister performance which is similar for all types of geological formations, again the solubility of rock salt leads to some specific aspects of this formation which are investigated by in situ-investigations, especially in the Federal Republic of Germany :

- . Water or brine intrusion :

    - measurements in flooded shafts

    - flooded drift experiment in Asse

    - flooded mine experiment in Hope.

3. Site investigations for salt repositories

In situ investigations in salt which were and are performed successfully in a number of countries around the world have led to such positive results that actual sites for the construction of repositories for several types of radioactive wastes are presently, or will be in the near future, under investigation. Thses are :

- investigation of the Mors Salt Dome, Denmark, by deep drillings

- Investigation of the Gorleben Salt Dome, Federal Republic of Germany, by shallow and deep drillings

- geologic and hydrologic characteristics established for seven salt sites in the US :
. Deaf Smith County, Texas, Permian Salt Basin
. Swisher County, Texas, Permian Salt Basin
. Davis Canyon, Utah, Paradox Salt Basin
. Lavender Canyon, Utah, Paradox Salt Basin
. Vacherie Salt Dome, Louisiana
. Richton Salt Dome, Mississippi
. Cypress Creek Salt Dome, Mississippi

- Investigation of salt in the USSR

- (Investigation of the Asse Salt Mine, Federal Republic of Germany)

- (Investigation of the Bartensleben salt mine, German Democratic Republic, for ERAM = Endlager für radioaktive Abfälle Morsleben)

The latter two were, are and will be used for the disposal of low- and intermediate-level radioactive wastes only.

Summary of discussion

This discussion, led by Mr. Feates, can be split according to the three
main rock types considered. The sensitive question of waste retrievability was also considered.

A.   Hard rocks (granite, basalt)

M. de Marsily pointed out that an underground cavern in fractured, water-
saturated rock, tends to act as a drain for water, i.e. in situ migration
tests are in the reverse direction compared to the future situation. There
should therefore be a need for migration tests towards the outside of the
excavations.

According to Mr. Simmons, the groundwater flow around a cavern largely
depends upon the fracture networks. At the URL site, fractures were found
parallel to the direction of the future caverns. It is therefore planned
to perform "point-to-point" migration tests in these fractures, hopefully
not affected by the presence of the excavations.

Mr. Lake stressed the fact that there will be some drainage of water in the
construction phase anyway. After the repository is backfilled, it might
very well be that the only cause for gradients be the heat emission from
the wastes.

Mr. Nataraja mentioned that measurements of heat-induced displacements in
hard jointed rocks rarely match the theoretical calculations. According to
Mr. Simmons, most of the discrepancies can be explained by some lack of
details on the structural geology of the testing place: most of the
displacements are known to occur along or across joints, which are
difficult to incorporate in computer models.

Mr. Dietz pointed out that similar discrepancies were observed at the
near-surface test facility in Hanford. The location of the extensometers
very close to the heater could be an explanation for this fact.

B.   Clay

Mr. Rometsch pointed out that marly clays are also considered in
Switzerland as potential host medium for LLW and MLW disposal. Such a
medium can already be examined in existing civil works such as the See-
lisberg tunnel.

111

C. Salt

Creep of rock salt will be enhanced by heating. Therefore, any void (e.g. annulus between canister and disposal hole, or porosity of crushed salt backfill) will be closed by the accelerated convergence of the rock salt (Mr. Kühn). Furthermore, the heat emission will generate pressures higher that the purely lithostatic ones. Therefore, no water could penetrate the repository area. Mr. Joshi expressed the opinion that leaving a repository open after waste disposal in salt would cancel this favourable property of salt rock, and should be avoided.

Mr. Rothfuchs pointed out that crystalline water can also migrate towards the heat sources. Mr. Jockwer added that, if this water is taken off by corrosion of the containers, then there could be some movement (mainly by diffusion and not by flow strictly speaking) of water towards the hot canisters. If no corrosion takes place, the water pressure increases close to the canisters and water movement is stopped.

Mr. Hunsche underlined the fact that experiments are planned in salt to determine whether or not cooling down of salt could induce cracking. Mr. Prij added that ECN's analyses of stresses in the salt dome having a repository with heat generating waste show no tensile stresses at any time and therefore the cracking could be thought of as highly unlikely. The analyses of the stresses in their heater tests however have shown that subtantial tensile stresses and therefore cracks can develop directly after shut down of a heater. This is also confirmed by experimental observation.

D. General comments about waste retrievability

This question was brought about by Mr. Rometsch's presentation (see Session 1), although retrievability as such was not considered there. (It was suggested to leave a repository "open" after waste emplacement for monitoring purposes, i.e. to use the repository as a full-scale underground laboratory with actual waste for some decades.)

Mr. Nataraja recalled the reasons for which US regulation foresees retrievability:

(a) better disposal methods may be found in the future;

(b) if faulty canisters are detected, or if accidents occur during waste emplacement, or if the geology is locally found to be unsuitable, then wastes can be retrieved if provision is kept for such an operation from the very beginning.

Retrievability, therefore, should be considered as a useful option.

Mr. Plodinec recalled that tests are already being performed with radioactive materials in hard rock (e.g. the Climax spent fuel tests). Similar tests with fully radioactive waste glass are planned to begin in 1989 in the salt at the WIPP site. As it is necessary to retrieve them, this operation must be possible right now.

Mr. Nataraja specified that NRC requires retrievability for up to 50 years after placement of wastes is initiated in a repository. Therefore, the equiment, techniques, procedures, etc., for retrievability should be in place prior to license for receiving and possessing waste can be obtained.

Session 3/Séance 3
Design and instrumentation for in situ experiments
in granite and basalt
Conception et instrumentation d'expériences in situ
dans le granite et le basalte

Chairman/Président:
R.ROBINSON
Battelle Project Management Division, Columbus, OH, USA

Hydraulic testing within the cross-hole investigation programme at Stripa

J.H.BLACK & D.C.HOLMES
Fluid Processes Research Group, British Geological Survey, Keyworth, Nottingham, UK

ABSTRACT

The intended programme of hydraulic testing is described in the context of an array of boreholes to be used for geophysical and hydrogeological testing of three million cubic metres of rock. Hydraulic testing is centred on the use of sinusoidal pressure tests which involve a source signal of sinusoidally fluctuating pressure. The likely propagation of this signal in the light of previously reported results of rock properties is assessed. The effect of the measurement zones themselves is also calculated and it is concluded that the frequencies likely to be used are in the range of a cycle per 10 minutes to a cycle per day. This is based on experiment geometry up to 500 m maximum distance between extremities. The equipment being constructed for the programme is described and is based upon automatic control using micro-computers. There is a source-zone system and a receiver borehole system (comprising 6 individual zones) which are inserted into the boreholes and measure automatically for long periods. Analysis and interpretation procedures are directly compatible with the data acquisition system and are semi-automatic based on a user-defined geometry.

1. INTRODUCTION

Underground scientific experiments concerning aspects of nuclear waste disposal have been underway at the worked-out Stripa iron-ore mine in Sweden since 1977. A new series of experiments, known as Stripa Phase II, were initiated during 1983. Amongst these is the so-called "Cross-hole Programme" which is a multi-disciplinary approach to the detection and characterisation of fracture zones. Within this programme it is broadly intended to detect and define fractures and fracture zones in the vicinity of exploratory boreholes and subsequently assess their significance. The fractures are to be detected using geophysics, more specifically borehole radar and borehole seismics, and are to be asessed using hydraulic techniques, particularly the sinusoidally varying pressure method. As well as being multi-disciplinary the programme also involves a number of organisations with Swedish Geological AB being responsible for the preparation of the site and the radar investigations, the Swedish National Defence Research Institute for the seismics and the British Geological Survey for the hydraulics. This paper concerns only the hydraulic aspects of the Crosshole Programme.

The Cross-hole Programme consists of two related investigations; one at the Stripa mine and another larger scale investigation of Gidea in northern

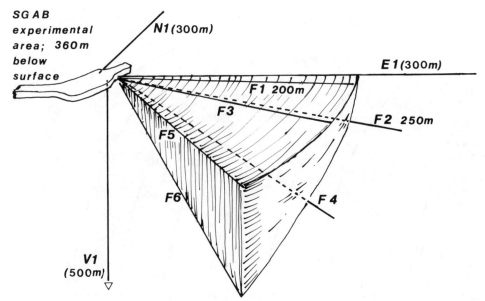

SGAB experimental area; 360m below surface

N1 (300m)

E1 (300m)

F1 200m

F3

F2 250m

F5

F6

F4

V1 (500m)

Figure 1   General perspective of the Crosshole Programme borehole array

Sweden. The full range of techniques is only being applied at the mine site. The intention is to investigate a volume of fractured rock [about 3 million cubic metres] using an array of boreholes emanating from the end of a drift 360 metres below surface.   The fan-shaped array (Figure 1) of six purpose-drilled boreholes (denoted F1, F2 ... F6) lies within an orthogonal array of three existing boreholes [denoted E1 (East), N1 (North) and V1 (Vertical)].

The purpose-drilled boreholes are between 200 and 250 m in length whereas the pre-existing boreholes are between 300 and 500 m long. Within this volume of rock it is already known that there is a variety of conditions including at least one major fracture zone.   The rock will be investigated by methods which are based on both single boreholes and those which are inter-borehole.   For these purposes, therefore, the distances between the boreholes range from practically zero at the drift wall to upwards of 500 m between the extremities of the existing boreholes.

This then is the environment in which cross-hole hydraulic testing is to be carried out. It is constrained by the dimensions of the site, the layout of the boreholes, the mine environment and the overall aims of the programme. Within these constraints the aim of hydraulic testing is to provide a hydraulic basis for calibrating the properties measured in the crosshole geophysics sub-programmes and, to assess the effectiveness of a method based on sinusoidally varying pressure to measure the rock and fracture properties relevant to subsequent safety analysis.   The crosshole hydraulics sub-programme is based primarily on sinusoidal pressure testing.

2.   TESTING PROGRAMME

The hydraulic cross-hole testing will be performed in the same fan array of boreholes as are used for the geophysical measurements.   It will be a

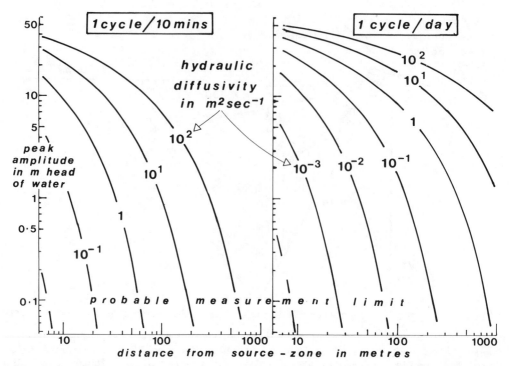

Figure 2   Attenuation of the peak amplitude of a sinusoidal signal with distance from a source zone for two frequencies   a) 1 cycle/10 mins b) 1 cycle/day

combination of existing "standard" interference techniques with the more novel sinusoidal method. It is envisaged that most of the field testing will rely on measuring directional hydraulic properties between "packed-off" intervals in separate boreholes. The method will use a sinusoidal variation of pressure applied in a limited zone (the source zone) of an excitation borehole by a carefully controlled regime of injection and abstraction. Receiver zones consisting of pressure transducers in short packed-off zones in adjacent boreholes will detect a sinusoidally varying pressure which has a smaller amplitude and lags behind the source zone signal. The decrease in amplitude and the retardation (phase lag) of the received signal provides the basis for an interpretation in terms of the hydraulic properties of the rock between the source and the receiver. Both the excitation and observation points can be moved within the available boreholes, so building up a series of hydraulic diffusivity vectors within the available boreholes. The movement of the observation point whilst maintaining the excitation signal is possible since there is no net flow of water onto, or out of, the excitation zone.

The main advantages of this method compared to more conventional step-change testing are perceived as the ability to move receiver zones without stopping the excitation, the ability to detect the signal against the variable background of the mine and the ability to tailor the test frequency to measure either fracture or bulk-rock properties. The main dis- advantages of the method are seen to be signal penetration and equipment complication.

3.   EXPERIMENT DESIGN

The critical steps in designing the Stripa experiment were to calculate the likely penetration of the hydraulic signal through the rock mass and to ensure that it would not be obliterated by receiver zone characteristics.

3.1 Penetration of the sinusoidal signal

The factors which control the penetration of the signal are the magnitude of the peak source-zone amplitude, the frequency of the signal and the hydraulic diffusivity of the rock.   These are related to the maximum observed response by equations which depend upon the geometric system envisaged.   Hence the signal observed at some distance from a source zone would depend on the presence or absence of fractures and whether or not the source and receiver zones were in the same fracture system.

The simplest geometrical configuration, a point source in an isotropic homogeneous porous medium, is also the most strongly attenuating.   However, given that the programme concerns the evaluation of fractures, perhaps a more relevant configuration is that of a source and receiver both within a fracture bounded by permeable rock.   In this configuration the maximum attenuation is calculated by assuming that the fissure has the measured fissure "hydraulic conductivity" combined with measured total specific storage (ie. the specific storage of the fissures and the matrix combined.   The equation applying to this configuration is given in Black and Kipp [1].   The results of calculating the amplitude variation for two different frequencies of source zone excitation are shown in Figures 2a) and 2b).

The major unknown is the variation of hydraulic diffusivity (hydraulic conductivity/specific storage) likely to be encountered within the volume under test.   This can be quite closely estimated from other investigations which have been carried out in the mine.   The results are summarised in Table I.   All the results are in fact reported as hydraulic conductivities and have been divided by an "average matrix specific storage" for granite to yield hydraulic diffusivities.

Table I Hydraulic conductivities measured in the Stripa mine

Title of Test or area	range of calculated hydraulic diffusivities ($m^2 sec^{-1}$)
Swedish)pulse tests	6×10^{-3}
American)injection tests [2]	$1 \times 10^{-5} - 1 \times 10^{-2}$
Co-operation)ventilation test [2]	1×10^{-4}
Stripa) Neretnicks 2D tracer test maximum	1×10^{-1}
Phase I) SGAB Borehole V1 [2]	$5 \times 10^{-4} - 7 \times 10^{-1}$
) SGAB Borehole E1 [2]	$5 \times 10^{-5} - 4 \times 10^{-1}$
Stripa) Site investigation for Crosshole	
Phase II) Programme - borehole F1*	$2 \times 10^{-5} - 1 \times 10^{-2}$
borehole F2*	$1 \times 10^{-4} - 1 \times 10^{-1}$

* preliminary results

119

It can be seen that apart from the SAC pulse tests all the results are relatively consistent with maximum hydraulic diffusivities in the 0.1 m^2 sec^{-1} range and minima around 2×10^{-5}. It can probably be correctly assumed that the lowermost values apply to the hydraulic diffusivity of rock matrix and that the higher values refer to fissures.

Returning to the problem of defining the likely observable signal from a sinusoidally varying source it only remains to set values for the maximum workable source-zone pressure and the detection limit of the receiver zone. These are set at 100 m of applied water head and a detection limit of 0.1 m resulting in an amplitude ratio of 1×10^{-3}. Referring to Figure 2a) it can be seen that a 10 minute cycle would be observable in the more conductive fractures up to 25 m from the source. Likewise a daily cycle (Figure 2b) should be detectable up to 250 m from the source.

3.2 Effect of receiver zone response lag and attenuation

Receiver zones contain a finite volume of water and compliant packers such that they do not respond instantly to pressure fluctuations in the adjacent rock and diminish the observed maximum amplitude. It is possible using a formula given in Black [3] to calculate the zone response time of a zone given the geometry of the zone and the hydraulic conductivity of the surrounding rock. In order for the receiver zone not to reduce the received amplitude to less than 80% of that in the surrounding rock then the ratio of zone response time to wave period should not exceed 0.1 [4]. Given a receiver zone length of one metre, an effective zone compliance equivalent to an open water level in a 1 mm diameter tube, and a borehole diameter of 76 mm, then the zone response time (a) will be $1 \times 10^{-6}/K$ (where K is the hydraulic conductivity of the surrounding rock in m sec^{-1}). It can be seen, therefore, that, for a sinusoidal signal of 1 cycle per ten minutes (ie. a wave period of 600 secs), the value of a should not be greater than 60 and the value of K should be greater than 1×10^{-8} m sec^{-1}. When daily cycles are considered rock hydraulic conductivities can be as low as 1×10^{-10} m sec^{-1} before significant attenuation of the signal is induced by receiver zone response time. It is the intention during the work programme to test the response time of each zone by carrying out a pulse test. It is also possible that the compliance effects may be less than the estimate above which is based on experience with larger packers and 100 mm diameter boreholes. However, it is unlikely to be reduced by more than one order of magnitude.

3.3 Test design summary

It is clear from the above considerations that signal penetration and signal detection should be possible for major fracture zones over distances of several hundred metres. If the rock matrix proves to be less permeable than has been estimated then short-period signals should be detectable at distances up to 50 m or more from the source zone. For the lower end of the range of hydraulic conductivities and diffusivities the penetration of any signal will be meagre and its detection impossible. It is in the fractures whose hydraulic conductivity lies in the range 1×10^{-10} to 1×10^{-9} m sec^{-1} where care in the design of the equipment should make meaurements feasible.

4. EQUIPMENT DESIGN

4.1 Basic criteria

The equipment must be capable of operating within the mine to produce a well defined source signal and possibly receive quite small hydraulic signals. It should be sufficiently automated and rugged to run for extended continuous periods, without human intervention, and be capable of sensing any test problems and reacting to abort a test safely.

4.2 Problems influencing the design

Most of the design problems occur in attempting to make the equipment sufficiently adaptable to operate in the hydraulic pressure field of the mine. The pressure field, within the plane of the array of test boreholes, results from a modification of the natural regional gradients by drainage of groundwater into the mine cavities. Any of the array boreholes will, therefore, experience different pressures at various distances into the rock mass. As a further complication, the boreholes themselves, by connecting zones of different pressure will further modify the field.

The pressure field will also fluctuate with time either as a general trend if groundwater drainage to the mine is not at equilibrium or periodically in response to earth tides, rainfall, etc. Also other testing procedures operating in the mine may cause "hydraulic noise". Fluctuations from these sources will be small, in the order of several centimetres, but may influence the measurement of low amplitude test signals.

During the cyclic tests, selected zones are isolated in a source and receiver borehole using packers. These will block the movement of groundwater between zones of different hydraulic pressure. Each separated length of borehole will attempt to attain a new equilibrium pressure further modifying the pressure field. To minimise this effect the boreholes will be allowed to equalise pressure along their full length for some months before testing begins, so that groundwater within the rock mass will reach a pseudo-equilibrium pressure.

5. GENERAL EQUIPMENT DESCRIPTION

5.1 Source borehole and signal generation

The source zone (Fig. 3) borehole contains two packers, usually one metre apart, although the distance can be varied. The packers are positioned by a rod string which also connects the source zone to a pump in the mine to perform water injection or abstraction. Located adjacent to, and connected with, this isolated zone is an absolute pressure transducer to measure the source zone pressure. A bypass tube through the packer assembly allows hydraulic connection between the two lengths of borehole either side of the packers.

To create the cyclic test pressure pattern, usually a sine waveform, water is pumped into or out of the source zone. The rate of water movement is determined by measuring the pressure in the source zone and comparing it to a predicted value. If the pressure is too high the injection rate is reduced or vice versa. A micro-computer controls these operations as part of its overall control function.

121

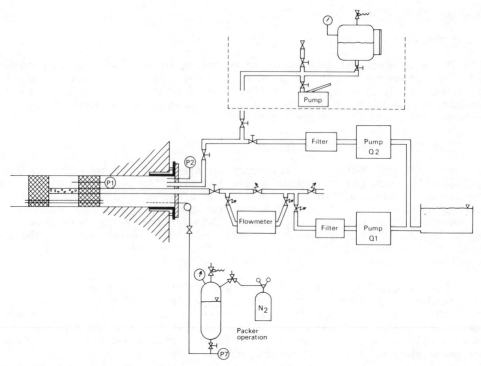

Figure 3 Schematic of equipment associated with the source-zone borehole

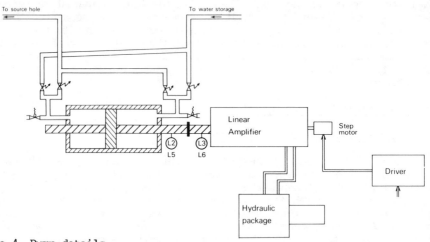

Figure 4 Pump details

The pumps (Fig. 4) which provide the water movement are located in the mine test working area. Each pump comprises a finely machined cylinder with a tightly fitting double acting ram. This is actuated by a hydraulic-augmented-mechanical driving ram, the exact linear position of which is controlled by a stepping motor. Solenoid values are fitted to the exit and

122

entrance ports of the water cylinder. A micro-processor on the pump accepts commands from the main control micro-computer to increase or decrease flow rate. It interprets these to vary the stepping rate and solenoid status of the pump which provides the necessary flow rate and direction. The motor stepping rate is transmitted to, and recorded by, the control computer as this is a direct measure of flow rate, assuming there is no back-leakage around the water-cylinder ram. During the initial stages of the testing program the flow rate will be measured by a sensitive flow meter to evaluate the cylinder leakage rate. Periodic flow measurements will   also be made to ensure that the cylinder is functioning correctly.

Under certain rock conditions water injected into or abstracted from the source zone will leak around the packers and produce a hydraulic signal in the remaining length of the borehole. This signal could be transmitted to the receiver borehole and incorrectly interpreted as originating from the point source zone. To obviate this problem a second pump in the mine area holds the pressure of the borehole water constant. The bypass tube in the   packer usually ensures that the entire borehole, excluding the source zone, is affected. Variations in the stepping rate of this second pump indicate the magnitude of any leakage.

5.2 Receiver borehole and signal detection

In the receiver borehole (Fig. 5) the minimum requirement is for two packers to isolate a single section. However, to increase the volume of rock under examination during one test, a series of six packers are used isolating five short zones, approximately 1 m in length. Additionally, the two long portions of open borehole may also be considered as zones, giving seven in

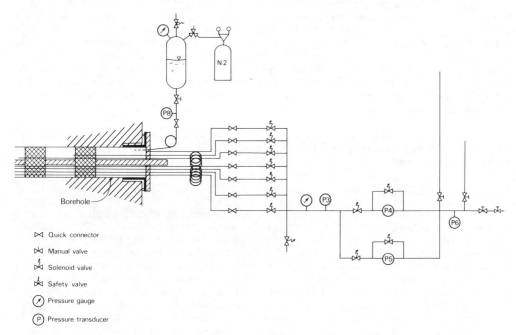

Figure 5 Schematic of equipment associated with receiver zone borehole

123

total. Each zone is connected, by a narrow diameter flexible tube, to the pressure measuring system located in the mine test area. The packer assembly is positioned in the boreholes by a blank rod string.

Pressure sensors for the receiver zones must operate at absolute values from 100 to 250 metres water equivalent. However, to measure hydraulic signals generated during the cyclic tests, they must discriminate pressure to 0.10 m or less. There are absolute pressure transducers capable of this but they tend to be rather expensive. The selected alternative is to use differential pressure transducers supplied with a back pressure through a tube in a mine shaft. The column of water in this tube, with a free upper surface, can be varied in height to match the pressure in a specific receiver zone.

Each receiver zone does not have its own transducer as this results in calibration differences between the separate sensors. Instead a single measuring system is used and each zone pressure is accessed to it in turn by a series of solenoid valves, operated by the main microcomputer. The measuring system comprises two absolute and two differential pressure trans- ducers. The differential pressure transducers are of different ranges, one from 0-70 metres and the other 0-10 m water equivalent. The two absolute pressure transducers measure water pressure in the test section and in the back pressure tube. For each pressure measurement the main computer dictates the same pattern. Initially the differential pressure transducers are isolated by solenoid valves from all pressure changes. The microcomputer reads and compares the pressure from the absolute transducers and if the difference is small will open the solenoids to one and/or both the differential transducers. These sensitive transducers are thus protected from overpressure.

Any packers used in the borehole array are inflated hydraulically by a tube from the mine using compressed gas pressure. Once inflated the packer inflation pressure is monitored by the control microcomputer. If it falls below a pre-set value an alarm is given and the packers can be reinflated. If, however, no action is taken and the inflation pressure continues to drop to such a level that the packers become detached from the borehole walls, then the test is stopped automatically and all solenoid valves are closed to isolate the boreholes from water inflow or outflow. Similar action is taken by the system if there is a power failure in the mine.

Pressure data will be collected from those boreholes not directly involved in the cyclic testing to provide qualitative information on transmissive zones within the rock mass.

5.3 Control system

The electronic control system comprises a central microcomputer driving a series of intelligent and non-intelligent peripherals. The source- borehole pumps contain their own processors which react to command statements from the central micro. Similarly all the transducer analogue outputs are fed to a processor-based unit which, on receiving a command, interrogates a transducer, converts the signal to digital format and presents it to the central micro. These two intelligent peripherals free the central processor from time consuming control functions. Collected data is stored on a floppy disc unit and some of it is presented "real time" on a printer so that the trend of a test can be examined directly. The switching of solenoid valves is performed directly through the central microcomputer output ports and relays.

Programming of the system is in Pascal/MT to facilitate the handling of data. The main control programme has been constructed to enable the test system to perform various hydrogeological tests. These include sinusoidal and square wave cyclic tests and continuous abstraction or injection tests based on either constant flow or pressure. Passive data collection is also possible to obtain information on the natural and artificial small scale pressure fluctuations. This programme also contains the option of performing instantaneous pressure-drop pulse tests in the source and receiver zones. This test allows a response time for each zone to be calculated for inclusion in the calculation of hydrogeological parameters from the cyclic tests.

Data sampling for the cyclic tests is based on the wave length of the hydraulic signal. One hundred scans are made and stored on diskette to characterise each cycle. Each test is programmed to last the time equivalent to seven cycles. During the time equivalent to the first two cycles the source-zone pressure is maintained at a selected near-equilibrium pressure. During the next three parts the source zone hydraulic pressure is made to follow the cyclic pattern, to the selected amplitude. In the two remaining periods the source zone pressure is again held at equilibrium whilst transmitted hydraulic signals are monitored in the receiver zones. During pulse tests and continuous flow tests scans are made at increasing logarithmic time periods to facilitate plotting. In passive data collection mode regularly spaced scans are taken.

6. ANALYSIS AND INTERPRETATION OF RESULTS

6.1 Background
The sinusoidally varying pressure test was chosen as the basis for this subprogramme because of some advantages which it offers over conventional step-change methods [5] under certain circumstances such as in mines. However, unlike conventional methods there is not a wealth of literature covering all possible approaches to testing and interpretation. This has meant that it has been necessary to develop a complete analysis and interpretation procedure from scratch. Naturally many of the ideas underlying conventional analysis are reapplied to the analysis of sinusoidal tests but in the process many concepts are examined for their applicability.

6.2 Analysis

The method was introduced by considering signal propagation in homogeneous, isotropic porous media from lines in cylindrical regions and from points in spherical regions [1]. This limited analysis has recently been extended to include point and line sources in fissured anisotropic porous media and to single fractures bounded by permeable rock. The likening of fractured granite to a fissured porous medium is specially relevant to sinusoidal pressure tests. This is because under certain circumstances it should be possible to vary the frequency of the signal and allow or disallow interaction with water in the rock matrix [5]. However, with the Crosshole Programme in mind, particular consideration is being given to the ranges of properties relevant to the fissured granites of the Stripa mine. Also further analysis is due to concentrate on fissure geometry (size and range of fissures, distance to boundaries), fissure and matrix interaction and the effect of scale and connections.

125

6.3 Interpretation

An interpretation technique is being developed so that interpretation of the test results is comparatively automatic. It involves a relatively complex numerical code which is based on the mathematical analysis described above. It allows some operator freedom so as to be able to take into account the results emanating from the geophysical sub-programme. Sensitivity analyses will be carried out to ensure that the most effective testing techniques are employed.

7. CONCLUSIONS

A hydraulic testing programme has been designed for Phase II of the Stripa Programme which is relevant to the needs of a complex interdisciplinary approach to bulk rock characterization. As far as it is possible to predict, rock properties are such that sinusoidal signals with periods up to a day will be detectable over the intended scale of the investigation programme. Under some circumstances more than one frequency will be measureable. Attenuation resulting from storage in receiver zones should not pose problems since the type of rock which induces receiver zone response problems will also rapidly attenuate the signal.

The equipment design takes into acount these theoretical considerations but also includes practical aspects such as background noise and the expected trends adjacent to mines. It is conceived and built as an "intelligent" testing system with considerable versatility. It relies upon current micro-computer technology combined with active peripherals and sensors. The data acquisition system is designed with the interpretation procedure in mind and will also provide data at the time of acquisition. Mathematical analysis is now quite far advanced and forms the basis for a semi-automatic interpretation procedure.

It is hoped that the design of this sub-programme has demonstrated an awareness of the practical and theoretical constraints of in-situ experiments in mines. Further progress awaits the inevitable developments associated withh actual field testing due to start in the summer of 1984.

8. ACKNOWLEDGEMENTS

The authors are indebted to their co-workers in the SGAB, especially M Sehlstedt and O Olsson. This work is published by permission of the Director of the British Geological Survey (NERC).

REFERENCES

[1] Black, J.H. and Kipp, K.L. Jr., 1981. Determination of hydrogeological parameters using sinusoidal pressure tests: a theoretical appraisal. Water Resources Res. Vol. 17. Pt. 3. pp 686-692.

[2] Carlsson, L., Norlander, H. and Olsson, T. 1983. Hydrogeological investigations in boreholes. Proc. of the Workshop on Geological Disposal of Radioactive Waste:- in situ experiments in granite. OECD/NEA and SKBF/KBS Stockholm 25-270 et 1982 pp 109-120.

[3]   Black, J.H. and Kipp, K.L. Jr., 1977. The significance and prediction of observation well response delay in semi-confined aquifer test analysis. Groundwater 15: 446-451.

[4]   Hvorslev, M.J., 1951.   Time lag and soil permeability in groundwater observations.   Ball. 36, US. Corps. of Eng., Waterways Exp. Sta. Vicksburg, Miss.

[5] Black, J.H. and Barker, J.A. 1983. Application of the sinusoidal pressure test to the measurement of hydraulic parameters.   Proc. of the Workshop on Geological Disposal of radioactive waste:- in-situ experiments in granite. OECD/NEA and SKBF/KBS Stockholm 25-27 Oct. 1982 pp 121-130.

Le laboratoire souterrain de Fanay-Augères

A.BARBREAU
Commissariat à l'Energie Atomique, IPSN-DPT, Fontenay aux Roses, France

RESUME

L'influence de l'effet d'échelle sur la valeur mesurée de la perméabilité et
du coefficient de dispersion est actuellement en cours d'étude dans une galerie
d'une mine d'uranium, dans un massif granitique. Cette mine est bien connue
du point de vue géologique et a déjà fait l'objet, dans le passé, d'études
structurales, géomécaniques et hydrogéologiques. Après une phase d'étude en
vue de la qualification du site, qui a consisté principalement à vérifier au
moyen de forages autour de la galerie que la roche était bien saturée en eau,
l'expérience sur l'effet d'échelle doit comprendre des mesures de perméabilité
par observation du débit drainé dans la galerie, par injection dans des forages, entre obturateurs, à distances croissantes et la détermination du coefficient de dispersion au moyen de traceurs injectés à distances croissantes.

ABSTRACT

The influence of the scale effect on the measured value of the permeability
and dispersion coefficients is presently underway in a drift of an uranium
mine, in a granitic massive. This mine is geologically well known and structural, geomechanical and hydrogeological studies have been carried out in the
past. After a first phase of qualification studies consisting in controlling
with boreholes drilled in the drift the saturation of the rock around the
gallery, the sole effect experiment will consist in a permeability measurement
by observing the drained flow rate in the drift itself and by injection
between packers in boreholes at increasing distances. Dispersion coefficient
will be determined by injection of tracers at increasing distances in
boreholes.

1. INTRODUCTION

L'Institut de Protection et de Sûreté Nucléaire ayant décidé de lancer
une série d'études concernant les propriétés du milieu fissuré, études qui
doivent être réalisées in situ, il a été nécessaire de rechercher au début de
1983 un site qui, sans correspondre aux conditions qui seraient celles d'un
véritable stockage géologique, présenterait des propriétés favorables à la
réalisation des expériences. Il s'agissait de trouver, à une certaine profondeur, dans une formation granitique, un terrain saturé en eau et suffisamment
fissuré pour fournir des résultats aisément interprétables.

En pratique, il fallait trouver une galerie déjà existante et l'on a procédé à l'examen des mines en formation granitique actuellement en exploitation.

Le choix final s'est porté sur la mine de Fanay-Augères, située à une vingtaine de kilomètres de Limoges, qui appartient à la Division Minière de la Crouzille, exploitée par la COGEMA du groupe CEA.

Cette mine se trouve dans le massif granitique de Saint Sylvestre (cf. fig. 1) qui est un leucogranite à deux micas, à grains moyens à grossiers, de 20 km sur 30 km, d'âge namurien, soit environ 320 millions d'années. Ce massif est intrusif dans des formations métamorphiques. Le massif est traversé par de grandes failles NW-SE et c'est dans ces failles que l'on trouve l'uranium.

Cette mine présentait à priori plusieurs avantages. Tout d'abord, la géologie de la région est assez bien connue grâce aux travaux de reconnaissance effectués par la COGEMA et grâce aussi à l'exploitation minière.

De plus, le BRGM a déjà, dans le passé, effectué dans cette mine, avec l'Ecole des Mines de Paris, une étude comparative de la fracturation profonde et superficielle avec analyse statistique et une étude hydrogéologique qui a consisté essentiellement à déterminer le tenseur de perméabilité en surface et en profondeur, cette étude étant toutefois restée assez sommaire. Par ailleurs, le BRGM y a également effectué des études de mécanique des roches concernant l'étude des contraintes in situ. On a ainsi mis au point et comparé différentes techniques de mesure des contraintes en galerie et en sondage (méthode du vérin plat, surcarottage et fracturation hydraulique). Ces études s'étaient déroulées dans le cadre d'un contrat avec la Délégation Générale à la Recherche Scientifique et Technique (DGRST) et étaient donc étrangères aux programmes d'études concernant les déchets radioactifs.

Par contre, le BRGM a effectué, pour le compte du CEA-IPSN et dans le cadre d'un contrat à frais partagés avec la CCE, l'étude comparative de la fracturation superficielle et de la fracturation profonde du massif grâce à des observations dans les galeries. Le granite présente une première phase de compression (Stéphanien, Autunien) suivie de distension (Saxonien, Crétacé et Jurassique), puis de nouvelles phases de compression (phase Pyrénéenne, puis phase Alpine). De nombreuses mesures ont porté sur les fractures : direction, pendage, épaisseur et venues d'eau. On a aussi procédé à des études photo-interprétatives. On n'a pas noté de variation dans l'orientation des fractures entre la surface et le fond et la fracturation en profondeur a une densité à peu près analogue à celle observée en surface.

La zone expérimentale finalement retenue, d'environ 100 m de long, se trouve dans une galerie située au niveau dit 320, en réalité à environ 170 m sous la surface du sol, très irrégulière dans cette région (Cf. fig. 2, 3, 4). Cette portion de galerie est apparue à priori favorable par suite des venues d'eau que l'on peut y observer et qui laissaient espérer un milieu saturé.

2. ETUDES DE QUALIFICATION DU SITE

Il a été alors décidé de procéder à un certain nombre d'études afin de procéder à la qualification définitive du site. Ces études ont compris :

- une étude structurale détaillée,

- des mesures dans des sondages,

- une mesure du débit dans la galerie.

2.1 Etude structurale

Elle a consisté à un relevé systématique de toutes les discontinuités
sur le parement Est de la galerie, sur une hauteur de 2 m, qui ont fait l'objet
d'une étude très précise. Toutes les fractures supérieures à 0,20 m de long
ont ainsi été relevées et l'ensemble des données (plus de 1000 mesures) a fait
l'objet d'une série de traitements statistiques fournissant le tracé de dia-
grammes de Schmidt, la délimitation de familles structurales, le tracé des
histogrammes de divers paramètres. Quatorze familles de fractures ont ainsi
été mises en évidence. Les fractures sont, en moyenne, de faible longueur
(6 % seulement dépassent 3,5 m) et les fractures uniques sont les plus nombreuses ; les 2/3 sont sans ouverture observable, 44 % paraissent humides à
l'observation, 5 des 14 familles paraissent susceptibles de jouer un rôle
hydraulique.

2.2 Etude par sondages (Fig. 4)

Six sondages carottés d'une longueur de 8 m à 16,70 m ont été réalisés
à partir de la galerie afin de déterminer la position des accidents tectoni-
ques majeurs au voisinage de celle-ci, les caractéristiques de fracturation
naturelle et l'état hydraulique du milieu.

On a ainsi pu déterminer la position de la faille dite "de la Recette"
et la densité de fracturation sur les carottes (Fig. 5 et 6). On a également
effectué la mesure des débits d'eau dans les différents sondages au moyen d'un
obturateur mobile, débits qui se sont révélés très variables d'un forage à
l'autre.

La faille de la Recette qui est le principal accident tectonique au
voisinage de la zone expérimentale ne semble jouer ni un rôle de drain, ni
un rôle de barrière étanche.

Les six sondages ont été équipés chacun en piézomètre, afin de déterminer la pression hydraulique existant dans le massif au niveau de leur partie
terminale. La connaissance de cette pression était essentielle pour la qualification du site, puisqu'il est nécessaire de disposer d'une galerie entourée
entièrement de terrains saturés en eau.

Dans trois sondages, la chambre de mesure a été isolée au moyen d'un
obturateur et le trou de sondage entre l'obturateur gonflé et la galerie a
été cimenté. La pression d'eau dans la chambre peut être lue en permanence
sur un manomètre installé dans la galerie tandis qu'un second manomètre permet
le contrôle permanent de la pression s'exerçant dans l'obturateur. Les trois
autres sondages ont leur chambre de mesure isolée par un train d'obturateurs.
On dispose aussi de deux manomètres pour le contrôle permanent des pressions.

Les mesures de pression très constantes dans le temps qui varient
de quelques décibars à environ 2 bars, montrent que le massif autour de la
galerie est bien saturé quoique la pression soit peu élevée par suite des
perturbations apportées par les travaux miniers. La répartition des pressions
est toutefois très hétérogène.

2.3 Mesure de débit en galerie (Fig. 7)

La galerie a été équipée afin que l'on puisse mesurer le débit d'eau drainé par le tronçon expérimental ; on a donc, à cette fin, aménagé le radier et installé un barrage amont et un barrage aval, distants de 97,60 m et hauts de 0,40 m. Les eaux provenant de la partie amont du tronçon expérimental sont collectées et rejetées en aval de celui-ci, après le deuxième barrage. Le débit obtenu (5,75 l/mn) conduit, avec les observations dans les forages, à une estimation de la perméabilité du massif granitique autour de la galerie de l'ordre de 10^{-7} à 10^{-8} m/s.

3. PROGRAMME D'ETUDE "EFFET D'ECHELLE"

A la suite de ces travaux, il a été décidé que le site était qualifié pour la phase d'étude à proprement parler. Celle-ci comprend dans l'immédiat l'étude de l'effet d'échelle, en milieu fracturé, sur les valeurs de la perméabilité et du coefficient de dispersion, qui a été confiée au BRGM avec la collaboration du Laboratoire d'Informatique Géologique de l'Ecole Nationale Supérieure des Mines de Paris.

Cette étude, dont le déroulement est actuellement en cours, a pour objectif d'améliorer la détermination de la perméabilité à grande échelle des milieux fissurés et d'améliorer nos connaissances sur les modes de cheminement des éléments en solution. Or, beaucoup de techniques de mesure utilisées en hydrogéologie ne permettent pas d'accéder facilement à la connaissance du comportement d'un massif à grande échelle en ce qui concerne les milieux hétérogènes et peu perméables. Il est donc nécessaire de préciser l'effet d'échelle.

Le principe de l'étude est de mesurer les débits et les pressions à distances croissantes de la galerie au moyen de forages radiaux partant de la galerie et équipés de plusieurs obturateurs successifs isolant plusieurs sections de chaque forage en chambres indépendantes reliées par des tubes à la galerie. Les mesures, effectuées en un grand nombre de points devraient permettre d'accéder au tenseur de perméabilité.

Le même dispositif sera utilisé pour l'étude des propriétés hydrodispersives du milieu, par l'injection dans les forages, à distances croissantes, de traceurs appropriés, qui seront ensuite recueillis dans la galerie.

Par ailleurs, des corrélations seront établies avec les caractéristiques du réseau de fracture, l'état de contrainte du massif et les perméabilités locales déterminées par essais d'eau dans les mêmes forages.

Le programme de travaux en cours de réalisation comporte les opérations suivantes :

3.1 Etude des perméabilités (phase 2A du programme)

- L'exécution de 10 forages carottés de 50 m de longueur chacun, dans des directions radiales par rapport à la galerie 320 : ces sondages seront répartis sur trois profils dont un profil central de quatre sondages et deux profils latéraux de trois sondages (Fig. 8, 9, 10, 11).

- Le relevé structural de toutes les discontinuités existantes sur les carottes orientées en cours de forage.

- La détermination des zones de venues d'eau en sondage à l'aide d'un obturateur simple déplacé en forage et permettant des mesures globales de venue d'eau dans la portion de sondage situé entre le fond du trou et l'obturateur. Le pas de mesure sera d'environ 2 m. Les venues d'eau partielles, parvenant dans chaque tronçon de forage, seront déduites par différence.

- La détermination des perméabilités du massif par injection d'eau en forage entre double obturateur. Les essais seront conduits en deux étapes successives, d'abord, sur des chambres de mesure d'environ 10 m, puis sur des chambres d'environ 2 m.

Parallèlement à ces travaux de forages et de mesures, on concevra et fabriquera les équipements des obturateurs multiples nécessaires à la mesure des pressions et à la réalisation des injections de traceurs. Le dispositif qui sera mis en place comportera pour chaque forage sept obturateurs, individualisant sept chambres de mesure. Chacune de ces dernières sera en communication avec la galerie par deux conduits en rilsan 6/4 mm permettant la purge, la mesure des pressions et l'injection des traceurs.

Après mise en place du dispositif à obturateurs multiples dans chaque sondage, suivi de la montée en pression dans les 70 chambres de mesure, on procèdera à la mesure des pressions stabilisées et des débits.

On interprétera l'ensemble des mesures sur un modèle tridimensionnel représentant l'écoulement autour de la galerie. On déduira la perméabilité en grand à partir des mesures de pression et du débit global. Le tenseur de perméabilité sera relié aux mesures locales de perméabilité entre obturateurs et aux relevés de fracturation (effet d'échelle sur la perméabilité).

3.2   Etude de l'effet d'échelle sur la dispersion (phase 2B du programme)

Cette expérience consistera à étudier et à caractériser l'écoulement hydraulique autour de la galerie en utilisant des traceurs. Ceux-ci seront injectés dans les chambres individualisées lors de la phase 2A. Les opérations suivantes seront effectuées :

- Des essais de traçage par injection de traceurs dans des chambres situées de plus en plus loin de la galerie (effet d'échelle). On utilisera simultanément des traceurs différents pour limiter la durée de l'expérience. Les traceurs seront collectés directement en galerie, sans chercher à localiser ou échantillonner directement les fractures individuelles puisque c'est justement l'effet global intégré que l'on cherche à connaître.

- La détermination de l'état de contrainte dans le massif (par fracturation hydraulique dans les forages réalisés lors de la phase de qualification de la galerie ou par surcarottage).

- L'interprétation des mesures de traçage sur modèle tridimensionnel continu représentant le transfert autour de la galerie et la détermination de la variation du coefficient de dispersion longitudinal en fonction de la distance à laquelle a été injecté le traceur.

4. <u>AVENIR DU LABORATOIRE SOUTERRAIN DE FANAY-AUGERES</u>

Ce programme concernant l'effet d'échelle sera vraisemblablement suivi d'autres programmes d'études qui seront réalisés, au moins en ce qui concerne certains d'entre eux, également dans la mine de Fanay-Augères, soit dans le même tronçon de galerie, soit à d'autres endroits appropriés de la mine. Parmi les études envisagées, citons :

- l'effet de la chaleur dégagée par une source thermique sur la perméabilité en grand d'un massif fissuré ;

- l'étude de l'évolution des propriétés mécaniques du granite sous l'effet de la chaleur ;

- l'étude in situ des phénomènes de migration en milieu fissuré ;

- le développement de techniques de mesures in situ des propriétés du milieu fissuré.

5. <u>CONCLUSION</u>

Le laboratoire souterrain de Fanay-Augères ne doit pas être comparé aux laboratoires souterrains tels qu'ils sont actuellement conçus et en cours de réalisation dans certains pays (Belgique, Canada, par exemple). Il ne s'agit en effet nullement d'étudier un milieu potentiellement favorable pour l'évacuation de déchets radioactifs, ni par conséquent de chercher à valider la sûreté du concept d'évacuation en formation géologique de ces déchets. Le site de Fanay-Augères, défavorable de ce point de vue, ne recevra jamais de déchets d'aucune sorte.

Il s'agit d'un site expérimental destiné à effectuer en profondeur, dans des conditions représentatives, des études sur les propriétés fondamentales des milieux fissurés encore mal connus afin d'améliorer nos connaissances sur ces milieux, de développer les méthodes et les techniques qu'il sera nécessaire de maîtriser lors de la réalisation d'un véritable laboratoire souterrain et, à plus forte raison, d'un dépôt de déchets radioactifs en formation géologique.

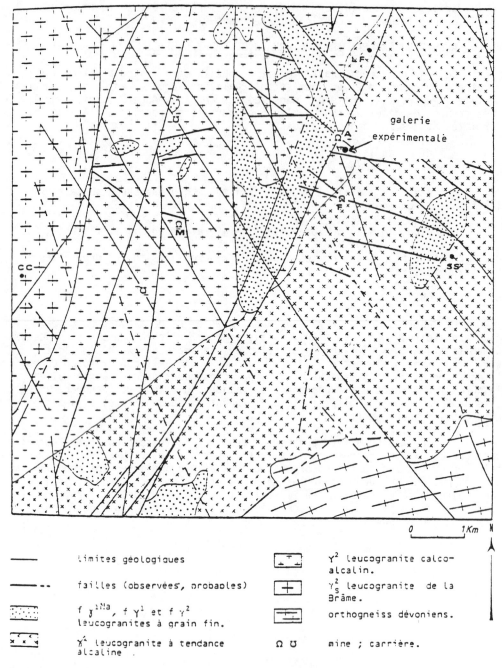

galerie
expérimentale

——	Limites géologiques	

Limites géologiques

faiLles (observées, probables)

f γ^{1Na}, f γ^{1} et f γ^{2}
leucogranites à grain fin.

γ$^{?}$ leucogranite à tendance
alcaline .

γ^{2} leucogranite calco-
alcalin.

γ$^{2}_{s}$ leucogranite de la
Brâme.

orthogneiss dévoniens.

Ω ℧   mine ; carrière.

A = puits d'Augeres ; F = puits de Fanay ; M = puits de Margnac ; CC = Compreignac ;
LF = Le Fraisse ; SS = Saint-Sylvestre.

Fig 1 - Situation géologique de la zone expérimentale :
extrait de la carte géologique au 1/50 000 d'AMBAZAC.

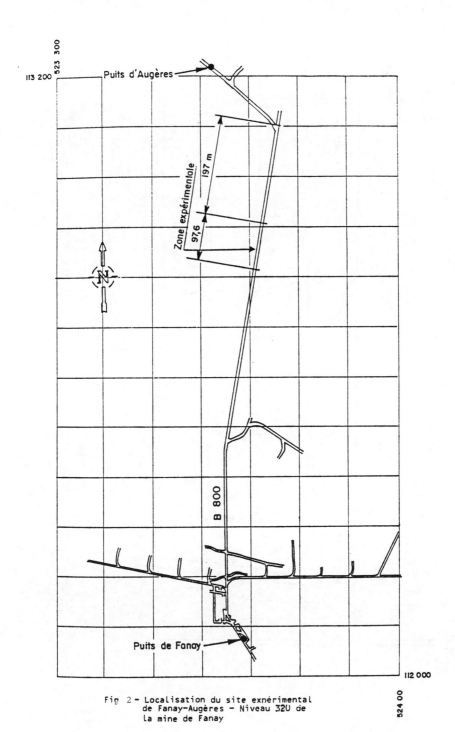

Fig 2 – Localisation du site exnérimental
de Fanay-Augères – Niveau 320 de
la mine de Fanay

135

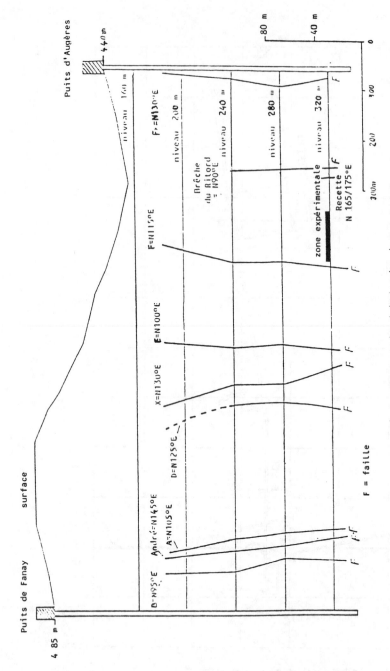

Fig 3 - Coupe nord-sud de la mine de
Fanay-Augères passant par la zone expérimentale :
localisation des principaux accidents géologiques.

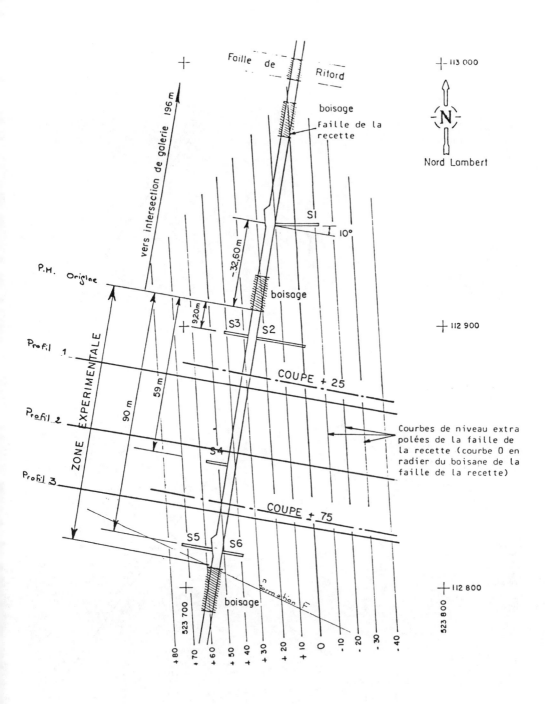

Figure 4

Echelle.1/1000

VUE EN PLAN DU SITE EXPERIMENTAL
(MINE DE FANAY-AUGERES- NIVEAU - 320)

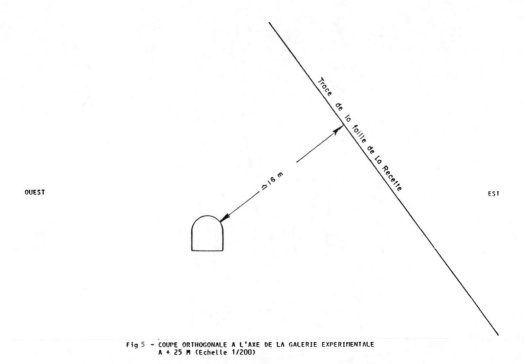

Fig 5 - COUPE ORTHOGONALE A L'AXE DE LA GALERIE EXPERIMENTALE
A + 25 M (Echelle 1/200)

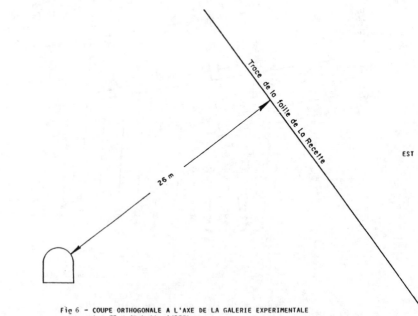

Fig 6 - COUPE ORTHOGONALE A L'AXE DE LA GALERIE EXPERIMENTALE
A + 75 M (Echelle 1/200)

138

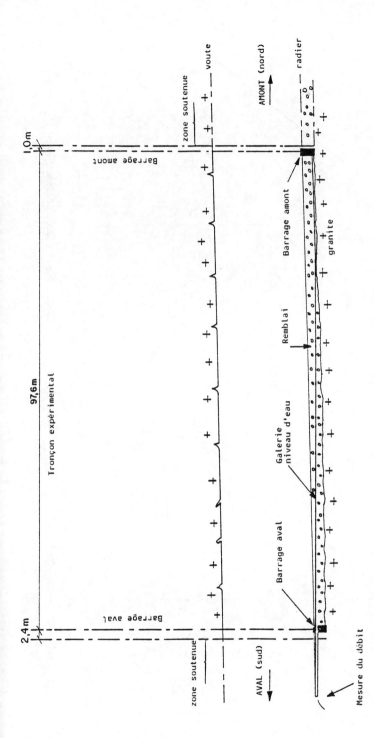

Fig 7 – COUPE LONGITUDINALE SCHEMATIQUE DANS L'AXE DE LA GALERIE EXPERIMENTALE

139

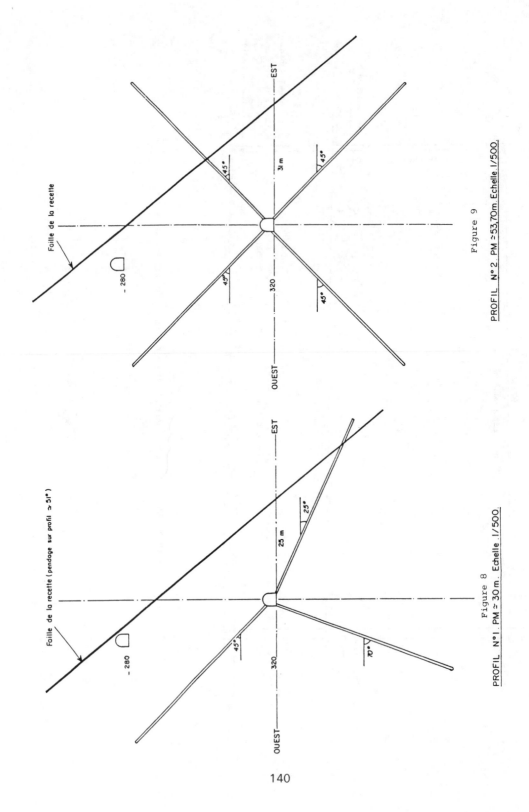

Foille de la recette

- 280

45°

45°

45°

320

31 m

OUEST — EST

Figure 9

PROFIL N° 2 PM ≃ 53,70m Echelle 1/500

Foille de la recette (pendage sur profil ≃ 51°)

- 280

45°

25°

70°

320

25 m

OUEST — EST

Figure 8

PROFIL N°1. PM ≃ 30 m. Echelle 1/500.

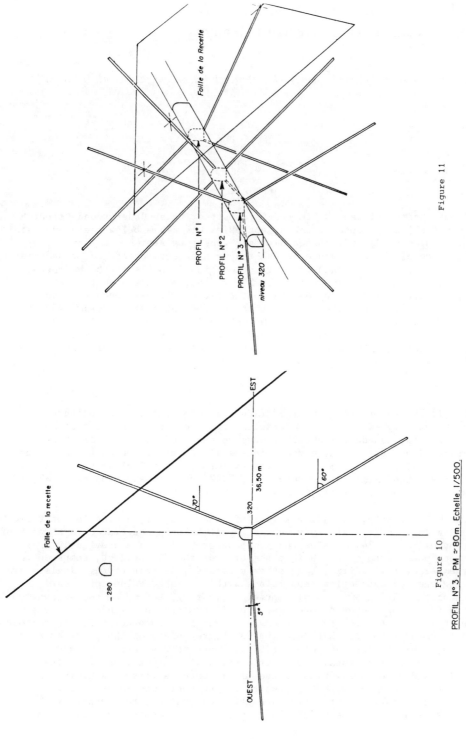

Figure 11

Figure 10

PROFIL N°3 . PM ≃ 80m. Echelle. 1/500.

141

Far-field hydrogeological monitoring at the site
of Canada's Underground Research Laboratory

C.C.DAVISON
Atomic Energy of Canada Ltd, Whiteshell Nuclear Research Establishment, Pinawa, Manitoba

ABSTRACT

Atomic Energy of Canada Limited is constructing an Underground Research Laboratory (URL) at a depth of 250 m, approximately 245 m below the groundwater table, in a previously undisturbed plutonic rock body near Lac du Bonnet, Manitoba. One of the main geotechnical research objectives of the project is to develop and validate comprehensive three-dimensional models of the hydrogeology of a volume of rock, 4.8 km^2 in area by 500 m deep, encompassing the URL excavation site. Prior to any excavation, these models have been used to predict the far-field hydrogeological perturbation (piezometric drawdown) that will be created by the excavation of the URL facility. As a model validation exercise, these predictions are being compared with the actual drawdown conditions, monitored by means of an extensive network of specially instrumented boreholes.

INTRODUCTION

Atomic Energy of Canada Limited (AECL) is constructing an Underground Research Laboratory (URL) within a granitic rock body near the town of Lac du Bonnet, Manitoba, Canada (Figure 1). Once constructed, the URL will provide researchers with representative geological environments in which to carry out a variety of in situ geotechnical experiments for Canada's Nuclear Fuel Waste Management Program (Simmons and Soonawala, 1982; Davison and Simmons, 1983; Simmons, 1984; Thompson et al., 1984).

The URL facility, including a vertical access shaft, a ventilation raise bore, and a main horizontal experimental level, is being excavated to a depth of 250 m, approximately 245 m below the groundwater table, in a previously undisturbed rock mass. One of the unique features of the URL project, compared to similar experimental facilities in other countries, is that hydrogeologists are monitoring the hydrogeological conditions within a large volume of rock surrounding the URL excavation site prior to, during, and after the excavation of the shaft and underground workings. The information obtained from investigations carried out prior to any excavation has been incorporated into various numerical computer models that describe the hydrogeology of the study area (Guvanasen, 1984; Lafleur and Lanz, 1984). These models have been used to predict the three-dimensional piezometric drawdown that will occur in the rock mass during and after excavation of the underground facility. When these predictions are compared with the results of the continuous hydrogeological monitoring being carried out during and after excavation, researchers will be able to assess how well the models have actually represented the three-dimen-

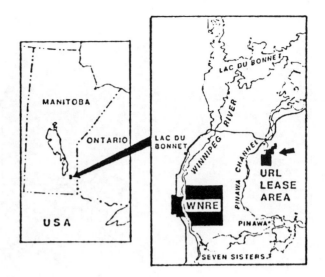

FIGURE 1:
LOCATION OF STUDY AREA

sional hydrogeological conditions of the rock mass surrounding the URL excavation site. This validation exercise is a major step toward the development of reliable models with which to predict solute transport through large volumes of plutonic rocks.

HYDROGEOLOGICAL CONDITIONS

Detailed investigations have been underway since 1980 to determine the three-dimensional physical and chemical hydrogeological characteristics to depths of 500 m within a study area 4.8 km^2 in size encompassing the URL excavation site (Davison, 1984). Over 130 boreholes have been drilled, logged, tested and instrumented at the study area to accomplish this. The locations of all these boreholes, as well as the URL shaft location, are shown on Figure 2. Fifty-nine shallow boreholes, referred to as the O-series boreholes, were drilled into the thin deposits of unconsolidated sediment at the site to determine their hydrogeological characteristics. A series of 40 boreholes was drilled to investigate the shallow bedrock conditions, both in the areas of bedrock exposure at surface as well as in the areas overlain by the unconsolidated deposits (B-series boreholes). Twenty-five boreholes were drilled to depths ranging from 160 m to 1090 m to investigate the deeper portions of the granitic rock mass. Eleven of the deep boreholes were of 76-mm diameter and completely cored using diamond drilling equipment (URL-Series). Fourteen were of 152-mm to 158-mm diameter and drilled using percussion techniques (M-Series). Many of the deep URL- and M-series boreholes were drilled close to the site chosen for the construction of the URL facility to provide a comprehensive network of boreholes with which to accurately monitor the groundwater perturbations created by the excavation.

Assimilation of all the fracture information obtained from the entire network of boreholes revealed that three major extensive subhorizontal fracture zones were present within the rock mass. It has been found that these zones largely control the movement of groundwater at the study area. Except for these distinct extensive fracture zones, the rock mass is relatively unfractured, with fracturing being slightly more pronounced near the ground surface. Figure 3 is

143

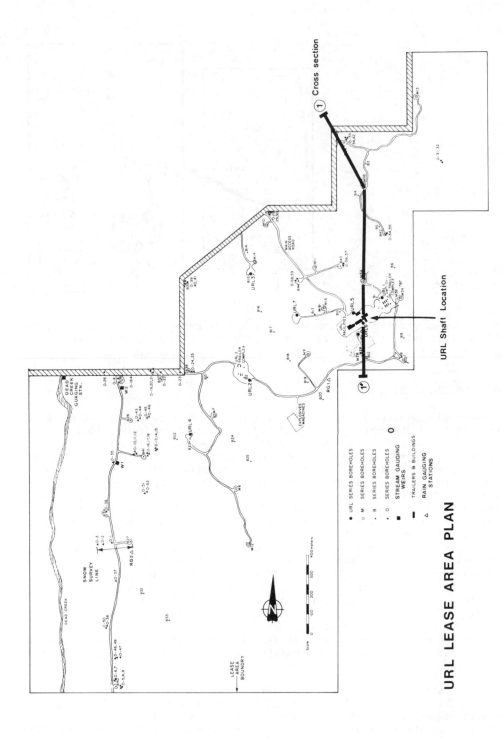

URL LEASE AREA PLAN

FIGURE 2: LOCATION OF BOREHOLES – URL SITE

144

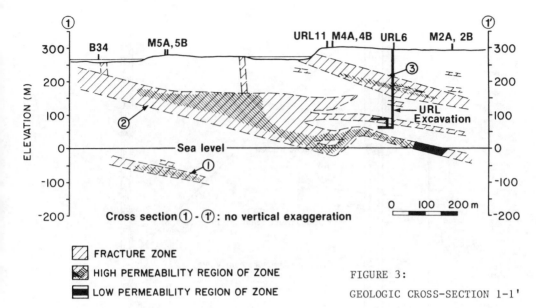

FRACTURE ZONE

HIGH PERMEABILITY REGION OF ZONE

LOW PERMEABILITY REGION OF ZONE

FIGURE 3:

GEOLOGIC CROSS-SECTION 1-1'

a geological cross section (along line 1-1' of Figure 2) that illustrates the location and attitude of the main subhorizontal fracture zones. These zones have been referred to as fracture zone 1, fracture zone 2, and fracture zone 3, in order of increasing elevation (Davison, 1984). The URL shaft will be excavated through fracture zone 3, but will stop well above a region of intense fracturing in fracture zone 2. However, the shaft and some portion of the horizontal underground level will be excavated through an off-branching, low permeability limb of fracture zone 2 (Figure 3). Therefore, the hydrogeological characteristics of fracture zones 2 and 3 will control most of the piezometric disturbance that will occur in the rock mass when the underground facility is excavated. The hydrogeology of these two fracture zones has been studied in considerable detail using a wide variety of borehole-testing techniques, including single-borehole straddle-packer tests and large-scale multiple-borehole hydraulic pressure interference tests (Davison, 1984). The results of this work have revealed that a complex pattern of permeability exists within each of the two fracture zones, and that these permeability distributions control the patterns of hydraulic head and groundwater chemistry. A hydro-geological monitoring system has been designed on the basis of the system has been designed on the basis of the hydrogeological conditions that have been found at the site. Monitoring points have been isolated in the main fracture zones as well as in the less-fractured zones to provide a comprehensive three-dimensional array with which to record the disturbance that will be created by the excavation of the shaft, the ventilation raise bore and the horizontal underground levels.

BOREHOLE EQUIPMENT FOR HYDROGEOLOGICAL MONITORING

Several types of multiple-interval completion systems have been installed in the boreholes at the URL site to monitor the three-dimensional, physico-chemical hydrogeological conditions. These include piezometer nests and water-table wells that have been installed in shallow holes (less than 30 m deep) and multiple-interval casing systems and multiple-packer/multiple-standpipe

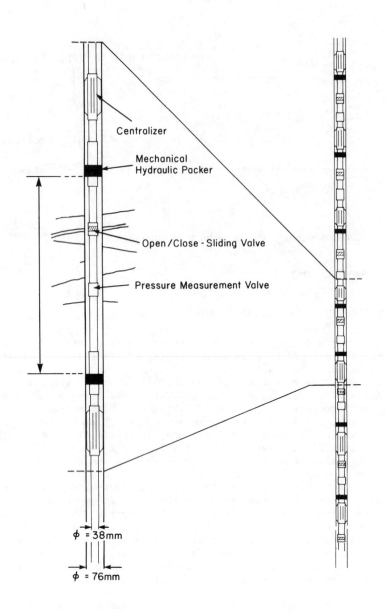

FIGURE 4: SCHEMATIC OF MULTIPLE-INTERVAL CASING SYSTEM
USED TO COMPLETE 76-mm DIAMETER BOREHOLES

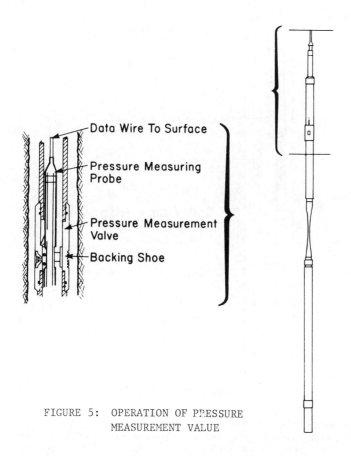

FIGURE 5: OPERATION OF PRESSURE
MEASUREMENT VALUE

piezometers that have been installed in the deeper holes (30 m to 1090 m deep).
Some of the equipment is conventional and has been used in a wide variety of
groundwater-monitoring applications. Much of the equipment is innovative and
was designed and fabricated to meet the particular requirements of this project.
Special consideration was given to such factors as the expected range of
piezometric fluctuation, borehole diameter and long-term reliability, in the
design and installation of the hydrogeological monitoring equipment.

Conventional bottom-hole standpipe piezometers and fully slotted water-
table wells were installed in the boreholes drilled into the unconsolidated
overburden deposits (O-series of Figure 2). The shallow (10 m to 15 m deep)
core-holes of 76-mm diameter that were drilled into the bedrock at outcrop areas
(B-1 thru B-26, B-32, and B-33 of Figure 2) were completed as either single,
fully slotted water-table wells or as dual-standpipe, bottom-hole piezometer/
upper water-table well installations depending on the location of fractures and
the groundwater level in each borehole.

Special multiple-interval instrumentation was designed, developed and
fabricated to complete the deep, 76-mm diameter coreholes and the deep, 152-mm
to 158-mm diameter percussion-drilled boreholes. Multiple-interval casing
systems were installed in all the 76-mm diameter coreholes that were deeper than
15 m (B-26, B-27, URL-1 thru URL-11 of Figure 2). These casing systems

147

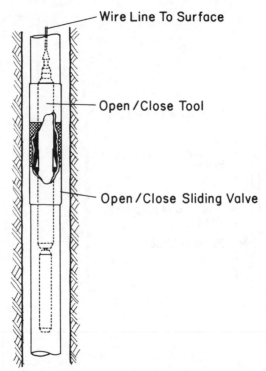

Wire Line To Surface

Open/Close Tool

Open/Close Sliding Valve

FIGURE 6: OPERATION OF OPEN/CLOSE SLIDING VALVE

consisted of a single-liner casing string (38-mm I.D.) containing external packer elements and through-the-casing access valves, as shown on Figure 4. Numerous monitoring zones were isolated in each 76-mm diameter borehole by means of the multiple-interval casing system.

Two types of access valve were placed within each packer-isolated interval: a pressure measurement valve and an open/close sliding valve. The pressure measurement valves are activated by means of a winch-operated wireline pressure measurement probe. The pressure probe is lowered down inside the casing string to the bottom, and it stops sequentially at each pressure measurement valve as it is raised to the surface (Figure 5). The pressure probe activates the pressure measurement valve, measures the hydraulic pressure conditions outside the casing, transmits the pressure measurement via the wire-line cable to a digital readout and control unit located at ground surface, and then deactivates the pressure measurement valve before being raised up the casing to the next valve. In this manner, a vertical profile of hydraulic pressure conditions can be obtained quickly from all the packer-isolated zones in a single borehole. The open/close sliding valves (Figure 4) provide the option of also having intermittent long-term access to any packer-isolated monitoring zone in the casing string. The sliding valves are normally installed closed; however, any valve can be opened or closed by means of a special tool that is lowered inside the casing by means of a wire-line cable (Figure 6). Once a sliding valve is opened, a direct hydraulic communication is established through the casing string, between the packer-isolated interval and the inside of the casing, which enables long-term continuous water-level monitoring, fluid

148

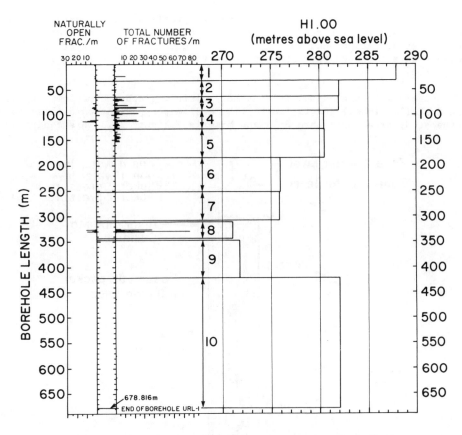

FIGURE 7: PRESSURE MEASUREMENT PROFILE FROM BOREHOLE
URL-1; DATA OF 1984 JANUARY

sampling or hydraulic tests to be performed as required. A schedule-80
polyvinyl chloride (pvc) plastic version of this multiple-interval casing
system, which was commercially available from Westbay Instruments Ltd. (Westbay)
of Vancouver, B.C., Canada (Rehtlane and Patton, 1982), was installed in all of
the 76-mm diameter boreholes shallower than 200 m. Hydraulically inflated
external-gland packers were utilized on the plastic casing system to provide the
isolation between the various monitoring intervals. A stainless-steel version
of the multiple-interval casing system, jointly designed and fabricated by AECL
and Westbay, was installed in the 76-mm diameter bore-holes deeper than 200 m.
Permanent-locking mechanical packers were used on the stainless-steel casing
system to isolate the various intervals (see Figure 4). These packers were set
individually using a hydraulic pressure injection tool. Because the multiple-
interval casing systems consist of modular components, the locations of packers
and access valves could be tailored to meet the specific requirements of each
individual borehole. Monitoring intervals were selected on the basis of the
detailed fracture logs and other information available from previous hydrogeological testing programs. As an example, Figure 7 shows the multiple-interval
casing completion for borehole URL-1, along with the fracture log of the
borehole. This figure also presents the hydraulic pressure data obtained at

149

each monitoring interval when the casing was profiled 1984 January 3. The hydraulic pressure measurements have all been converted to the equivalent fresh-water elevation head ($H^{1.00}$), to allow comparison with hydraulic pressure data obtained from other piezometric monitoring installations at the study area.

Specially designed multiple-packer/multiple-standpipe piezometer systems were installed in all the 152-mm to 158-mm diameter percussion-drilled boreholes at the study area (B-34 thru B-44 and M-series, Figure 2). This equipment is

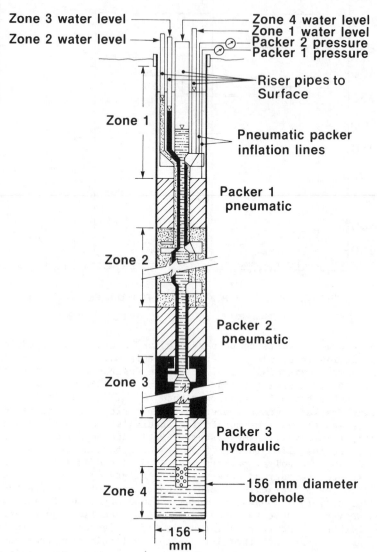

FIGURE 8:   SCHEMATIC OF MULTIPLE-PACKER/-STANDPIPE PIEZOMETER SYSTEM

USED TO COMPLETE 152 mm TO 158 mm DIAMETER BOREHOLES

somewhat similar in concept to conventional multiple-piezometer installations, in that a separate riser standpipe connects each piezometer interval to ground surface for monitoring purposes. Figure 8 illustrates the main features of the multiple-packer/multiple-standpipe piezometers used in this study. Up to four intervals can be isolated in each borehole by means of packer elements placed at any desired location along a tubing string. The bottom packer element is a retrievable self-locking hydraulic packer that is inflated by pressurizing the fluid column within the annulus of the tubing string. The other packers on the string are pneumatic packers which are each inflated from ground surface by means of individual inflation lines. After installation, the inflation pressure within each of the pneumatic packers is continuously monitored at surface to ensure packer integrity. A separate tube connects each packer-isolated monitoring zone to ground surface: the bottom zone is accessed through the annulus of the central tubing string (35-mm I.D.) and the other zones are accessed through individual 25-mm diameter riser pipes. A complex feed-through system was designed and fabricated to pass the various piezometer lines and packer inflation lines through the packer elements. After installation of the multiple-packer/multiple-standpipe piezometer systems, water was bailed, swabbed or pumped from each of the riser pipes, to ensure complete homogenization of the fluid chemistry within the piezometer column. This was necessary because of the variations in groundwater salinity that occur at the study area; these variations have a large effect on the hydraulic head values that are calculated from the standpipe measurements. Because a relatively simple pattern of large-scale fracturing was found to exist within the rock mass at the study area, adequate monitoring of the main hydrogeologic units could be achieved in the 152-mm to 158-mm diameter boreholes using this multiple-packer/multiple-standpipe piezometer equipment. A typical installation of this equipment is illustrated in Figure 9, along with the corresponding $H^{1.00}$ measurements.

CONTINUOUS MONITORING OF HYDROGEOLOGICAL CONDITIONS

Continuous monitoring has been underway to establish the natural hydraulic head fluctuations that occur at each piezometer location. This information has been used to establish baseline trends in the hydraulic head conditions at the study area, prior to any excavation of the URL facility, so that perturbations that are caused by the excavation can be discriminated. The continuous monitoring will extend through the construction and operation phases of the facility to record the influence that excavation during various phases of the project has on the surrounding groundwater regimes. It is hoped that this monitoring can continue until the year 2000, at which time it is planned to decommission the facility.

In each of the multiple-interval casing systems, a single, sliding-sleeve valve has been opened so that the water level within the casing annulus represents the hydraulic head of the single interval adjacent to the open valve. The water levels in all the casings, and the water levels in all those piezometers equipped with standpipes to surface, are being measured weekly using standard manual water-level tapes. These manual readings are entered into computer data files that facilitate the rapid production of up-to-date hydrographs for all piezometers.

Hydraulic head profile measurements are being made in each of the multiple-interval casing systems on a regular schedule, using the pressure measurement probe equipment described earlier. Table 1 presents an example of sequential profile measurements obtained from the multiple-interval casing

151

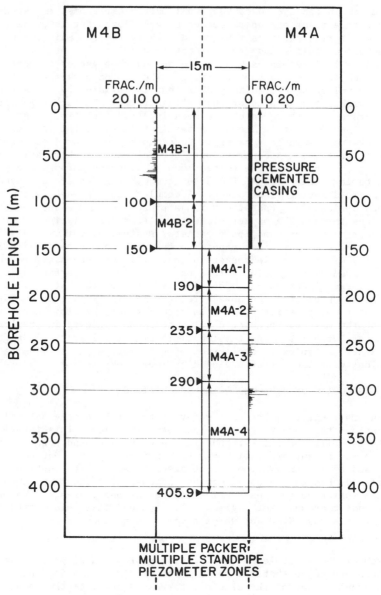

FIGURE 9:   MULTIPLE-PACKER/MULTIPLE-STANDPIPE
PIEZOMETER COMPLETION OF BOREHOLE M 4

systems. Profile measurements in the casing systems are currently being made either monthly or bi-weekly to monitor the effects of shaft excavation.

In addition, a computer-controlled water-level measuring system has been designed and installed to continuously monitor 75 selected piezometers at the study area. The water level in each piezometer is monitored by means of an

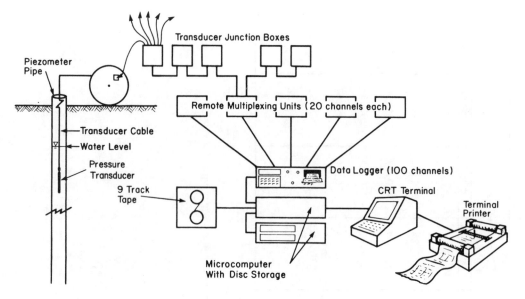

FIGURE 10: AUTOMATIC PIEZOMATIC MONITORING SYSTEM FOR THE URL SITE

18-mm diameter, 0 to 100 psi-output pressure transducer (installed) below the water level. The slim-diameter transducer fits easily into the casing and riser pipes of the multiple-interval casing systems and the multiple-packer/multiple-standpipe piezometer systems installed in many of the boreholes. When the transducers are in place, the piezometric water levels can also be monitored manually from surface by means of a water-level tape, to provide an independent check of the transducer output. The pressure transducers are connected to a central data acquisition system by means of a network of wire links. The data-logging equipment can automatically record the water levels at all 75 piezo-meters as frequently as every 15 seconds, if necessary. This automatic monitor-ing system will record the rapid water-level changes expected to occur during excavation of the shaft and underground workings, particularly when major fracture zones, such as fracture zone 3, are excavated. Figure 10 is a schem-atic that illustrates the various components of the monitoring system. This figure also shows those piezometers being monitored by the automatic system. Multiconductor data lines transmit the transducer output signals to remote multiplex units (RMU) and the RMUs relay the information along data wires to the central microcomputer-controlled data-logging equipment. The RMUs scan 20 pressure transducer channels, and are located up to 1 km from the main data-logging unit. Signals from transducers located as much as 1.8 km from the main data logger are being recorded using this equipment. Currently, automatic measurements are being gathered from all 75 transducers at a frequency of once every hour, and these are stored in data files. The automatic measurements are complemented by a complete set of manual measurements made weekly, using a water-level tape, to ensure the accuracy and define the resolution of the data recorded by the automatic data acquisition system. Small diurnal piezometric fluctuations such as those caused by earth tide effects, which are of the order of 1 to 2 cm, are easily identified on the automatic records.

Monitoring Interval #	Borehole Interval (m)	Hydraulic Head as H Date							
		29-6-82	8-7-82	21-9-82	11-4-83	16-6-83	10-8-83	15-11-8
1	9.1 – 15.2	272.65	272.54	272.16	269.80	271.66	271.23	272.65
2	16.8 – 24.4	272.33	272.42	271.62	270.25	271.22	270.67	271.95
3	25.9 – 35.1	272.02	272.11	270.31	269.89	270.77	269.62	260.70
4	36.6 – 45.7	271.98	272.10	270.41	269.90	270.74	269.66	270.50
5	53.3 – 73.2	271.77	271.90	270.28	269.73	270.64	269.47	270.50
6	74.7 – 94.5	271.63	272.10	270.83	270.60	271.03	269.91	271.20
7	100.6 – 157.6	271.84	272.14	271.64	271.47	271.85	271.69	272.07

TABLE 1

BOREHOLE URL-4:   MULTIPLE-INTERVAL CASING PRESSURE PROFILES

Any variations in groundwater chemistry that might accompany the piezo-metric drawdown are also being monitored by sampling sequentially from selected piezometers. The zones and sampling techniques have been selected to ensure that minimal hydraulic disturbance is associated with the collection of the water samples, so as not to compromise the piezometric pressure monitoring program.

SUMMARY

A wide variety of monitoring instrumentation has been installed in the rock mass surrounding the URL excavation site to observe the three-dimensional physical and chemical hydrogeological perturbations that will be associated with the excavation of the facility some 245 m below the groundwater table. These include conventional piezometers and wells installed in shallow boreholes, and specially designed multiple-interval casing systems and multiple-packer/multiple-standpipe piezometers installed in deep boreholes. An automatic system has been installed to monitor 75 piezometers and to record the rapid piezometric responses that will be created when the facility is excavated through permeable fractures and fracture zones.

Detailed studies have characterized the main hydrogeological conditions of the rock mass encompassing the excavation site, a volume of approximately 2.5×10^9 m^3, and these have been incorporated into several numerical models to predict the three-dimensional hydrogeological effects of excavation. These predictions will be compared to the actual piezometric drawdowns that are being measured at the site by the monitoring instrumentation throughout the construction and operating phases of the project. This comparison will assess how well the various numerical models have described the hydrogeological conditions in the rock mass. The model validation exercise provided by the URL hydrogeological monitoring program is a major component of AECL's overall hydrogeological research, and it will lead toward the development of reliable models with which to predict groundwater movement and solute transport through large volumes of plutonic rock.

REFERENCES

Davison, C.C. and G.R. Simmons (1983). The research program at the Canadian Underground Research Laboratory; in proceedings of the Nuclear Energy Agency workshop on geological disposal of radioactive waste and in situ experiments in granite. Stockholm, Sweden, 1982, October 25-27, p.p. 197-219.

Davison, C.C. (1984). Hydrogeological characterization at the site of Canada's underground resarch laboratory; in proceedings of the International Associat-ion of Hydrogeologists international symposium on groundwater resource utilization and contaminant hydrogeology, Montreal.

Guvanasen D. (1984). Flow simulation in a fractured rock mass; in proceedings of the International Association of Hydrogeologists international symposium on groundwater resource utilization and contaminant hydrogeology. Montreal, Canada, 1984 May 21-24.

Lafleur, D.W. and R. Lanz (1984). Numerical Modeling of the URL: Calibration to field hydraulic tests and prediction of URL effects; in proceedings of the International Association of Hydrogeologists international symposium on groundwater resource utilization and contaminant hydrogeology, Montreal, Canada, 1984 May 21-23.

Rahtlane E. and F.D. Patton (1983). Multiple port piezometers versus standpipe piezometers: an economic comparison in proceedings of 2nd national symposium on aquifer restoration and groundwater monitoring, Columbus, Ohio p.p. 287-295.

Simmons, G.R. and N. Soonawala (1982). Underground research laboratory experimental program. Atomic Energy of Canada Limited Technical Record, TR-153*, 261 p.

Simmons, G.R. (1984). In situ experiments in granite in underground laboratories - a review; in Proceedings of the CEC/NEA Workshop on the Design and Instrumentation of In-Situ Experiments in Underground Laboratories for Radioactive Waste Disposal, Brussels, 15-17 May 1984.

Thompson, P., P. Lang and P. Baumgartner (1984). Planned construction phase geo- mechanics experiments at the underground research laboratory; in Proceedings of the CEC/NEA Workshop on the Design and Instrumentation of In-Situ Experiments in Underground Laboratories for Radioactive Waste Disposal, Brussels, 15-17 May 1984.

---

* Technical Records are unrestricted unpublished reports available from SDDO, Atomic Energy of Canada Limited Research Company, Chalk River, Ontario KOJ 1J0.

Water injection test and finite element calculations
of water percolation through fissured granite

LUTZ LIEDTKE & ARNO PAHL
Federal Institute for Geosciences and Natural Resources, Hannover, FR Germany

1. Introduction

Within the framework of the German/Swiss Cooperation agreed in
1983 it is intended to test and further develop engineering geological rock-mechanical investigative methods for use in crystalline rock. Partners involved are the Nationale Genossenschaft für
die Lagerung radioaktiver Abfälle (NAGRA), the Institut für Tieflagerung der Gesellschaft für Strahlen- und Umweltforschung (GSF)
and the Bundesanstalt für Geowissenschaften und Rohstoffe
(Federal Institute for Geosciences and Natural Resources - BGR).

The NAGRA Rock Laboratory at Grimsel is situated in the Aare and
Gotthardt massiv in the Swiss Alps, in the vicinity of the Grimsel Pass.

The main access tunnel to the control centre of Grimsel II, of
the Kraftwerke Oberhasli AG (Electricity Generating Company), was
investigated by NAGRA and an area below the Juchlistock at a
depth of approx 450 - 500 m was chosen for the rock laboratory
(NAGRA 1983 [1]). A review of the investigations proposed in the
NAGRA rock laboratory, Grimsel, has been published by PFISTER, E.
1983 [2] .

The laboratory tunnel and the test sites were cut in 1983/84.
Fig. 1 shows the layout of the rock laboratory after [1] and [2].

The BGR shall carry out the following research projects:

- A modified water injection test (Bohrlochkranzversuch BK),

- rock stress overcoring techniques, (Gebirgsspannungen) and

- geophysical high frequency techniques.

The following report describes the planning of the modified water
injection test and the accompanying investigations proposed.

2. Purpose of the Modified Water Injection Tests

The test site is situated in central Aare granite, which is fissured and also in part cut by faults. The purpose of the tests is

NAGRA
FELSLABOR GRIMSEL

BK   Water Injection Test
GS   Rock Stresses
ZB   Central Area
WT   Thermal Test
AU   Damage Assessment
VE   Ventilation Test

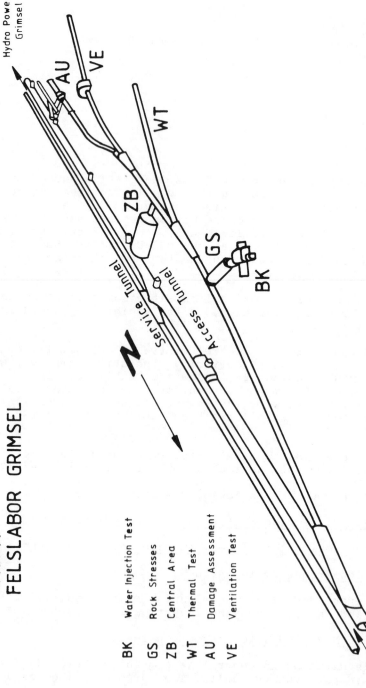

Fig. 1 Grimsel Rock Laboratory

to further develop equipment for the determination of water conductivity and water permeability in crystalline rocks along discontinuities, e.g. cleavage, fissures and faults. In solid rock
these are the major factors influencing the permeability of the
rock, whereas the permeability of intact rock without discontinuities is of secondary, or even negligible, importance. The
hydraulic conductivity depends for the most part on the discontinuity surfaces, their width and filling material as well as
their origins and intersections. For this reason the BGR
started, within the framework of investigations carried out for
the underground siting of nuclear power stations, to develop a
flow test apparatus. This apparatus and the work carried out in
the studies for the underground siting of nuclear power stations
is contained in the BGR reports 1981/82 [4] and in other
publications, e.g. SCHNEIDER, H.J. 1982 [5] and PAHL, A. &
SCHNEIDER, H.J. 1982 [6]. This first apparatus and measurements
taken in varieagated sandstones and granites provide the basis
for a research project intended to comprehensively improve and
further develop the investigative methods available.

The modified water injection test may comprise of the following
methods:

Water percolation through fissured rock

direct methods

- in-situ percolation tests with
  a control injection borehole
  and surrounding observation
  boreholes, located according
  to the geological structure

- block model tests in a laboratory or in the rock laboratory

indirect methods

- simulation using analytical and numerical methods,
  taking rock discontinuities
  into consideration
- special geological structure
  modelling, e.g. degree of separation in parts of the rock
- back-up analyses

The tests are designed to quantify the direction dependant permeability of the rock with respect to rock pressure and water
pressure. The problems of investigating rock pressure are being
studied in the BGR research project on rock stress, "Gebirgsspannungen GS". In-situ tests for this project will be started in
1984 such that the first results may be taken into account in the
modified water injection tests.

3. Description of the test site BK

The position of the test site is shown in Fig. 2. The dominant
factor in the selection of this site was the presence of a water
bearing fracture zone in which a test drilling was made in 1980
by NAGRA and which was evaluated by GEOTEST (Switzerland). In the
deeper section of the borehole prior to the excavation of the
test site the water pressure had reached 13 bar for relatively
short periods. During the excavation using blasting the drill
hole was intersected and the deeper section reclosed.

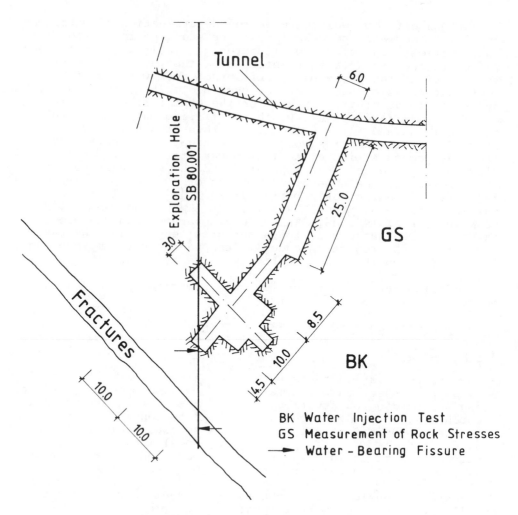

Tunnel

6.0

Exploration Hole
SB 80.001

3.0

25.0

GS

8.5

Fractures

10.0

4.5 10.0

BK

10.0

10.0

BK Water Injection Test
GS Measurement of Rock Stresses
→ Water-Bearing Fissure

Fig. 2   Versuchsort

The front section of the test site (GS) is characterised by a
stress zone, caused by shear surfaces, which strikes through the
tunnel whereas the rear section (the BK test site) is cut by
individual fissures and cleavage surfaces which can be attributed
to definable systems (Fig. 3, structure diagram, recorded by
Dipl. Geologist D. Segond von Banchet): In the deepest section
of the test site water is seeping in from fissures. In a special
mapping programme the roof, side walls and floor were mapped in
detail for fissures, faults and cleavage surfaces. Fig. 4, (drawn
up by Dipl. Geologist Bräuer) illustrates a section of the floor
mapping in the area of the modified water injection test. The
diagram clearly shows that the discontinuities in the rock are
fairly limited. Only a few fissures are of larger extension. Many

fissures grade into others or have a visibly small extension. The structural evaluation, which is still in progress, is intended to provide a basis for siting the first boreholes of the modified water injection tests. The following procedure is planned: A fissure system is selected which is regarded as being a good conductor of water. Two holes, one injection and one observation, will then be drilled into the fissure system, which will then be tested.

Following the results of this test other boreholes will be made, either in the same system or in a different system. In the final phase it is proposed that approximately 15 boreholes be made, aimed downwards into the water-bearing rock such that investigations are made of the two-phase water/solid rock system.

The geological records of the boreholes will be made up from data stemming from the cores and with the help of a video camera. This system allows fissures to be measured or also fissure openings to be determined as appropriate. The results are extremely important for the positioning of the packer and injection sections in the boreholes.

The TV camera and the system for geological recordings is fully described by KREBS, E. 1969 [7] .

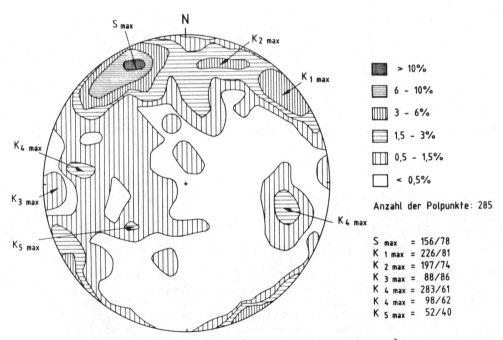

Fig. 3: Cleavage and fissuring in the BK test tunnel.
Schmidt's network, projection in the lower hemisphere.
Engineering geological structure recording:
Dipl. Geologist SEGOND von BANCHET.

161

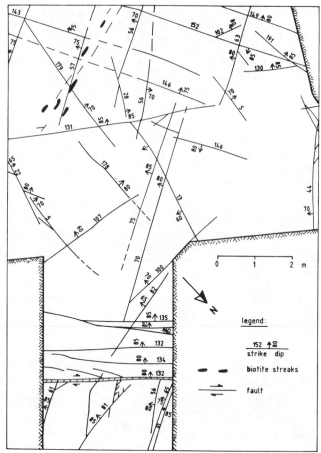

Fig. 4: Section of the floor mapping by Dipl.-Geologist BRÄUER

4. Equipment Description

The equipment used may be divided into three groups

1. Probes and packers in the boreholes
2. Pumps
3. EDV equipment

As shown in system diagrams Figs. 5 and 6 the probes in the bore-holes may show either a natural or an artificial water potential of the test area. Measurements will be taken of the pressure using pressure sensors (5-100 bar), temperature and conductivity will also be recorded. A probe consists therefore of three sensors and is enclosed in the observation borehole (86 mm dia.) by two pneumatic packers (length approx. 1.0 m). In general anything up to 4 probes, each separated by a packer, will be introduced one after the other. In exceptional cases, however,

162

15 probes may be connected serially in one borehole. The number of drill holes depend on the local hydrological conditions, which may be very varied in the test area. Initially boreholes with a regular spacing of 20 - 50 m are intended.

Since the individual sensors are scanned in series a 5 core cable suffices for all probes in one hole. The electrical circuit shown in Fig. 7 is then required. The data transmission cable runs between the probes and the data collection box in a watertight stainless steel pipe (length 1-2 metres).

The processing computer sends a pulse to the circuit shown, which is contained in each probe, which starts a comparator. This then compares a rising electrical voltage (0 start) with the voltage present across the sensor under test and when the two voltages are equal a pulse is returned to the computer. The elapsed time (5-55 s) is measured in the multiprogrammer (in s) and stored. A pause of approx. 10 ms is programmed between the testing of each sensor. After the first probe, with its 3 sensors, has been tested, the next probe in the same borehole is initiated by an electric pulse. This procedure is repeated until all probes in all boreholes have been tested. The time period between the individual measurement cycles is determined in the measurement programme by the computer.

A second independent measurement system is installed for the compilation of measurement values in the central injection hole. In this system the measurement values are decoded by separate electrical circuits prior to storage on a floppy disc.

The third measurement system allows the scanning of individual temperature, pressure and conductivity sensors. This individual scanning is mainly required for the regulation of the pumping equipment. The system diagram of the pump units, packers, probes and steel pipes is shown in Fig. 8. The steel pipes are designed to allow calibration of the equipment when not on the test site. Three high pressure hydraulically driven piston pumps are used.

Power for the hydraulics is provided by either a 55 Kw electric or 55 PS diesel motor. The pumps are laid out according to capacity required, i.e. according to pumping volume and pump head. Nominal values may be as follows:

	Pump Volume	Pump Head
1. Pump	Q max = 340 l/min	H = 400 m
2. Pump	Q max = 70 l/min	H = 400 m
3. Pump	Q max = 12 l/min	H = 7000 m

The pump head or pump volume may be regulated according to test requirements. This regulation is achieved either by regulating pump revolutions or with by-pass valves. The measurement of water volume is carried out electronically using flow volume measurement equipment of various accuracies.

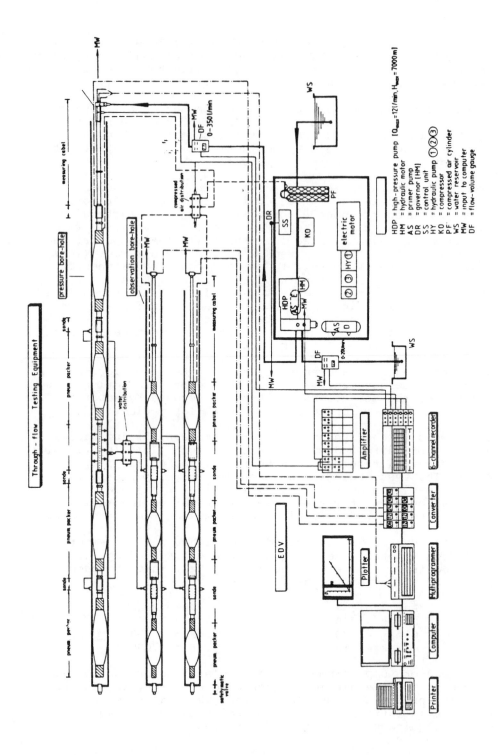

Fig. 5: Through-Flow Testing Equipment

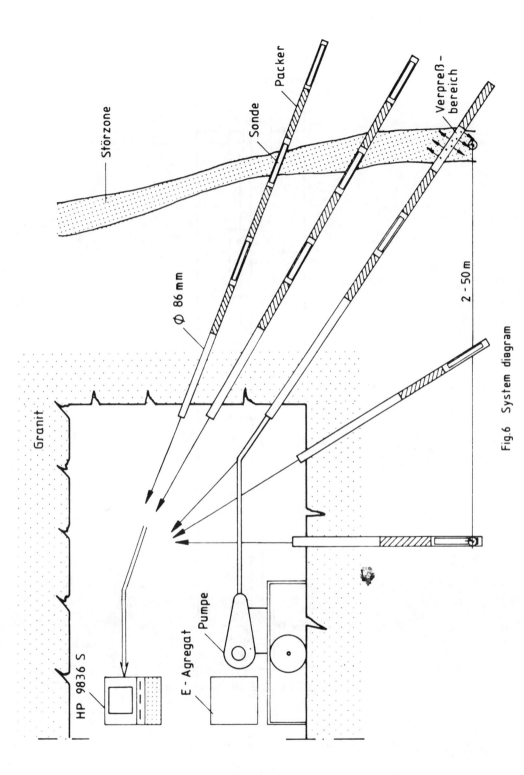

Störzone

Packer

Sonde

Verpreß –
bereich

Ø 86 mm

2 - 50 m

Granit

HP 9836 S

E - Agregat

Pumpe

Fig.6 System diagram

165

Sonden:

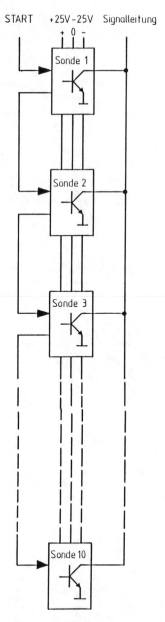

Fig. 7: The sondes are connected with 5-poles jacks

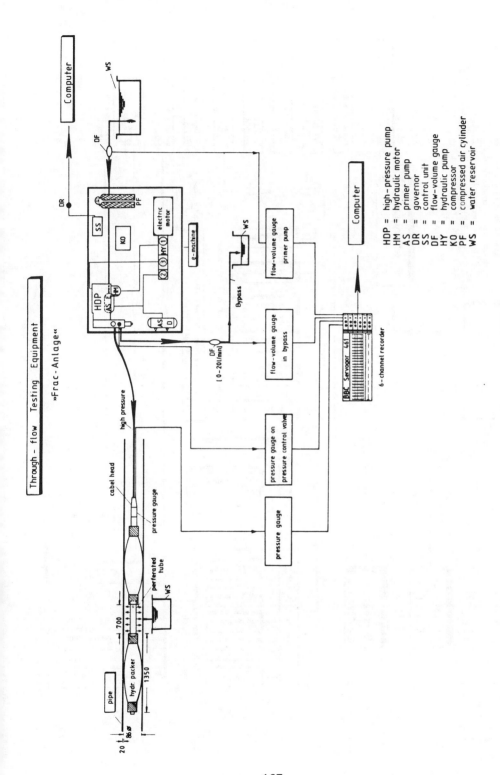

Fig. 8: Through-flow Testing Equipment "Frac-Anlage"

HDP = high-pressure pump
HM = hydraulic motor
AS = primer pump
DR = governor
SS = control unit
DF = flow-volume gauge
HY = hydraulic pump
KO = compressor
PF = compressed air cylinder
WS = water reservoir

167

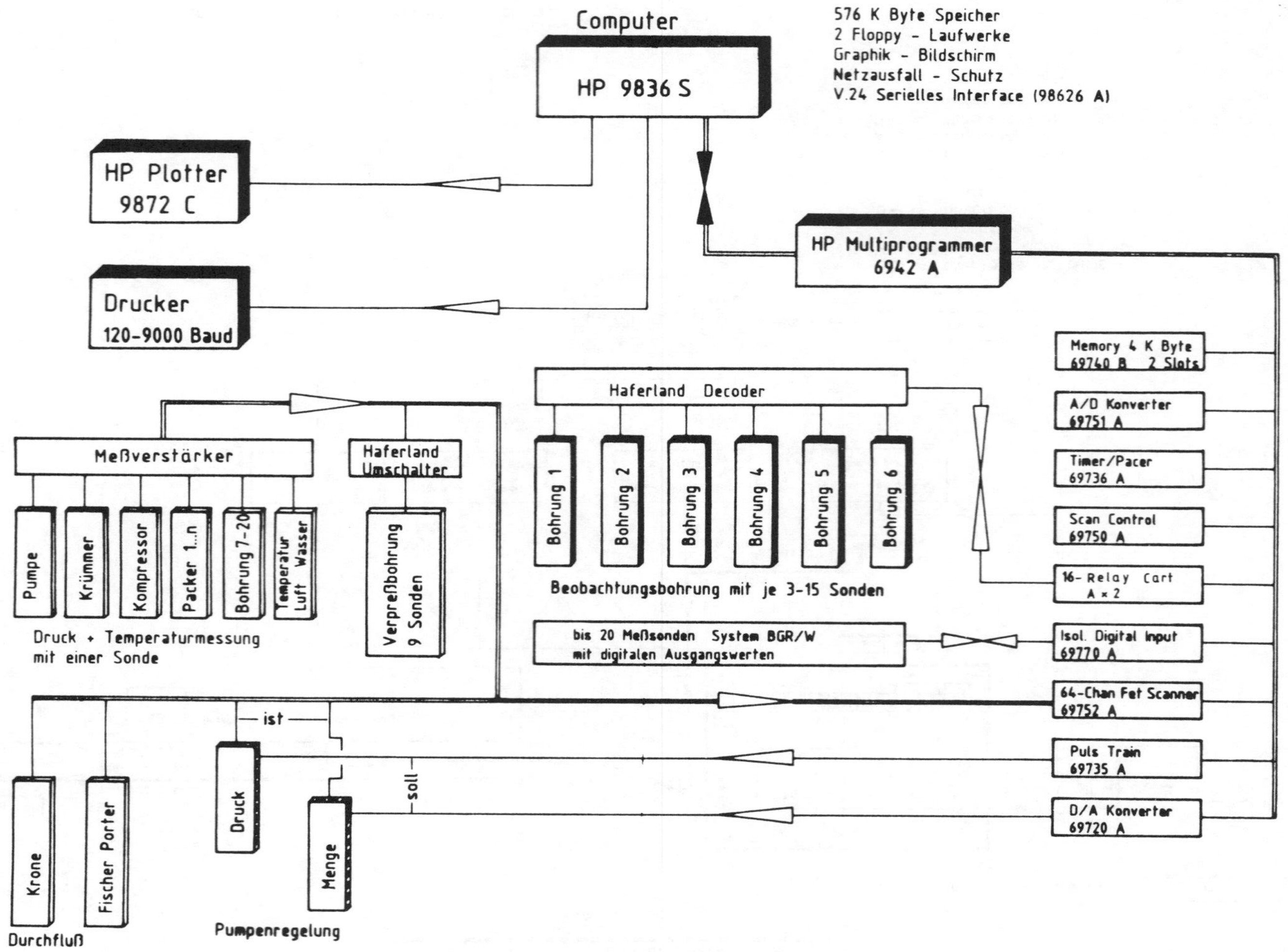

Fig. 9: Computer

The high pressure pump (pump head of 7000 m) allows not only permeability measurements but also rock fracturing (Fig. 8) and stress measurement within the rock. For safety reasons hydraulic packers are used. It is intended to carry out measurements of the absorbable frac pressure of the rock and stress measurement in, amongst other sites, the rock stress tunnel (GS) at a depth of around 150 to 200 m below the tunnel.

The third equipment group consists mainly of one older computer and one newer computer system and two loggers. The older computer, a Compulog system, is no longer capable of meeting requirements because of its low performance capacity and unreliability.

The new computer is, as shown in Fig. 9, also divided into the following three subgroups:

a. Computer, HP 9836S, douple floppy disc unit, with graphic facility, power failure protection and a V24 serial interface.

b. The second subgroup consists of a plotter (DIN A 3) and a printer. Both units are connected to the HP 9836S as is the multi-programmer.

c. The multi-programmer has 9 circuits, which, because of requirements, are in part independent of one another. The regulation of the measurement value scanning is carried out by a computer programme aided by the "Scan Control" card which, depending upon requirements, uses either the 64 channel Fet Scanner, 16 Relay card or Isol. Digital Input card, as shown in Fig. 9. The analoque measurement values are converted into digital values using the A/D Con erter Card and then stored in the Memory Card.

   The timer periods between the previously described pulses from the probes in the borehole are determined by the timer/pacer card. Short-term storage of all measurement values on the memory card of the multi-programmer serves to considerably reduce time taken for the scanning cycle. The data may then be smoothed by a seperate program and stored on a floppy disc and/or be plotted on the screen or the plotter. The pulse train and the D/A converter card allow the flow volume to be controlled by a stepping motor. The pressure head is controlled by a potentiometer. The desired accuracy is achieved by comparing nominal and actual values.

5. Test Execution

The actual test procedure may first be decided after the two above-mentioned boreholes have been sunk and simple flow tests carried out. These two boreholes will allow a rough estimation of the extension of the geologically-inferred water bearing fissures.

The pilot tests and the information thus gained will determine the actual execution of the flow tests using either pressure or injection volume regulation. For these tests the central injection hole may be used as observation hole and one of the observation holes used as injection hole. It is intended that the pressure or water volume during the testing are either constant, rise or fall continuously, or rise or fall in steps. The duration of the steps will also be varied. The single holes will be so set as to allow if possible a closed water system to be evaluated. It is feasible that circumstances may require bore-holes in the far laboratory tunnel and the main access tunnel, e.g. if during injection tests a pressure decrease or increase were observed in the far observation points.

The permeability of the rock is dependent upon the excavation method used, at least in the close vicinity of the gallery. Should it be established that an increased permeability is present in the excavation area then this must be taken into account when planning the further tests.

6. Finite Element Calculations

Natural water pressure may rise in isolated areas in an open system up to 40 bar, this is approximately equivalent to the overlying height of the rock with respect to the test site.

However, in a closed fissure system heat expansion or possible creep in certain rock sections may cause the pressure to reach rock pressure. Thermal expansion of the rock is, for example, a possibility where highly radioactive wastes are stored. For these reasons a stress expansion calculation using the finite element programme ADINAT 8 was carried out. An 8 point iso-parametric element using the modified material law of Drucker Prager was used. Fig. 10 shows a section from a network with 140 elements and the fundamental rock mechanical parameters. For reasons of symmetry only one half of a 13 m long vertical fissure, 5 m below a tunnel, was simulated.

In evaluating the calculation results two major systems were distinguished:

1. Fissure surfaces having contact pressure (fully relaxed water pressure in the fissure) equivalent to petrostatic pressure.

2. Fissure surfaces having no contact, even without fissure wa-ter pressure.

Results for Case 1 are shown in Figs. 11 and 12. Two cases were assumed, one with rock pressure of 25 MPa (Fig. 11) and one with a rock pressure of 10 MPa (Fig. 12), corresponding to an overlying rock height of 1000 m and 400 m respectively.

During the calculations fissure water pressure was continuously raised. Only as it reached approx. 50 % of rock pressure was a

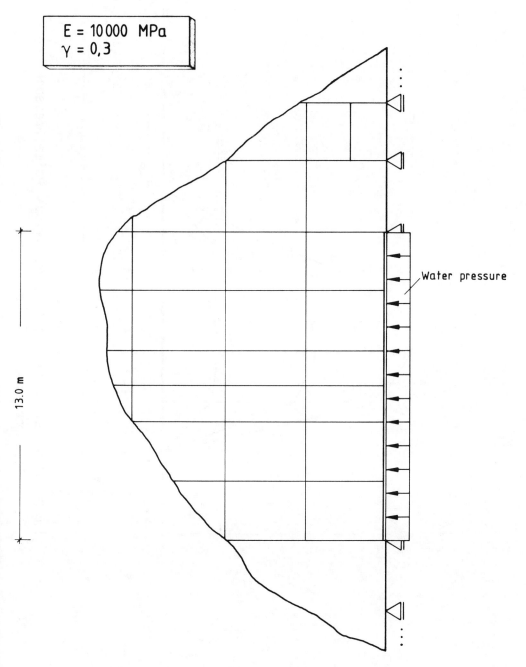

Material properties

E = 10 000 MPa
γ = 0,3

13.0 m

Water pressure

Fig. 10 Central portion of finite element mesh

171

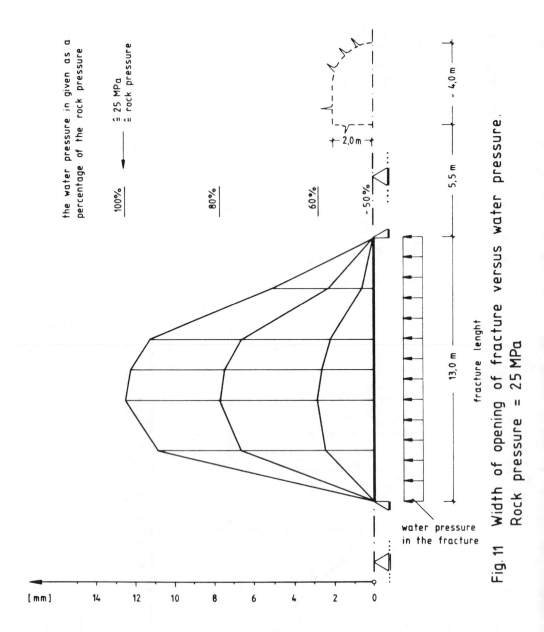

the water pressure in given as a percentage of the rock pressure

≙ 25 MPa
≙ rock pressure

100%

80%

60%

- 50 %

2,0 m

5,5 m

- 4,0 m

13,0 m

fracture lenght

[mm]

14

12

10

8

6

4

2

0

water pressure in the fracture

Fig. 11 Width of opening of fracture versus water pressure. Rock pressure = 25 MPa

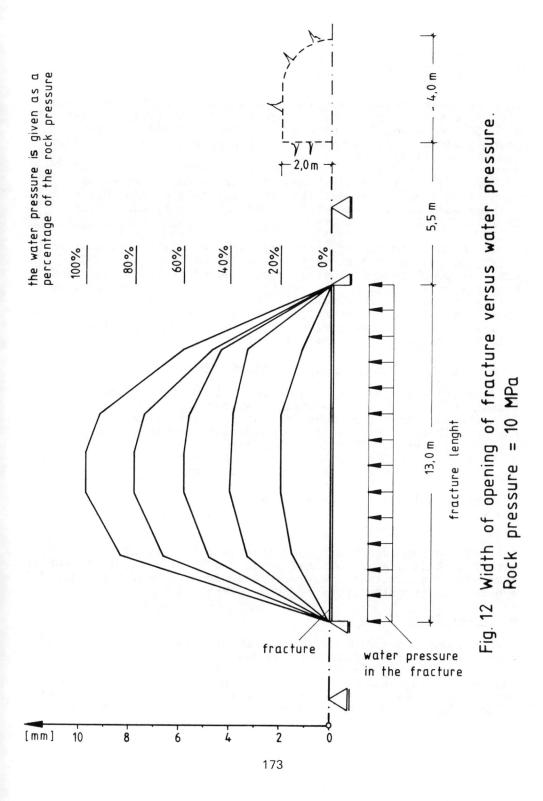

Fig. 12 Width of opening of fracture versus water pressure.
Rock pressure = 10 MPa

173

fissure opening in evidence. The width of opening depends amongst other factors on both the rock rigidity and fissure dimensions. In the system considered maximum unilate ral fissure openings of 12.5 mm and 4,6 mm were calculated. Fissure water pressure was 25 MPa and 10 MPa respectively.

In cases where fissure surfaces are not in contact even without fissure water pressure then a rock pressure of 10 MPa results in a maximum opening of approx. 9.8 mm. The form and individual opening width in dependance upon water pressure in the fissure can be seen in Fig. 13.

The opening sizes dependent upon fissure water pressure so found give a basis for calculating the possible permeability changes when carrying out future estimates of fissure water flow.

It is intended to carry out the fissure water flow calculations using the finite element method. The FE programme ADINAT [8] [9] will be used. The hydrological recording of the test field will show the areas which may be described as being either continuum or discontinuum. Initial calculation results have shown that spatial models are necessary to describe the flow processes. The verification and presentation of the results in a publication will have to wait because of lack of examples.

7. Comparison with other investigations

Small tests are possible in-situ or in laboratory using a granite block with edge lengths of approx. 1 m. Such tests have the advantage that they present a closed system and thus any artificial or natural fissures present may be measured exactly. The clear parameters then also allow the best possible graduation of the FE model. The extrapolation then neces sary up to full size rock areas is thus more reliable.

A block or slice model may also be used to simulate, under suitable parameters, single fissures of preset opening width, roughness and fissure material.

The tests under laboratory conditions may be carried out in either rock laboratory Grimsel or in the laboratory of the BGR in Hannover. Similar tests have been described by BARTON (1982) [6] and for smaller dimensions by EVANS (1983) [7].

8. Tracer Tests

The previously described probes allow not only the measurement of pressure but also temperature and salinity.

Raising the temperature of the injection water is carried out using a flow heater. Temperature increases of up to 40 °C provide a basis for considering the storage capacity of the near field.

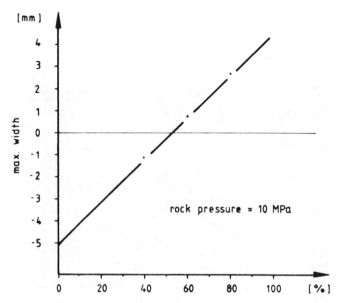

Fig. 13: Maximum width of opening of fracture versus water
pressure

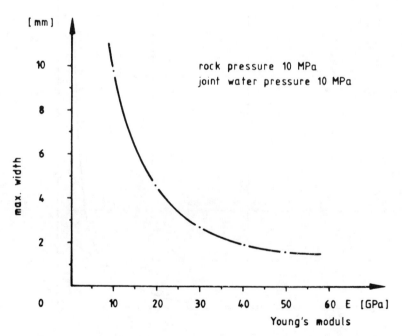

Fig. 13a: Maximum width of opening of fracture versus rock
stability

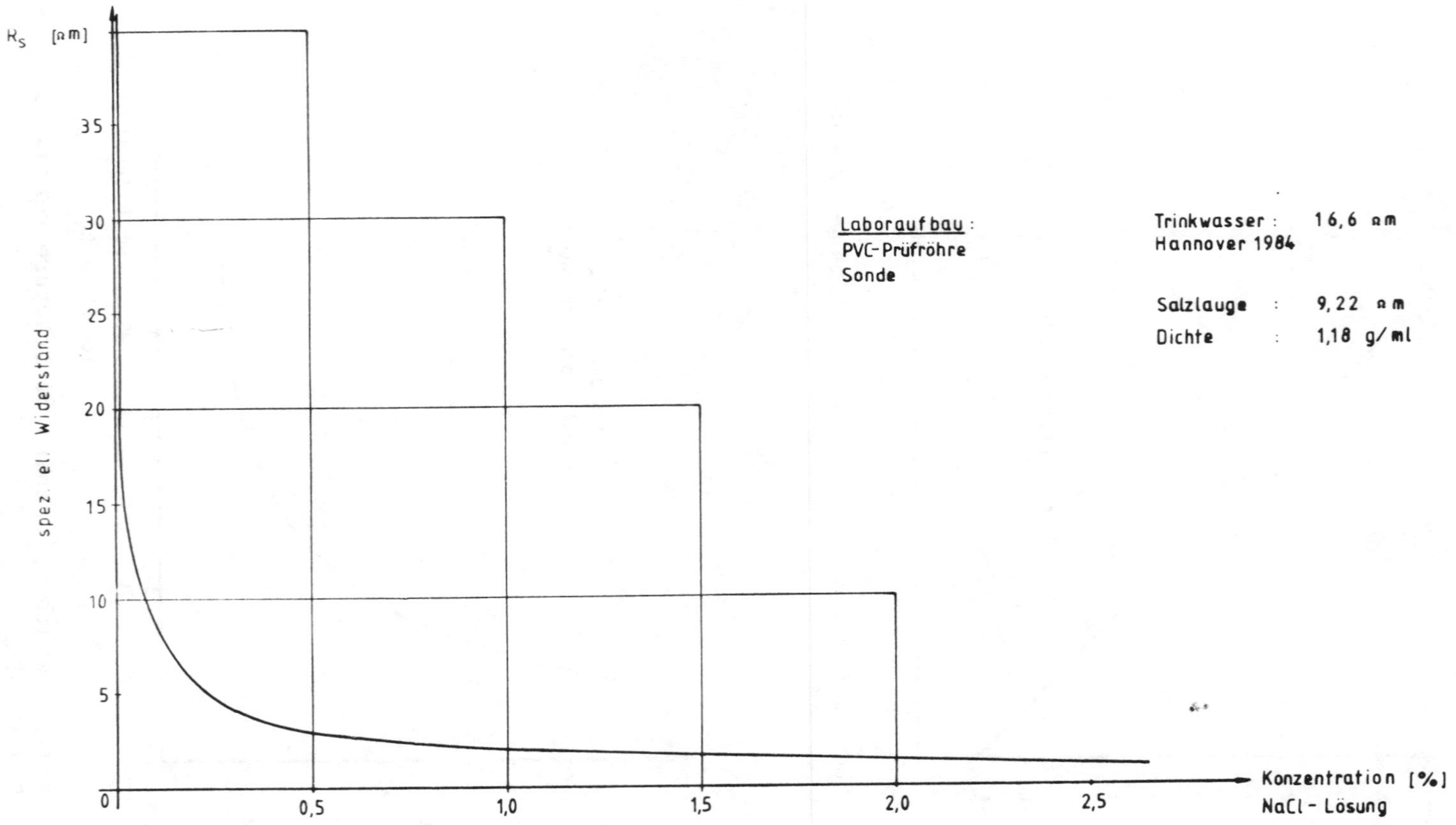

Abb 14 Abhängigkeit des spez. el. Widerstandes vom Sättigungsgrad einer NaCl-Lösung

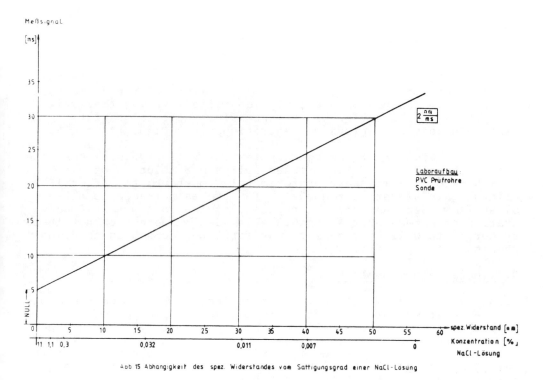

Abb 15 Abhangigkeit des spez. Widerstandes vom Sattigungsgrad einer NaCl-Losung

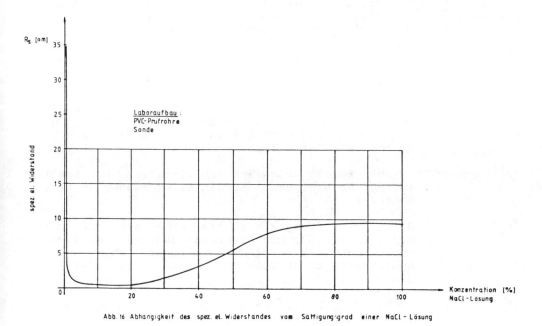

Abb. 16 Abhangigkeit des spez. el. Widerstandes vom Sattigungsgrad einer NaCl-Lösung

This type of tracer is only useful for tests of short-term duration because of the very short distances and fast cooling
involved.

These problems may be avoided by using brine solutions. Concentrations of approx. 0.3 %, i.e. approx. 1/10th of North Sea water
salinity are clearly and reliably identifiable using the available probes. For a review of the specific electrical resistance of
salt solutions of various concentrations, see Figs. 14 and 15.

From these Figures it can be seen that as salinity increases, the
specific resistance falls exponentially to a solution of 5 %,
from 5 % to 15 % it stays approximately constant, and after 16 %
climbs again almost continuously. For use as a tracer to establish flow speed and storage capacity of the rock a salt concentration of between 0.2 % and 0.3 % is ideal. Whether or not these
tracer tests will be necessary may first be decided in the final
phases of the project.

9. Timetable and outlook

Finally, I would like to briefly describe the projected timetable
of the modified water injection tests. The completion of the
comprehensive programme will require approximately 5 years. The
preparatory work, as for example simulation tests for testing of
test apparatus, was started in 1983 in the test department of the
BGR. The geological recording with evaluation is in progress. The
in-situ tests in the NAGRA Rock Laboratory will be carried out
during 1985/1986 at the sites described. Following this a second
test site is being considered in order to test a rock mass
complex of lower hydraulic conductivity because at the present
test site the majority of the jointed rock mass is expected to
have a relatively high hydraulic conductivity.

References

[1] NAGRA   Nagra informs, Baden, July 1983

[2] PFISTER, E., Proposals for in-situ research in the proposed la-
                boratory at Grimsel in Switzerland, Proc.
                Workshop on Geological Disposal of Radioactive
                Waste, NEA, Paris 1983

[3] BUNDESANSTALT FÜR GEOWISSENSCHAFTEN UND ROHSTOFFE, Untersuchun-
                gen zur Ausbreitung kontaminierter Wässer im Fels,
                Hannover 1982

[4] SCHNEIDER, H.J., Ausbreitung kontaminierter Wässer im Fels und
                Konsequenzen für die Kavernenbauweise, Symposium
                Underground Siting of Nuclear Power Plants
                Bundesanstalt für Geowissenschaften und Rohstoffe
                Hannover 1982

[5] PAHL, A. u. SCHNEIDER, H.J., Standortmöglichkeiten für unterir-
                dische Kernkraftwerke im Fels aus ingenieur-

geologisch-felsmechanischer Sicht
Symposium Underground Siting of Nuclear Power Plants
Bundesanstalt für Geowissenschaften und Rohstoffe
16. - 20. März 1981, Hannover 1982

[6] BARTON, N.,Modelling Rock Joint Behaviour for In Situ Block
      Tests; Implications for Nuclear Waste Repository
      Design, ONWI-308, Terra-Tek Distribution Category
      UC-70 Salt Lake City 1982

[7] EVANS, D.D., Unsaturated Flow and Transport Through Fractured
      Rock-Related to High-Level Waste Repositories,
      Nuregier-3206 Department of Hydrology and Water
      Resources, University of Arizona, Tucson 1983

[8] BATHE, K.-J., Finite Element Procedures in Engineering
      Analysis, Prentice-Hall International, Inc., London
      1982

[9] BATHE, K.-J., Sonnad and P, Domigan, Some Experiences using
      Finite Element Methods for Fluid Flow Problems.
      Finite Elements in Water Resources, Proceedings of
      the 4th International Conference, Hannover 1982

Migration experiments in the Stripa Mine, design and instrumentation

H.ABELIN & L.BIRGERSSON
Royal Institute of Technology, Stockholm, Sweden

ABSTRACT

Migration modelling in safety analysis for nuclear waste repositories in crystalline rock are based on the assumption that leached radionuclides (RN) will interact with the bedrock.

This paper presents two in-situ tracer experiments which objectives are to determine important parameters for retardation of RN due to the interaction.

Methods for water sampling at the face of a drift are presented. Equipment for automatic pressure pulse tests and for tracer injection with a constant overpressure are described.

1.   BACKGROUND

In a final repository for radioactive waste in crystalline rock water flowing in fissures may transport the radionuclides leached from the waste. The migration modelling in the safety analysis for this repository is based on the assumption that if and when any radionuclides are leached from the waste, practically all of the important radionuclides will interact chemically or physically with the bedrock and will thereby be considerably retarded. The magnitude of this retardation depends upon the flow rate of the water, the uptake rates and equilibria of reactions as well as the surface area in contact with the flowing water.

Most studies concerning the water flow in bedrock are based on the assumption that the water flow can be described as a porous media flow. This might be true for very large distances where the flow would encounter a multitude of channels and some averaging may be conceivable on the scale considered.

Transport over short distances, i.e. in the vicinity of the canister, most probably occur in individual fissures. On an intermediate scale where more than a few fissures conduct the flow, well type tracer tests alone cannot give the detailed information needed to understand dispersion and sorption phenomena in fissured rock.

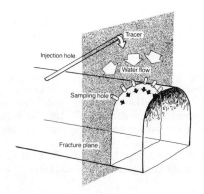

Figure 1. Schematic view of injection and sampling

Two migration experiments are underway in the Stripa mine:

o   "Migration in a single fracture", distances up to 10 meters
o   "3-D migration", distances up to 50 meters

The experiment "Migration in a single fracture" will end in jan 1985. The 3-D migration experiment is planned to end mid 1986.

2.   MIGRATION IN A SINGLE FRACTURE

2.1   Purpose

The main purpose of this investigation is to see if it is possible to extend results, on sorption and retardation of radionuclides in granitic rock, obtained from laboratory experiments to a real environment with migration distances up to 10 m.

It was also of interest to try to determine if there is any channeling within fractures. This channeling would reduce the surface that is in contact with the flowing water and thereby effect the retardation of radionuclides.

The experience gained from this 2-dimensional i.e. single fracture experiment should be used when designing and running an experiment over a larger volume of rock to study 3-dimensional dispersion and channeling.

2.2 Design of experiment

Tracers have been injected with a small constant over pressure into a single fracture which has a natural water flow towards the drift. All water coming out of the fracture at the face of the drift has been collected. Figure 1 shows a schematic view of a fracture with an injection hole and several sampling holes.

In the actual experiment <u>five injection holes instead of one</u> have been used.

Nonsorbing (conservative) tracers have been used to characterize the water flow within the fracture.

181

The results from these tracer runs and data on sorption from laboratory experiments have been used to predict breakthrough curves for a sorbing tracer. The predicted curves will later be compared with the experimentally obtained results.

Most of the sorbing tracers would not reach the sampling points within the time of the experiment. They would be strongly sorbed in the close surroundings of the injection hole. To see how far out from the injection point these tracers had reached, parts of the fracture surface around the injection point have been excavated and analysed for sorbing tracers.

As no radioactive tracers could be used stable tracers with the same chemical behavior as the important radionuclides have been used.

2.3   Preparations

A suitable fracture for the experiment was selected by monitoring the water flow rates from different fractures seen at the face of the drift. The water coming out of the fractures was collected in plastic sheets, that were glued to the wall of the drifts covering these fractures. The plastic sheet method was further developed in the 3-D migration experiment, see 3.3. The major part of the preparations were the location of intersecting points between the fracture plane and the five injection holes. To locate these intersections pressure pulse tests were done. The results from these tests were compared with core logs and TV-logs.

2.4   Pressure pulse tests

The method used, was to seal off the injection holes with single packers, pressurize one of the injection holes and monitor the water loss rate and the pressure responses in the other injection holes and in some sampling holes in the fracture. The sampling holes chosen, where to monitor the pressure responses were those with the highest water flow rates. The inter- pretation of the results is difficult because of the presence of different fractures which are interconnected by the five injection holes.

The pressure pulse test equipment consists of two major parts:

o   Data acquisition system

o   Injection system

With this equipment it is possible to do five pressure pulse test with relaxation time in between, one in each injection hole. After such a set of tests one or more packers were moved and a new set of tests was performed. A schematic drawing of the pressure pulse system is given in figure 2.

For the data acquisition system equipment from HP (Hewlett Packard) were used. A HP 3497A was used for data acquisition and A/D conversion. This unit was controlled by a micro computer (HP85).

The system monitored the water loss rate as well as the pressure responses in the injection holes and in the sampling holes. The data were stored on tape and later processed by the HP85 to give plots on pressure

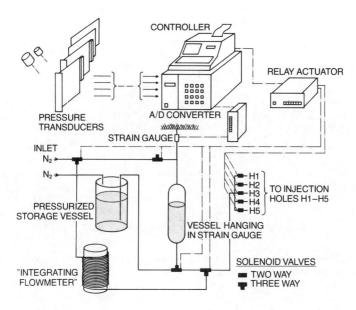

Figure 2.   Pressure pulse system

responses and water loss rate versus time. The built-in printer in the HP85 was used as a real time recorder for the pressure responses.

Piezo resistive pressure transducers were used to monitor the pressure responses at different points.

The water, contained in a metal flask which was hanging in a strain gauge, was injected by compressed nitrogen during the pulse testing. By using the data acquisition system it was possible to monitor the relief on the strain gauge caused by the emptying of the flask. The flask was automatically filled when its water content reached a certain low level. During the filling of the flask, water was injected from an other system consisting of a 500 m long nylon tube wound around a large (diameter 70 cm) card board cylinder. With the strain gauge equipment it was possible to measure water loss rates down to ml/h.

There was one problem with the strain gauge equipment. The amplifier, used to support the strain gauge, was sensitive to temperature variations. This temperature dependence was not due to the actual strain gauge but was located within the amplifier. A temperature change of a few degrees Celsius caused errors in calculated flow rates in the order of ml/h. It is possible to measure flow rates around ml/h by keeping the temperature more constant or by using a more sophisticated amplifier.

2.5 Equipment, injection of tracers

The overall requirements on the equipment were that it should be possible to do a step introduction of tracers into the fracture and that it should also be possible to overlay a nonsorbing (conservative) tracer pulse on the sorbing tracer injection. The possibility to take out small volume samples

183

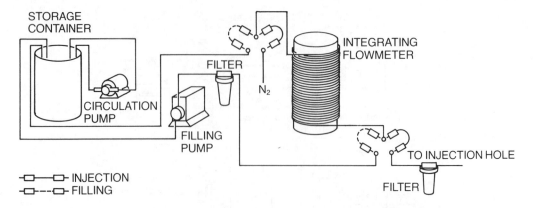

STORAGE
CONTAINER

INTEGRATING
FLOWMETER

FILTER

N₂

CIRCULATION
PUMP

FILLING
PUMP

TO INJECTION HOLE

INJECTION
FILLING

FILTER

Figure 3. Injection equipment, main parts

from the injection compartment should also exist. The main parts of the injection equipment is shown in figure 3.

The "integrating flowmeter" has a resolution of about 1 ml and a total volume of ~ 5 litre. This was also used as an intermediate tracer solution container. The flowmeter was filled by pumping tracer solution from the storage container through a 1 micron filter. The flowmeter was then pressurized by compressed nitrogen. Based on the experience from a preliminary investigation it was decided that quick connectors should be used, instead of 3- and 4-way valves, to direct the flow.

Straddled mechanical packers were used to seal off the injection point of the injection hole. The distance between the packers is 0.1 m and the sealed off volume 0.05 l. The mechanical packer is so designed that the risk that just one of the rubber sleeves will be compressed leaving the others uncompressed is reduced. It is not possible to compress anyone of the rubber sleeves more than 20 mm. The packers system is shown in figure 4.

The injection compartment is designed in such a way that it is possible to take out a small volume of the injection solution and transport it out by nitrogen, see figure 5.

All water coming out of the fracture was kept under nitrogen atmosphere from the sampling holes to the fractional collectors. The nitrogen that flushed the sampling holes were also used to speed up the transport of the water samples, from the sampling holes to the fractional collectors in the glove boxes.

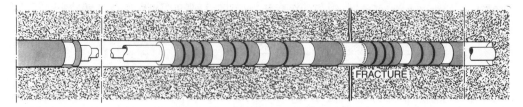

FRACTURE

Figure 4. Mechanical packer system

1. INLET VALVE N_2
2. OUTLET VALVE TRACER SAMPLE
3. CHECK VALVE
4. INLET TRACER SOLUTION

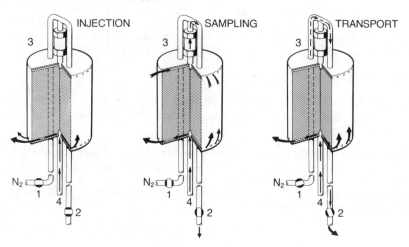

Figure 5. Injection compartment

A preparatory experiment has shown that the major problem with the water collecting system was to minimize the leakage of water past the sampling packer, due to capillarity in the fracture. By using two concentric holes when collecting water, see figure 6, this leakage could be reduced.

2.6 Analysing equipment

Two automatic analysing systems have been developed, one for ion selective electrodes and the other for multi component analysis with a spectrophotometer. The result from the multi component analysis is presented in such a way that one can compare the calculated spectrum with the actually measured one. All analysis have been done at the laboratory in Stockholm.

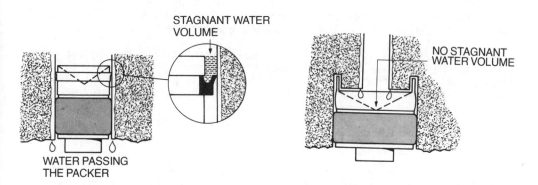

Figure 6. Design of sampling packer

185

2.7 Excavation of the fracture surface

As was mentioned earlier most of the sorbing tracers were predicted not to reach the sampling holes within the time used for the experiment. To get the possibility to determine how far from the injection point they have migrated, part of the fracture around the injection point has been excavated. This excavation was done by core drilling (diameter 200 mm) from one of the sampling holes up to the sorbing tracer injection point.

2.8 General conclusions and discussion

A problem when injecting with a small constant overpressure is that minor variations of the injection pressure causes large variations in the injection flow rate.

Using quick connectors instead of three and four way valves to direct the flow made the system tremendously flexible without any major piping work and it functioned very well. This might not be recommended when running "hot" experiments because now and then a drop of liquid emerges when connecting or disconnecting the male quick connectors.

The data acquisition system with the micro computer, all from HEWLETT PACKARD, was reliable and the autostart rutines worked well.

The "integrating flowmeter", that is the 500 m long nylon tubing wound around a large card board cylinder, is a very good instrument which gives direct information of how much fluid has been injected and knowing the time between two readings one can calculate the average injection flow rate. The flow rate can be calculated within 5 % and the total injected volume within 0.5 %.

In the pressure pulse equipment not only the output from the strain gauge amplifier should be monitored but also the feed to the actual strain gauge. This will make it possible to compensate for changes caused by temperature variations.

When working with small margins it has been noted that even small fluctuations of the surrounding temperature causes a lot of problems. More effort should be put into the possibility to keep the test site at constant temperature or else one has to use refined equipment. Even if measures have been taken to keep the temperature constant one should use the data acquisition system to also monitor the surrounding temperature.

The problem with treating all water samples in the same way was underestimated. Methods have to be developed that makes it possible to treat water samples, containing tracers that are light and/or temperature sensitive in the same way independent of weather conditions until analysis are performed.

3.   3-D MIGRATION

3.1 Purpose

The objective of this project is to get an understanding of the spacial distribution of water pathways in crystalline rock over long distance (up to 50 m) and at large depths (360 m).

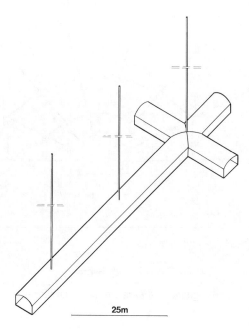

25m

Figure 7. Layout of the test site with the three vertical injection holes.

As channeling and transverse dispersion may have a profound influence on the arrival times and concentrations of radionuclides escaping from a repository, it is important that these effects are studied.

The results from the experiment will be used for water flow model verification and/or modification.

3.2   Design of experiment

The experiment will take place in the Stripa mine at the 360-m level. Water flows constantly into drifts at this level since it is located well below the water table. Conservative tracers will be injected into this water flow.

From the test site, which has a total length of 100 m, see figure 7, 3 vertical holes (length ~ 70 m) have been drilled.

Injection of tracers will be carried out from up to 3 separate high permeability zones (each zone ~ 2 m) in each of the 70 m holes. The injection zones will be placed approximately 20, 30 and 50 m above the drift.

Water will be collected at at least 350 sampling "points" located at the ceiling of the test site. The sampling arrangement covers a surface area of ~ 700 m^2.

187

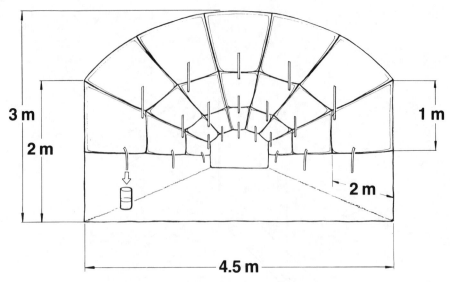

Figure 8. Sampling arrangement

3.3 Preparations

The major preparations of the test site, i.e. activities before start of tracer runs, are described below.

- Excavation of the test site

- Preparation of sampling "points"

- Location of the 2 m injection zones

Excavation of the test site

Since the distance from the injection zones to the ceiling of the drift will be up to 50 m, a "safety distance" to other major excavations at higher levels is required. By making an access drift (~ 135 m), before excavating the test site, the distance from the injection zones to other excavations at higher levels was extended to at least 100 m.

Preparation of sampling "points"

The upper part of the test site which has an area of ~ 700 m² has been covered with plastic sheets, each sheet with an area of about 2 m² (see figure 8).

This means a total number of 350-400 sheets, that will serve as sampling "points" for water emerging from the rock.

The advantages with covering the ceiling with plastic sheets instead of drilling sampling holes are:

188

-   A better water (and tracer) mass balance can be made.

-   A better sampling density can be obtained.

And the disadvantages:

-   Fissures with their orientation parallel to the drift tend to be closed
    near the face of the drift due to the release of rock stresses.

-   Methods for covering this large area (~ 700 m^2) had to be developed.

The problems with the sheet covering method is that a wall of a drift
is usually moist and that a dust layer remains from the excavation. To overcome these problems, the whole surface was first cleaned with a high
pressure cleaner and before covering with plastic sheets, the surface was
dryed with heated air.

Several glues and plastic materials were tested. The glue that worked
best, i.e. had a good adhesiveness to a moist surface was a one component
construction glue, GOODRICH PL 400. Among the different plastic sheets that
were tested, a thin (0.1 mm thich) PVC-film was found to be easy to handle and
worked well with the glue.

Location of the 2 m injection zones

As mentioned earlier injection will be carried out from up to three
separate high permeability zones in each injection hole.

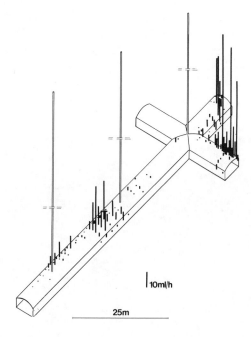

Figure 9. Water flow rates into sampling sheets.

189

In order to find these high permeability zones, the water inflow into 2 m sections of the injection holes will be monitored.

3.4   Discussion

The plastic sheet covering method seems to work well. The sheets has now (May -84) been up for 6 months without any problem. This sampling arrangement has made it possible to collect all water emerging from a 700 m^2 area. The results up to now from the water collection shows that there is a pronounced channeling within the crystalline rock, see figure 9.

Solute migration experiments in fractured granite, South West England

M.J.HEATH
Elcon Western (Electrical) Ltd, Research Services Section, Redruth, UK

ABSTRACT

Experimental techniques have been developed for the investigation in situ of solute migration in fractured granite. The location of suitable fractures providing long flow paths without intersection for such experiments may be difficult, especially if fracture frequency is high. A flow field involving pressurised injection and passive withdrawal has been necessitated by working above the water table. Steady state flow rates have been difficult to maintain for the long periods required as flow rates have progressively fallen. This is attributed to clogging by particulate matter of natural origin or introduced by pumping. The importance of selecting suitable solute concentrations, low enough to allow full interaction with fracture faces and high enough for subsequent detection, has been demonstrated.

1. OBJECTIVES AND CONSTRAINTS

The research being undertaken at the Troon experimental site in the Carnmenellis granite in Cornwall forms part of a research programme aimed at developing experimental and theoretical techniques that may be applied at prospective repository sites to investigation of the migration of radionuclides following their release from buried radioactive waste into the rock mass. The return of these hazardous substances to the biosphere will depend on the hydraulic flow characteristics of the rock mass involved, and on the behaviour of specific nuclides within that rock-water environment. A two-fold approach has therefore been adopted. The fracture permeability of the granite at the site has been investigated and mathematical techniques are being developed to enable the hydraulic flow characteristics of the rock mass to be modelled [1,2,3] . Experimental techniques whereby the migration of individual solutes along water-bearing fractures may be investigated are also being developed. It is this aspect of the work that is discussed in this paper.

Fracture networks are made up of individual fractures along which radionuclides may be transported. If radionuclide migration through such networks is to be understood, the behaviour of these solutes in individual fractures must be determined. If their results are to be applied to transport modelling, therefore, migration experiments must be carried out using single fractures without

191

intersection. However, if the retardation of a pulse of a solute by diffusion into the rock matrix and sorption is to be investigated, the time-dependency of these processes necessitates a long residence time for that solute in the fracture during the experiment. This can be provided by low flow rates and long flow paths. These constraints may be difficult to reconcile, particularly if, as at the Troon site, fracture frequency is high, as individual fractures without intersection that may be drilled to provide long flow paths may be difficult to locate.

A detailed description of the geology with a discussion of the permeability investigations carried out at the Troon site has been given by Heath [4]. A summary of other research into the structural properties of the granite at the site is being presented by Cooling and Tunbridge [5].

2. EXPERIMENTAL DESIGN

2.1 Selection of fracture

Early experiments have been carried out in a disused quarry, approximately 10m deep, using fractures made accessible by drilling into suitable rock faces. The fracture frequency at the site is high,

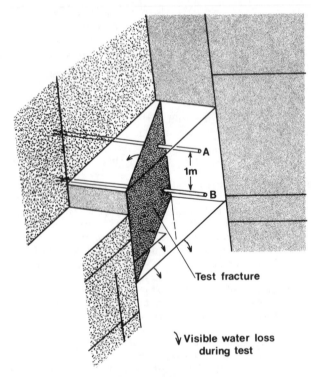

FIGURE 1. Arrangement of test fracture and boreholes.

192

as steeply-dipping fractures, two strongly-developed conjugate sets of which characterise the granite at depth, are supplemented near the surface by sub-horizontal stress-relief joints. A number of possible test fractures, satisfying the requirements mentioned in Section 1, have, however, been located and a suitable fracture selected for pilot experiments.

Two horizontal, 37mm diameter diamond-drilled boreholes, 1m apart, one vertically above the other, have been drilled to intersect the test fracture as shown in Figure 1. The fracture has been accurately located in each borehole by examination of the core which has also provided material for mineralogical and geochemical analysis and for diffusion investigations in the laboratory.

2.2   Hydraulic system

The layout of the hydraulic system employed in the Troon experiments is illustrated in Figure 2. The purpose of this system is to isolate an individual fracture and pass various solutes along it between two boreholes. Analysis of the results of solute migration experiments is often made difficult by the complex flow fields set up between the two boreholes. Breakthrough curves for non-sorbed and weakly-sorbed species from injection-withdrawal (dipole) experiments, for example, tend to be controlled largely by the effects of the many flow paths present: some short and some very long. Ideally, radial flow towards the actively pumped withdrawal hole, with passive injection into the other hole, is employed, in which case a direct flow path can be assumed and high solute recovery values may be anticipated. This preferred flow field has not been possible at the Troon site as most fractures accessible from the quarry faces are above the water table, locally lowered by the effects of the quarry itself, with its many boreholes, and of an adjacent mine. Active pumping from the withdrawal hole therefore results in air being drawn into the fracture. The flow field employed therefore involves active pumping under pressure into the injection hole with passive withdrawal from the other hole. The effect of opening the latter to atmospheric pressure is to produce a sink towards which water will tend to flow, thus producing a complex "semi-dipole" flow field with resultant difficulty in interpretation of experimental results.

Access to the experimental fracture has been facilitated by setting hydraulically inflated rubber packers, each 30cm long, in the borehole on either side of the fracture to isolate a test section, the volume of which has been minimised by inserting a large stainless steel spacer leaving the test zone in the form of an annulus. Around the central perimeter of the spacer, three ports have been machined to allow injection of solutes and access to a mixing loop (see below). The packers have been inflated to pressures of approximately 20m head above the injection pressure used in each test to ensure tight fitting in the borehole. These pressures have been recorded by pressure transducers linked both to a multi-channel pressure indicator and a chart recorder.

Pressurised water supply has been provided by a triplex ram pump via a double filtering system to remove particles of greater

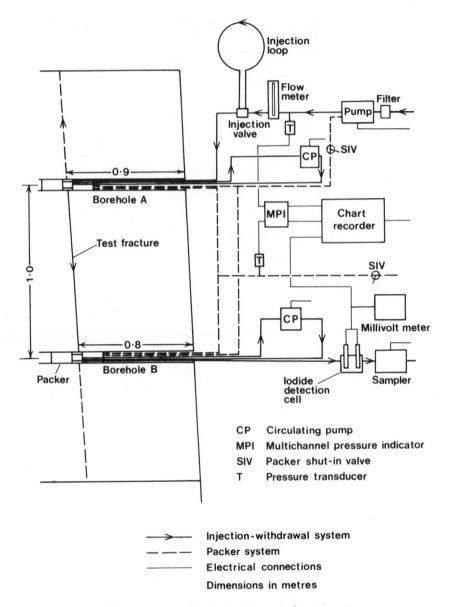

CP Circulating pump
MPI Multichannel pressure indicator
SIV Packer shut-in valve
T Pressure transducer

→ Injection-withdrawal system
--- Packer system
— Electrical connections
Dimensions in metres

FIGURE 2. Lay-out of experimental apparatus.

than 100 μm diameter. Injection flow rates have been measured using rotameters in which a steel ball rises in a tapered glass tube. Flow rates as low as 0.25ml/min. may be measured in this way but the rotameter is susceptible to calibration drift during long experiments due to particulate matter, especially of organic (algal) origin, being deposited in the glass tube. Withdrawal flow rates, as low as 0.1ml/min. have been measured manually.

Injection of solutes has been accomplished using an injection loop, of any desired volume, isolated from the main supply tube by a pair of valves which are turned at the required time to direct flow through the loop. Solutes have therefore been injected as pulses of short duration. When analysing the results of these experiments, it must be assumed that the entire volume of the injection system has been involved in the input of solute. To avoid areas of stagnant water, therefore, a mixing loop has been installed, around which water is pumped at a rate that is fast compared with the flow of water into the fracture. A similar mixing loop has been installed in the withdrawal system. The injection mixing loop is, of course, pressurised (at up to 50m head in experiments to date) so that the pump installed in it must operate at these pressures. Peristaltic pumps (as installed in the withdrawal mixing loop) do not operate at high pressures, while a piston pump installed initially produced unacceptable pressure perturbations in the injection system. A gear pump has now been produced for the purpose.

As the form of the solute input function and breakthrough curve will be modified by dispersion in the hydraulic system, especially when flow rates are very low (which is desirable), the volume of that system has been minimised by employing an annular test zone and by keeping hydraulic pipes as short and as narrow as possible. The volume of the system has been carefully measured in order to estimate the effect of dispersion.

As the experimental system must be applicable to studies not only of non-sorbed but also of strongly sorbed solutes, suitable materials have been used in its construction. Thus, all fittings with which solutes are brought into contact have been machined in stainless steel, and nylon hydraulic pipes have been used. The possibility that the rubber packer sleeves might sorb strontium or caesium has also been tested and discounted.

2.3   Analysis

Early experiments were carried out using the non-sorbing iodide ion injected as potassium iodide in aqueous solution and analysed in situ using ion specific electrodes. An iodide detection cell was prepared by housing the electrode pair in a block of clear perspex machined to minimise the volume of water in contact with the electrodes, thus making the detector respond quickly to changes in concetration despite the very low flow rates (Figure 3). Iodide ion specific electrodes are sensitive to concentrations as low as about 10^{-7} molar (M), though stabilisation is very slow to concentrations of less than $10^{-5}M$. The calibration was found to remain constant even after tests of many weeks duration. It was hoped to install an iodide detection cell in the mixing loop of the injection system to monitor the injection concentration as the pulse passed though the injection zone. However, while the iodide electrode is solid state, the sleeve-type reference electrode allowed water to enter around the inner cone under high injection pressures. A solid state combined electrode, such as that used for fluoride analysis, would be better suited to this purpose.

195

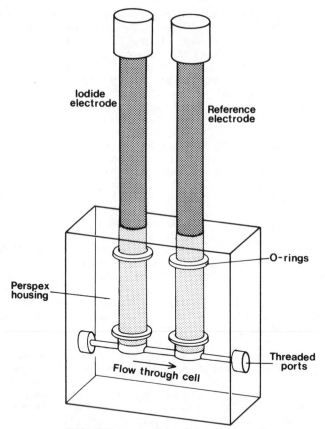

FIGURE 3. Iodide detection cell.

Later experiments have involved the simultaneous injection of iodide with either strontium (as strontium nitrate) or caesium (as caesium chloride). Analysis for both strontium and caesium has been carried out by atomic absorption spectrophotometry (AA) with detection limits of about 10^{-7}M and 10^{-6}M respectively. Samples have been collected downstream of the iodide detection cell, using an automatic sampler. Analysis by AA requires samples of several millilitres, which may take long periods to collect when flow rates are only around 0.1ml/min. Sampling periods of 30 minutes have had to be increased to 60 minutes or more as flow rates have fallen during tests (see Section 3).

Background concentrations of the solutes injected are very low (less than 10^{-7}M for iodide, 2.4×10^{-6}M for strontium and around the detection limit of 10^{-6}M for caesium).

3. HYDROLOGY OF SINGLE FRACTURES

Individual fractures are commonly modelled as planar and parallel-walled. In reality, such fractures rarely exist, most being

196

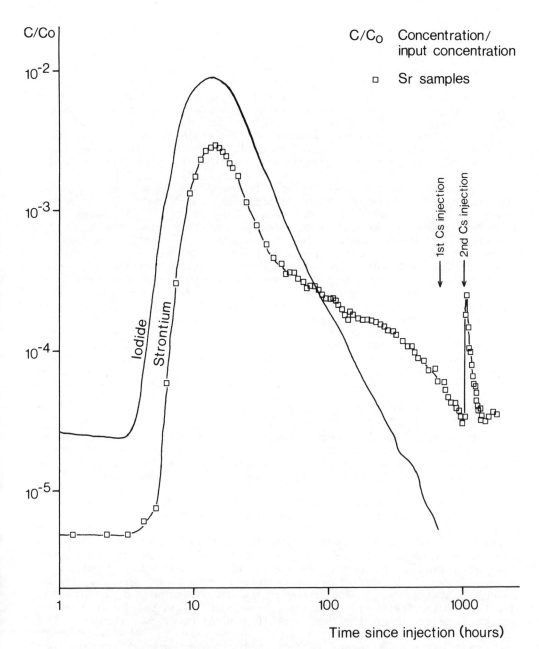

FIGURE 4. Breakthrough curves for iodide and strontium following simultaneous injection of a pulse of each solute.
(Initial iodide background elevated by previous experiment).

characterised by channelling. Where the experimental fracture at the Troon site is exposed along the top of the experimental block, for example (Figure 1), water has been observed flowing from the fracture along only one short length, 20cm or so long, during injection tests. Furthermore, the flow from that length has been found to be greater when pumping into hole B than into the much closer hole A. High iodide recovery values of around 30% are also consistent with a connection via a channel or channels between the two boreholes. Flow connections between two boreholes drilled to intersect a fracture characterised by channel flow will therefore vary considerably according to where in the fracture the holes are drilled.

Repeated pumping into the test fracture has been found to alter its hydraulic characteristics considerably in that its effective aperture has progressively decreased. This effect, also observed in tests at depth in vertical boreholes, is probably attributable to the blocking of narrow channels in the fracture by particulate matter. Naturally occurring alteration products such as clays, drilling debris, fine detritus, including algal debris, introduced during injection tests or the effects of chemical precipitation, especially if the water being injected differs geochemically from the water occurring naturally in the fracture, may be responsible for this clogging. Whatever the cause, flow rate variations present difficulties for the analysis of the results of solute migration experiments as steady state flow rates are required for modelling. Such flow rates have been achieved for long periods (up to several weeks) but have invariably given way to falling flow rates, often as the result of disruption to pumping due to power failure or equipment malfunction.

4.   MULTI-SOLUTE INJECTION EXPERIMENTS

A number of experiments involving the simultaneous injection of equal molecular concentrations of non-sorbing iodide ions with either sorbing strontium or strongly-sorbing caesium ions have been carried out. Typical results are shown in Figure 4. As the injection and withdrawal boreholes are only 1m apart, the residence time of each solute is short compared with the time required for retarding processes such as diffusion or sorption to take place. However, the breakthrough curves for iodide and strontium illustrated are significantly different. Initial breakthrough and peak arrival times are very similar, but the strontium peak concentration is lower than that of iodide by a factor of four, and the "tail" for strontium is much longer, the strontium concentration exceeding that of iodide after about 90 hours. That most of the strontium passed through the fracture with the iodide suggests that the strontium input concentration was too high for the short flow path involved. At about 1000 hours, the strontium breakthrough curve shows a secondary peak corresponding to the second of two caesium injections. It appears that the presence of caesium ions competing for sorption sites caused strontium to be eluted from the fracture. Experiments with individual solutes may not, therefore, give a true indication of their behaviour in groundwater systems in the presence of other solutes.

Two tests involving the simultaneous injection of iodide with strongly sorbing caesium have been carried out. The concentration of caesium injected in the first test was so low that none had been detected in water from the withdrawal hole two weeks after injection. The small amount of the element injected (5 x 10^{-4} moles) was perhaps entirely sorbed by the fracture surfaces. Alternatively, caesium may have come through at such a low concentration that it was below the 10^{-6}M detection limit of analysis. A higher caesium concentration was injected in the second test and a breakthrough curve for the element similar to that for strontium in Figure 4 began to develop. A falling flow rate after about 500 hours modified the curve by maintaining relatively high caesium concentrations and the test was abandoned after 750 hours when flow stopped.

5.   CONCLUSIONS

An in situ experimental system for the investigation of solute migration in fractured granite has been described. The processes by which radionuclides will be retarded as they are transported by groundwater along fractures (diffusion and sorption) are slow in operation. Flow paths and residence times for solutes within fractures used in migration experiments should, therefore, be long. However, analysis of the results of these experiments requires a single fracture without intersection. These requirements may be difficult to reconcile, especially if fracture frequency is high. Ideal flow fields involving radial flow to a withdrawal hole may not be possible if fractures have been drained; for example, when working above the water table or adjacent to mine workings.

In order to minimise the effects of dispersion of solutes within the hydraulic system, the volume of that system should be reduced to a minimum. This has been accomplished by reducing the packer-isolated test zones to annular form around central spacers, by using hydraulic pipes of small radius and by locating all service equipment as close to the fracture as possible.

Interpretation of the results of migration experiments is difficult if steady state flow rates cannot be achieved. This requirement is difficult to satisfy over the long periods involved as the effective aperture of a fracture has been found to be progressively reduced by prolonged testing, particularly if pumping is interrupted, an effect attributed to clogging of channels by particulate matter. This problem may be reduced by thorough flushing of boreholes before experiments commence to remove drilling debris, especially if the hole has been percussively drilled. Water injection into fractures should be carefully filtered to remove suspended detritus, and should be chemically similar to water naturally occurring within the fracture; ideally, it should be collected nearby and its chemistry should be known. Once established, flow fields should not be perturbed by interruption to pumping. To this end, protection against power failure is desirable.

The results of experiments with single solutes may differ from those involving simultaneous injection of several solutes, especially if there is competition for sorption sites. It is important to select the correct solute injection concentration as too high a concentration may allow much of the solute to pass through the fracture without interacting with the rock, while too low a concentration may escape detection.

ACKNOWLEDGEMENTS

The work described in this paper has formed part of the United Kingdom Atomic Energy Authority's research programme into radionuclide migration in fractured crystalline rock. The contribution of Mr. V. M. B. Watkins of the Research Services Section of Elcon Western (Electrical) Ltd. to all aspects of the field programme, and of Dr. E. M. Durrance of the University of Exeter, geological consultant to the project, is gratefully acknowledged.

REFERENCES

[1]   Bourke, P.J., Durrance, E.M., Hodgkinson, D.P., and Heath, M.J.: Harwell Report (in preparation).

[2]   Heath, M.J. and Durrance, E.M.: "Radionuclide migration in fractured rock: hydrological investigations at an experimental site in the Carnmenellis granite, Cornwall", Harwell Report (in preparation).

[3]   Hodgkinson, D.P.: "Analysis of steady state hydraulic tests in fractured rock", Harwell Report (in preparation).

[4]   Heath, M.J.: "Geological control of fracture permeability in the Carnmenellis granite, Cornwall: implications for radionuclide migration", Mineralog. Mag. (in press).

[5]   Cooling, C.M. and Tunbridge, L.W.: "Methods of rock mass structure assessment and in situ stress measurement carried out in Cornish granite", this volume.

Methods of rock mass structure assessment and in-situ stress measurement carried out in Cornish granite

C.M.COOLING
Building Research Establishment, Garston, Watford, UK
L.W.TUNBRIDGE
Golder Associates, UK

ABSTRACT

A series of experiments relating to the assessment of rock mass structure and the measurement of in-situ stress has been carried out within the context of a wider rock mechanics research programme aimed at developing a site assessment methodology in order to provide data for the design of excavations for the disposal of radioactive waste. Direct and indirect methods of assessing rock mass structure and techniques for analysing the data obtained are described. In-situ stress measurements using both hydrofracture and several overcoring techniques have been used on site in Cornwall and evaluated for near surface conditions.

RESUME

Une série d'expériences relative à l'évaluation de la structure des masses rocheuses et à la détermination des tensions in situ a été effectuée dans le cadre d'un programme de recherches de plus grande envergure sur la mécanique des roches, dont l'objet est d'établir une méthodologie d'évaluation des emplacements afin de fournir des données pour l'étude des excavations pour le déchargement des dechets radioactifs. Des méthodes directes et indirectes pour l'évaluation de la structure des masses rocheuses et des techniques pour l'analyse des données obtenues sont décrites. Les tensions in situ ont été déterminées à l'emplacement dans la Cornouailles, en se servant des méthodes d'hydrofracture et de plusieurs techniques de recouvrement, et les résultats ont été évalués pour des conditions près de la surface.

1. INTRODUCTION

In the United Kingdom, government research on the safe disposal of radio-active waste is the responsibility of the Department of the Environment (DoE). On its behalf the Building Research Establishment (BRE) is co-ordinating an interactive research programme on underground aspects of dis-posal drawing on the expertise of government research laboratories, universi-ties and consulting firms. An outline of this work during its formative stages is given by Hudson [1] and more recent developments by Cooling et al [2]. The experiments described below form part of this programme and are designed essentially to develop and validate a generic site assessment metho-dology. The choice of a granite site for the initial stages of the programme of research was based on its merits as an example of a strongly fissured medium, capable of holding high stresses and providing an area of homogeneous material appropriate to our tests. The experimental facility has never been regarded as a suitable candidate site for an actual repository.

The facility comprises a series of tunnels, roughly at right angles to each other, entered at ground level at the base of a quarry wall. A block of rock (see Fig 1) was selected for a series of experiments to assess the properties of the rock mass, especially those of its discontinuities since these control, amongst other things, anisotropy and inhomogeneity in the mechanical and hydrogeological properties of the rock mass. Since access to a rock mass is normally restricted to specific exposures or boreholes it is important that the validity of indirect techniques of assessment, such as seismic velocity and wave attenuation and the accurate estimation of fracture frequency in any direction through the rock mass, be explored. This paper focusses on the field assessment studies undertaken on rock mass structure and in-situ stress measurements in the granite adjacent to the tunnels and in-situ stress measurements by the hydrofracture method made down a borehole in an adjacent quarry.

2. ROCK MASS STRUCTURE

The accurate assessment of rock mass structure is vital both to the safe and efficient construction of the repository and in engineering the subsequent isolation of its contents. The present studies are directed towards developing a methodology for recording the characteristics of discontinuities and interpreting the data so that predictions of the discontinuity pattern at proximate locations may be made.

2.1 Current approach and techniques

Recording the discontinuity parameters on site has been carried out in two stages:

- A geological overview to determine the structural regions from the age relations of the discontinuity sets and variations across the site.

- An unbiased survey of discontinuity parameters to provide data for statistical analysis.

It is difficult to produce a survey generating unbiased discontinuity parameters. The scanline survey technique was considered the most appropriate to the mapping of discontinuities in tunnels and other man-made exposures, and data from vertical boreholes could supplement the predominantly horizontal scanlines. The scanline survey method used is based on recording the discontinuities that, when projected, cross a measuring tape laid along an exposure. The type of information recorded includes such key parameters as genetic type, dip, dip direction, trace length, etc. The 'type' and 'importance' parameters are defined from the geological overview. This successfully provided the neccessary data on discontinuity spacings and spatial distribution.

2.2 Interpretation of discontinuity data

Analysis of the information from the scanline surveys is aimed at providing statistical data on the distribution of discontinuities in the rock mass. The characteristics which are relevant to the problems of connectivity and rock mass stability are:

- Spatial distribution and orientation of the discontinuities
- Discontinuity trace length
- Discontinuity aperture

Data from the scanline surveys are analysed using a suite of computer programs being developed to run on a microcomputer [3]. The functions of the programs are as follows:

- Selection of subsets of the data based on any combination of values of
  the recorded parameters
- Factor analysis to determine relationships between parameters
- Plot of hemispherical projection of normals to discontinuity planes
- Plot of histogram of discontinuity spacings
- Plot of histogram of discontinuity trace length
- Calculation of discontinuity frequency in any desired direction
- Plot of rosette of discontinuity frequency in any desired plane
- Plot of semi-variogram of discontinuity frequency (Fig 2)
- Plot of semi-variogram of discontinuity orientation
- Plot of semi-variogram of discontinuity trace length

The selection of subsets may be based on the results of the geological
overview so that statistical information is determined for each set of discontinuities. The factor analysis determines the relation or degree of correlation between the various parameters and assists selection of subsets for
analysis. The plot of hemispherical projections of normals to the discontinuity planes helps identify the set according to orientation. Discontinuity
frequency may be calculated in any desired direction by resolving discontinuity frequencies for the angle of intersection with the sampling scanline
and with the required direction. A rosette of discontinuity frequencies on a
plane may be built up and used to determine the direction of maximum and
minimum frequencies, which are not necessarily orthogonal [4].

The variogram function is a basic tool in geostatistics [5] and describes
the spatial aspects of regionalised variables. The semi-variogram of discontinuity frequency is a graphing of the variance of discontinuity frequency in
discrete intervals against distance between the intervals. The concept is
similar to that of autocorrelation and may be used to evaluate spatial dependence in fracture set characteristics [6]. The semi-variograms of orientation
and trace length represent graphings of the variance of the parameter averaged
over discrete intervals. The application of geostatistics in estimating discontinuity frequencies from the semi-variogram data is being investigated [7].
At the current stage the semi-variogram is considered to be a useful tool in
quantifying patterns in the data.

2.3 Large scale anistropy and inhomogeneity

As a part of the integrated research programme it was considered important
to study the possibility of correlating the rock mass discontinuity characteristics with compressional wave velocity data. The velocity contour map
(Fig 3) was obtained by simple tomographic algorithms from the mean velocities
measured along lines between stations in roadways around the perimeter of the
selected test block of rock [8]. The velocities were measured, using a hammer
blow for the energy source, between two geophone stations with the aid of a
storage oscilloscope. The velocity map shows the effect of the free face and
possibly of damage in the rock adjacent to the boundaries of the tunnels.

3. IN-SITU STRESS

For the mechanical analysis and design of an underground excavation a knowledge of the in-situ stress values and their spatial consistency is essential.
The principal values of the stress tensor are rarely equal at the depths appropriate to engineering works; moreover the horizontal stress is frequently the
maximum principal stress. In order to evaluate different stress measurement
techniques a series of measurements using different instruments and techniques
was carried out at the experimental site. The techniques available to measure
the absolute value of the in-situ state of stress in a rock mass remote from
the influence of an excavation include: borehole overcoring and hydraulic
fracturing.

3.1 Borehole overcoring

The borehole overcoring stress relieving systems measure strains induced in
borehole core as it is released from the rock mass and the stress field. When
these strains are known, the state of stress may be calculated by assuming
homogeneous, isotropic and linear elastic behaviour of the rock.

The site work involved evaluating commercially available borehole overcoring instruments during a programme of in-situ measurement in the mine at a
depth of 34 m in granite. The following systems were evaluated:

- Borehole deformation gauge; [9]
- Hollow inclusion cell, [10]
- Solid inclusion cell; [11]
- Leeman-type triaxial cell. [12]

The overcoring instruments were installed in 38 mm diameter pilot holes at
the end of and coaxially with 101 mm diameter boreholes. The borehole deformation gauge, hollow inclusion cells and solid inclusion cells were monitored
during overcore by means of a cable passing through the drill string by way of
a modified water swivel. Analogue data signals from the strain gauges were
converted to digital records for display and storage on magnetic tape by an
automatic data logger at a rate of one scan of all gauges every 5 seconds. On
completion of a test the data was transferred to a microcomputer for processing
and plotting. The Leeman-type triaxial cell was not provided with a facility
for continuous monitoring and the strain gauge readings were recorded only
before and after overcoring.

In the borehole deformation gauge (Fig 4A) strain gauge cantilevers are
coupled to the pilot hole wall to measure deformations recorded across three
coplanar diameters. A plot of deformations during overcore is shown (Fig 5).
Following recovery, the core is tested biaxially in a pressure chamber to
determine the deformation modulus (Fig 6). The gauge measures strain in one
plane only, consequently tests in three boreholes in different orientations are
required to determine the complete three dimensional state of stress.

The hollow inclusion cell comprises nine strain gauges in rosettes encapsulated in a hollow epoxy cylindrical shell (Fig 4B) which is permanently glued
into the pilot hole. A plot of deformation recorded during overcore is shown
(Fig 7). Problems with setting of the epoxy glue in the low ambient temperature
of the mine (10°) which resulted in debonding of the cell when the rock core
was stress relieved and the sensitivity of the strain gauges to temperature
changes induced by the coring process and flushing fluid were revealed during

these tests. The formulation of an epoxy glue suitable for use at low temperatures was developed to overcome the first problem and monitoring of the cell
temperature by means of a thermistor during overcore permitted correction of
the strain recorded for the temperature effect. Following recovery, the core is
tested biaxially in a pressure chamber to determine the deformation modulus
(Fig 8). The strain gauges are arranged so that the complete state of stress
may be determined from one test.

The solid inclusion cell comprises strain gauges encapsulated in a solid
epoxy plug, which is permeanently glued into the pilot hole. The problems of
debonding the cell were more severe than with the hollow inclusion cell due to
the higher modulus of the solid cell. The strain gauges in the solid cell are
also sensitive to temperature variation.

The Leeman-type cell comprises strain gauge rosettes on moveable pistons
which are glued directly to the wall of the pilot hole. This cell is less
affected by debonding of the gauges as little tensile stress is developed at
the interface with the rock. A half bridge design utilizing a dummy gauge
attached to a chip of rock resulted in reduced sensitivity to temperature
variations.

3.2 Hydrofracture

The hydrofracture method [13] has become accepted as a reliable means of
measuring the in-situ stress at depth in a borehole. The technique relies upon
stress concentration around a borehole assuming an elastic response to the
in-situ stress field with one principal stress parallel to the axis of the
borehole. The lowest stress is predicted to occur at the same borehole azimuth
as the maximum principal stress perpendicular to the borehole.

Fluid penetration of a packed-off borehole interval results in rock failure
when the fluid pressure is sufficient to overcome the minimum stress on the
borehole wall and the tensile strength. The fracture is predicted to grow in an
orientation normal to the minimum principal stress.

A system to measure the in-situ stress by the hydrofracture method has been
developed and evaluated during a programme of tests in granite at depths down
to 130 m. A schematic arrangement for the system is illustrated (Fig 9). Pressures in the packers and test zone are monitored on the surface by means of
strain gauged diaphragm pressure transducers. Flow of water to the test sections is measured with a magnetic pickup turbine flowmeter. Analogue signals
from the pressure transducers are conditioned and flow pulses are cumulated
and recorded on magnetic tape by an automatic data logger at up to 10 scans per
second. In addition the flow pulses are converted to flow rates electronically
and both test pressure and flow rate are recorded on a chart recorder. Pressure
gauges and digital display of total flow are provided for manual recording of
the data. On completion of a test the data from the logger is transferred to a
microcomputer for processing and plotting (Fig 10).

The test is conducted by pressurising the test zone until a fracture is
created and opened against the in-situ stress normal to the plane of the fracture. Initiation of the fracture is identified by a rapid drop in pressure,
(Fig 10, (a)) the pump is then shut-off and further drop in pressure monitored
to determine the shut-in pressure when the fracture is closed by the in-situ
stress. This sequence is repeated several times to obtain values of the re-

fracture pressure and further values of shut-in pressure (Fig 10 (b)). A slow refracture test is conducted to determine the pressure at which the fracture begins to open against the in-situ stress (Fig 10, (c)). Constant pressures are maintained in the test zone and flow rates are monitored. The in-situ stress normal to the fracture plane is identified by the rapid change in slope of the graph of flow against pressure as the fracture opens. The in-situ stress parallel to the fracture plan is calculated as a function of pressure re-quired to initiate the fracture, in-situ stress normal to the plane of the fracture and the tensile strength.

4. CONCLUSIONS

A series of rock mass assessment and in-situ stress studies have been des-cribed. Emphasis has been placed on the design of the experiments, experience with and descriptions of the instruments used rather than detailed discussion of the results obtained; this is in accordance with the wishes of the organi-sers of this Workshop. This approach however accords well with the research described which is being conducted in order to develop a fundamental under-standing of the influence of discontinuities and stress on rock mass behaviour so that appropriate site assessment methodologies and instrumentation can be developed to assist in repository design.

ACKNOWLEDGEMENTS

The research described in this paper is supported by the Radioactive Waste Professional Division of the Department of the Environment, the project co-ordinated by the Building Research Establishment and the paper published with the permission of the Director of the Building Research Establishment. The results of this work will be used in the formulation of government policy but at this stage they do not necessarily represent government policy.

REFERENCES

[1]   Hudson, J A: 'UK Rock mechanics research for radioactive waste disposal' Proc 5th ISRM Congr. E161-165. Melbourne. (1983).

[2]   Cooling, C M, Tunbridge, L W and Hudson, J A: 'Some studies of rock mass structure and in-situ stress'. Proc ISRM Symp. Design and Performance of Underground Excavations, Cambridge 1984 (in press).

[3]   Tunbridge, L W and Richards, L R: 'Geotechnical Site Assessment Methodology', Report DOE/RW/83.063, (1983).

[4]   La Pointe, P E and Hudson, J A. 'Characterisation and interpretation of rock mass jointing patterns', Spec Paper Geol.Soc. of America (in press)

[5]   Matheron, G: 'The theory of regionalised variables and its applications' Les Cahiers due Centre de Morphologie Mathematique, Fontaineblue, 5. (1971).

[6]   Miller, S M: 'Geostatistical analyses for evaluating spatial dependence in fracture set characteristics'. Proc 16th APCOM Symp. SME/AIME Inc. (publ) 537-545 (1979).

[7]   Clark, J: Personal communication (1983).

[8]   New, B M. 'A seismic transmission tomography technique for rock quality
        evaluation'. Report DOE/RW/83.103. (1983).

[9]   Merrill, R H: 'Three-component borehole deformation gauge for determin-
        ing the stress in rock'. RI 7015 US Bureau of Mines, Washington (1967).

[10]   Worotniki, G and Walton, R J: 'Triaxial 'hollow inclusion' gauges for
        the determination of rock stress in-situ'. ISRM Symp. on Inv. of Stress
        in Rock and Advances in stress measurement. Sydney Supplement (1976).

[11]   Rocha, M and Silverio, A:'A new method for the complete determination of
        the state of stress in rock masses', Geotechnique, 19, 116-132 (1969).

[12]   Leeman, E R: 'The determination of the complete state of stress in rock
        in a single borehole - laboratory and underground measurements'. Int. J
        Rock Mech. Min. Sci., 5, 31-56 (1968).

[13]   Haimson, B C and Fairhurst, G: 'Initiation and extension of hydraulic
        fractures in rock'. Soc. Petrol Eng.J. 7 310-318 (1967).

[14]   Binnall, E P: 'Instrumentation and Computer Based Data Acquisition for
        in-situ Rock Property Measurements'. US Dept of Energy, Preprint
        LBL-10532 (1980).

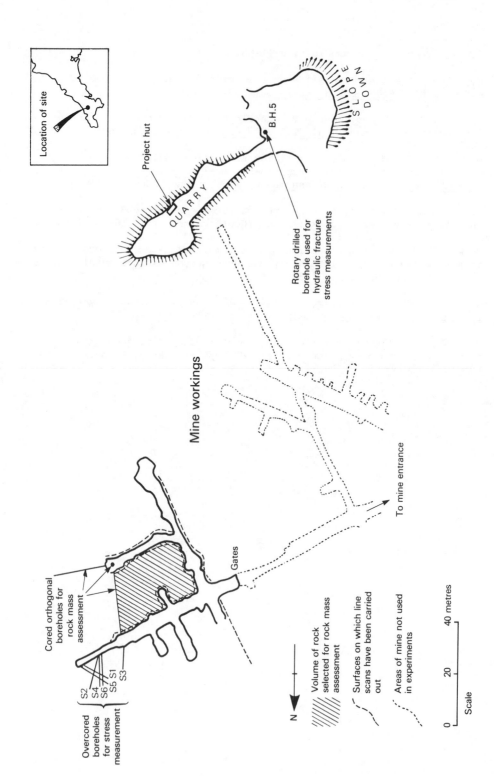

Figure 1 Location plan of experimental mine and quarry.

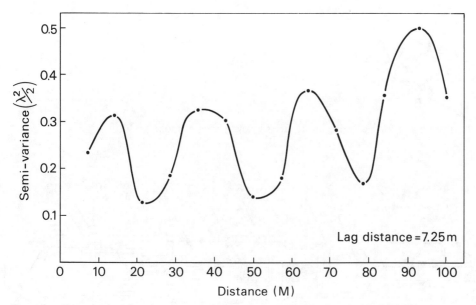

Figure 2 — Example plot of semi-variogram of discontinuity frequency from borehole data

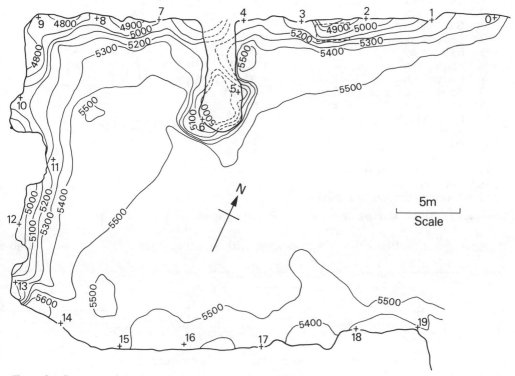

Figure 3 — Representative velocity contour map (From New, 1983)

209

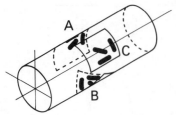

Figure 4   Instruments for measuring in-situ stress by overcoring method.

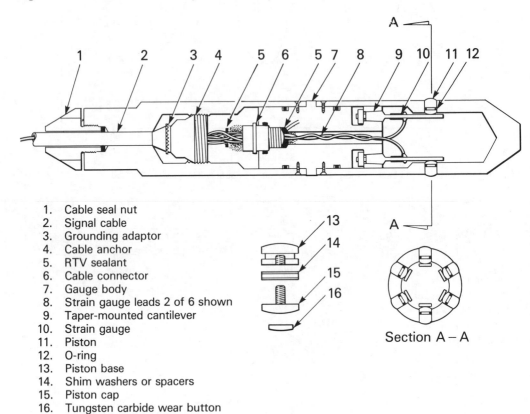

1. Cable seal nut
2. Signal cable
3. Grounding adaptor
4. Cable anchor
5. RTV sealant
6. Cable connector
7. Gauge body
8. Strain gauge leads 2 of 6 shown
9. Taper-mounted cantilever
10. Strain gauge
11. Piston
12. O-ring
13. Piston base
14. Shim washers or spacers
15. Piston cap
16. Tungsten carbide wear button

Section A – A

Figure 4a   The USBM borehole deformation gauge (after Binnall, 1980).

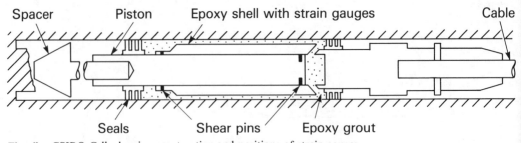

Spacer   Piston   Epoxy shell with strain gauges   Cable

Seals   Shear pins   Epoxy grout

Fig. 4b   CSIRO Cell, showing construction and positions of strain gauges

210

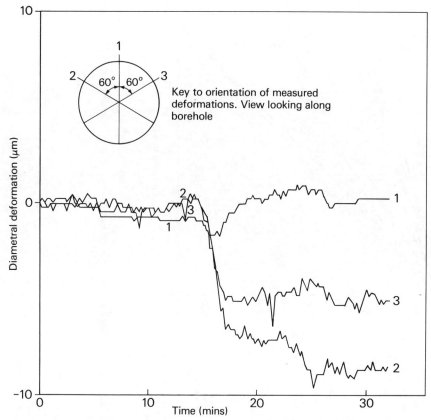

Figure 5 — Plot of deformations recorded from borehole deformation gauge during overcore

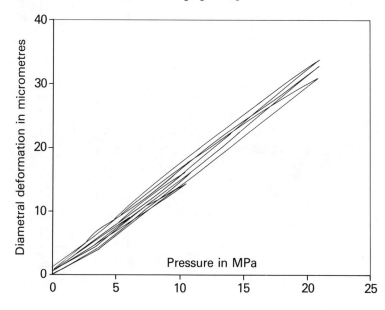

Figure 6 Plot of deformations recorded from borehole deformation gauge during biaxial test.

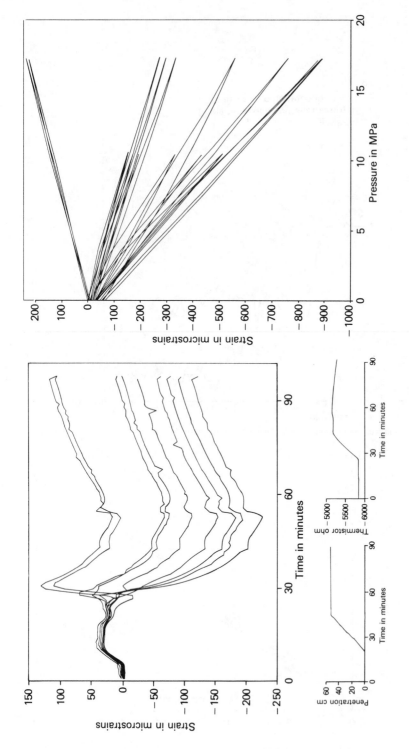

Figure 8 Plot of strain recorded from hollow inclusion cell during biaxial test.

Figure 7 Plots of strain, penetration and temperature recorded from hollow inclusion cells during overcore.

212

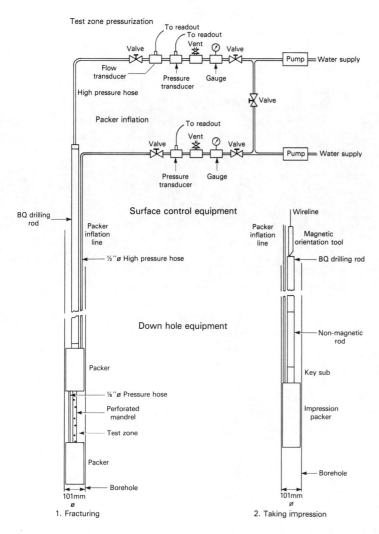

Test zone pressurization

To readout
To readout
Vent

Valve   Valve

Flow
transducer

Pressure
transducer   Gauge

High pressure hose

Pump — Water supply

Valve

Packer inflation

To readout
Vent

Valve   Valve

Pressure
transducer   Gauge

Pump — Water supply

BQ drilling
rod

Packer
inflation
line

½"ø High pressure hose

Surface control equipment

Down hole equipment

Packer

⅛"ø Pressure hose

Perforated
mandrel

Test zone

Packer

Borehole

101mm
ø

1. Fracturing

Wireline

Packer
inflation
line

Magnetic
orientation tool

BQ drilling rod

Non-magnetic
rod

Key sub

Impression
packer

Borehole

101mm
ø

2. Taking impression

Figure 9 Schematic arrangement of equipment for determining
the in situ rock stress by the hydrofracture method

213

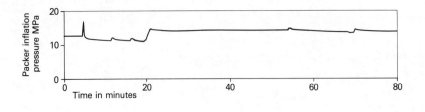

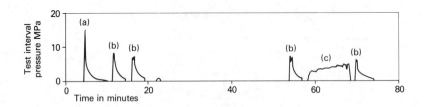

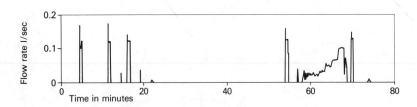

Figure 10   Records of packer pressure, test pressure and
flow from a hydrofracture test

Layout and instrumentation of in situ-experiments for determination of hydraulic and mechanic rock mass properties of granite with and without heat load in the underground laboratory Grimsel, Switzerland[1]

W.BREWITZ, W.SACHS, M.W.SCHMIDT, J.SCHNEEFUSS & W.TEBBE
GSF, Institut für Tieflagerung, Braunschweig, FR Germany

ABSTRACT

In the framework of a research contract with the Nationale Genossenschaft für die Lagerung radioaktiver Abfälle, the Gesellschaft für Strahlen- und Umweltforschung mbH carries out
in situ tests in the underground laboratory Grimsel. The
ventilation test is designed for the determination of the
macro-permeability of granite in particular of homogeneous
rock masses without distinct fractures. In a heater experiment
the thermomechanical behaviour of granite will be measured as
function of heat load and time. A set of 6 geophysical tiltmeters will be used in order to identify neotectonical movements as well as rock mechanical effects resulting from the
heater test. At present the underground testing sites are
almost ready for installation of the measuring equipment. The
first tests will be commenced by mid 1984.

1. INTRODUCTION

In May 1983 the Nationale Genossenschaft für die
Lagerung radioaktiver Abfälle (NAGRA), Switzerland,
entered into an agreement with the Bundesanstalt für
Geowissenschaften und Rohstoffe (BGR), Hannover, and the
Gesellschaft für Strahlen- und Umweltforschung mbH (GSF),
München, for the performance of a joint research and
development programme in the underground laboratory at
Grimsel/Switzerland. Main subjects of this programme are
in situ tests concerning the investigation of hydraulical
and mechanical rock mass properties of granite. The
underground laboratory has been developed offside the main
access tunnel to the hydro power Station Grimsel II of the
electricity supply company Oberhasli AG. The test sites
are placed in a section of the Juchlistock where the
Zentraler Aaregranite changes into Grimsel Granodiorit.
The rock mass as such is almost dry in a technical sense.

1) This paper was prepared under the research contract KWA
53154 with the Federal Minister for Research and Technology
(BMFT) of the Federal Republic of Germany

Only the various systems of faults can be regarded as water saturated.

Within the entire GSF-programme on radioactive waste disposal the in situ-tests planned at Grimsel mean a sensible completion to similar tests presently being performed in the disused iron ore mine Konrad. In the underground laboraty at Grimsel the following hydrogeological and rock mechanical aspects shall be investigated:

- macro-permeability of granitic rock masses,
- moisture release of faulted and non-faulted granite with low hydraulic potential, with and without heat load,
- mechanical behaviour of faulted granite under heat load,
- neotectonical movements and/or thermomechanical movements offside the heater tests.

An important task is the development of suitable investigation techniques and the gathering of sufficient experiences for later employment at actual repository sites.

This paper deals with the planned experiments in a more general manner. Specific informations about the underground laboratory, its geology and its installations are not subject of this paper.

2. VENTILATION TEST FOR MACRO-PERMEABILITY MEASUREMENT IN GRANITE

The ventilation test is an attempt to characterize the permeability of a large volume of granite by large scale field measurements. It was firstly developed and theoretically established by P.A. Witherspoon et al. [1] as part of the hydrogeological research program in the Stripa mine (Sweden).

By measuring groundwater seepage, water pressure and water temperature in several boreholes surrounding the test drift the hydraulic conductivity was determined by using the Thiem equation for steady, radial flow into a long cylindrical sink in a homogeneous, isotropic porous medium [2].

A similiar test is presently carried out at the Konrad iron-ore mine under different hydrogeological conditions [3], [4].

In the following discription emphasises the instrumentation of the ventilation test in the Grimsel underground laboratory (FLG) which will start in summer 1984.

2.1 LAYOUT AND INSTRUMENTION

The layout of the Grimsel ventilation test is a further development of the macro-permeability experiments performed at Stripa and in the Konrad mine.
Using a tunneling machine for the development of the ventilation test drift the rock mechanical effects on the drift wall could be minimized.
The circular cross-section of the drift provides simple inner boundaries for the hydrogeological system which will be useful for data interpretation.

The ventilation drift has been driven in the southern area of the FLG at a depth of about 500 m below surface (Fig. 1). The drift has a length of app. 70 m and a diameter of 3.5 m. Site specific is the occurence not only of dry but also of moist rock formations which is most important for a successful performance of the test [5], [6].

While the end of the drift is intersected only by a few waterbearing fractures a significant proportion of the water release from the granitic rock is connected to fractures and fissures in the front part of the drift wall. Thus for the purpose of better ventilation results the drift has to be subdivided into two sections and permeability data will be produced at first from the more fractured and moist section of the rockmass (Fig. 2).

Instead of a fixed brick wall as it is used in the Konrad test drift, a flexible and inflateble large size packer made of rubber-lined canvas will be installed in a way that there is no heat and humidity interchange between the two parts of the test drift or the access drift.

Determination of the macro-permeability of the rock mass around a drift requires the measurement of the ground water flow into the drift. Since the seepage is very small and most of the water will be lost to evaporation the water flow will be determined by measuring the moisture pickup in the air of the ventilation drift.
In contrast to the Stripa ventilation test a closed system will be used.

The ventilation unit consists of a fan, an air cooler and a heater (see Fig. 2).

The water vapour coming from the ventilation section of the drift will be partly condensed in the cooling trap. After heated up the ventilation air will be blown into the test drift again.

Measuring ventilation volume, air humidity and air temperature at the inlet and exhaust ducts as well as the water extracted by the cooling trap the volumetric eva-

217

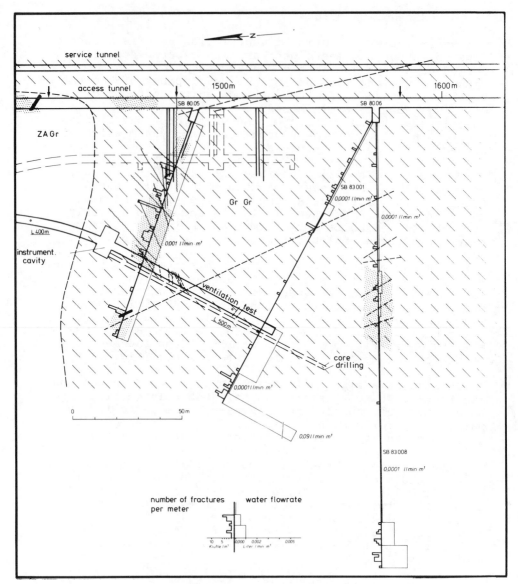

Figure 1: Location of the ventilation test gallery in the
underground laboratory Grimsel

poration rate into the drift can be calculated. The ventilation system is controlled by a humidity gauge mounted
in the exhaust duct, which will keep the relative air
humidity inside the gallery at an app. constant level.
In the drift air temperature and air humidity gauges will
be installed at different points.

218

Hydraulic gradients can be determined by measuring the pore water pressure in two app. 110 m long boreholes drilled parallel to the drift in a distance of 3.5 m and 5.25 m from its axis. The diameter of the boreholes measures 86 mm. For the instrumentation of the boreholes a new packer system had to be developed by which the pore water pressure and temperature can be monitored in at least about 15 sealed intervals of each borehole.

In addition the exploration holes SB 83.001 and SB 80.05 (see Fig. 1), which were driven from the access tunnel of FLG, will also be included in the macro-permeability experiment.

2.2 PERFORMANCE OF THE MACRO-PERMEABILITY EXPERIMENT

During the macro-permeability experiment the following measurements will be conducted.

- continuous control of the air humidity and air temperature in the drift
- continuous measurement of the air humidity and air temperature in the exhaust and inlet ducts of the ventilation unit
- continuous measurement of the extracted water
- continuous measurement of the water pressure and water temperature in the packer intervals of boreholes
- deformation measurements in the ventilation drift
- identification of water bearing fractures by periodic temperature measurements at the drift wall by an infra-red thermometer.

In phase I of the ventilation test no ventilation will by used in order to reach an app. equilibrium in the drift atmosphere and if possible water saturation in the surrounding rock mass. During phase II after connecting the ventilation pipes the ventilation test will be first performed in the front area, later also in the poorly fractured rock mass in the rear and of the drift. Phase II will consist of a series of tests to be performed at different air temperature and air humidity conditions. In phase III the drift will be locked again till an app. equilibrium of the drift atmosphere is reached again. Comparing phase I with phase III will allow conclusions on the hydraulic potential of the granite rock mass around the test drift.

2.3 HYDRAULIC CONDUCTIVITY OF LAMPROPHYRE VEINS AND FISSURES UNDER HEAT LOAD

In addition to the ventilation test which is aimed as the measurement of the macro-permeability of a large rock mass

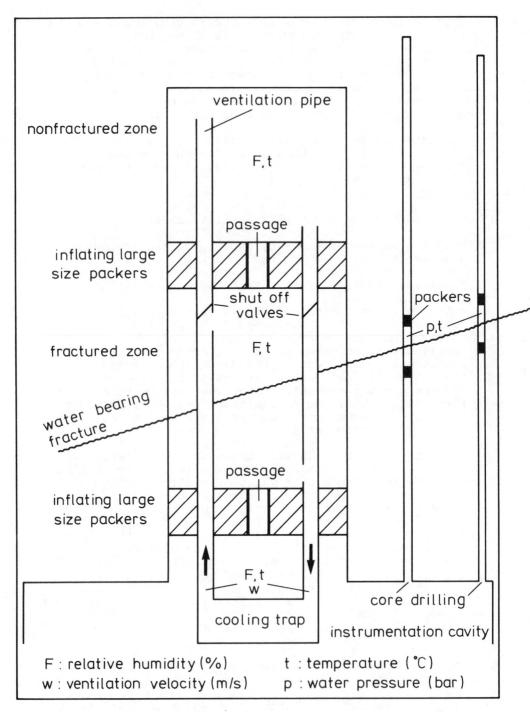

Fig. 2: Schematic diagram of the experimental setup of the ventilation test.

volume the dewatering system fitted in boreholes of the
heater test will be used for the determination of the
hydraulic conductivity of distinct geological water pathways in granite such as veins and fissures. A particular
goal is the investigation of the hydraulic behaviour of
these discontinuities under heat load to be simulated by
electrical heaters. A similar dewatering system was already used in heater tests at STRIPA [7]. The water flow
from the sourrounding rock mass into the boreholes was
pumped out and its volume was measured. It was proved that
the water flow into the borehole i.e. the volume of
pumping water was directly related to the heat take up of
granite. Expecially where the granite was heated up the
amount of intruding water increased rapidly due to thermomechanical dilatation and the closure of open fractures.
After the water had been pressed out the hydraulic conductivity of the rock mass was decisively reduced
permitting only very small quantities of water seeping
into the boreholes (Fig. 3). After the heating phase was
stopped and the granite had cooled off it lasted a
considerable period of time before the water flow started
again. Later the flow reached the same level as it was
before heating. Important result in respect of the
disposal of heat producing radioactive waste in granite is
that heat may close effective water pathways for some
time. These effects will be investigated in detail together with the heater tests in the underground laboratory
Grimsel.

According to GSF plannings the dewatering boreholes as
part of the heater test field will be seperated into
different measuring sections. The extension and depth of
these sections depend on the geological and hydrogeological conditions at the site. In particular thickness, dip,
and the existence of fractures and fissures are important
factors for the installation of packers in the boreholes
(Fig. 4). A specific geological feature at the heater test
site is a slaty lamprophyre vein which cuts through the
granite (Zentraler Aaregranit). Its width measures up to
2 m (Fig. 5).

By automatic pumps the intruding water will be lifted
from the bottom of each dewatering section. It will be
pumped into the water discharge system of the underground
laboratory after the volumetric flow rate has been
measured by a flow meter. The specific water volume from
each section will be registered as function of time and
heat load applied to the rock mass.

3. GEOTECHNICAL INVESTIGATIONS FOR THE DETERMINATION OF
THERMOMECANICAL BEHAVIOUR OF GRANITE

Basis for the heater test planned and designed for the
underground laboratory Grimsel are experiences gained by

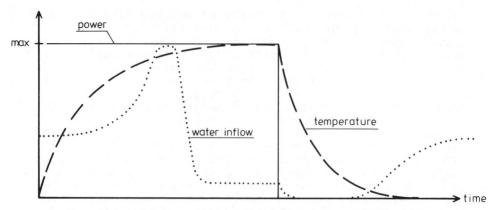

Fig. 3: Water flow into dewatering boreholes (schematic)

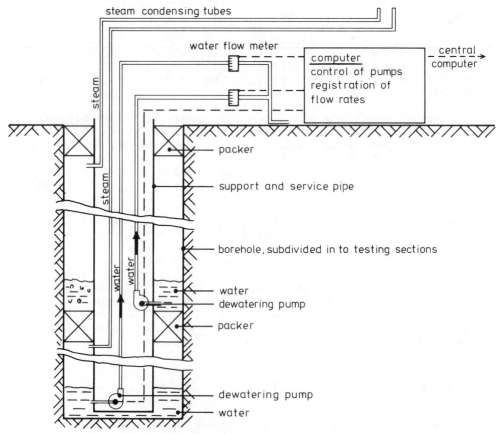

Fig. 4: Heater test, Dewatering system of boreholes (schematic)

GSF with heater tests in the Asse salt mine [8] as well as results achieved in the Stripa mine (Sweden) by Lawrence Berkely Laboratories (LBL) and other institutions [9, 10, 11]. Main subject of all these experiments was the determination of mechanical events and deformations in the rock mass affected by the heat load. Extent and distribution of temperature, mechanical stress and deformation were measured as a function of time and heat load in the near field of the heating sources. The heaters used in these experiments had a power output comparable to the power output of heat generated by radioactive waste canisters. The conventional layout provided an installation of these heaters in boreholes. In a pattern of peripheral boreholes the various measurements will be carried out. The observation of the temperature field as well as the heat wave spreading can be performed with measuring equipment already developed and successfully tested. The determination of the deformation rates in boreholes causes problems because of the elastic properties of granite. Even under heat load the expected convergence rates are very small that they can hardly be measured by conventional methods. In this concern the experiences gained by LBL in the Stripa Mine gave valuable assistance for selection of the right measuring equipment and installations. Though at Stripa the rock mass is intensely fractured and the prevented expansion is lower than normal the thermomechanical expansion of the rock mass was reduced. It is believed the open fracture volume acted as a buffer containing the thermomechanical strain to a very small area around the heater boreholes. The remaining deformations in the testing boreholes were so small that they could not be measured sufficiently with extensometers due to the wide spacing between the individual extensometer anchors.

For stress measurements as basic investigation of heater tests so far electrical gauges such as strain gauges, LVDT, vibrating wire have been used. The basic principle is the determination of borehole deformations and the computation of the actual rock mass stresses. The results depend very much on the overall knowledge of the rock mass properties and the numerical modelling as such. Another aspect is the way of instrumentation. Regarding the elastic behaviour of granite the fitting of the measuring gauges in the boreholes requires specific attention. In the Stripa mine the stress gauges at first supplied satisfying results being in good agreement with pre-calculations carried out. In the course of the test however it became obvious by the increasing number of defects i.e. faulty results that the electrical gauges are most sensitive to moisture. This effect has not been experienced in rock salt where the moisture content in the borehole atmosphere is almost nil.

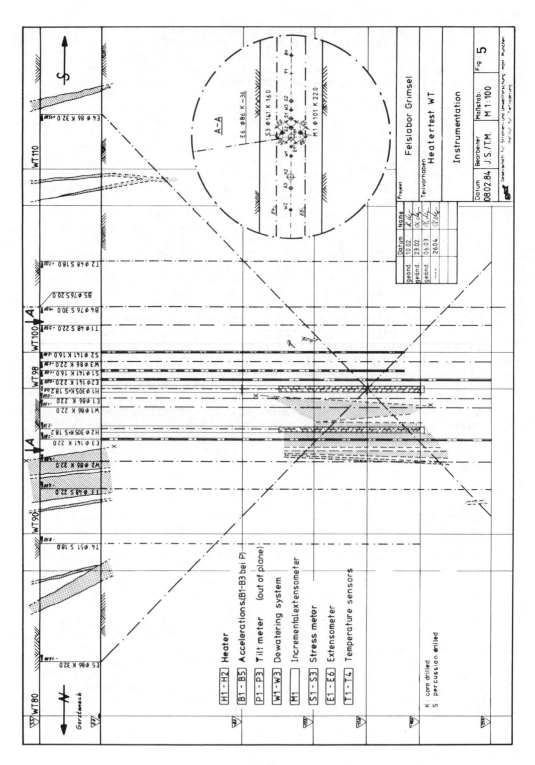

A-A

E6 Φ86 K~36

H1-H2 Heater
B1-B5 Accelerations(B1-B3 bei P)
P1-P3 Tilt meter (out of plane)
W1-W3 Dewatering system
M1 Incrementalextensometer
S1-S3 Stress meter
E1-E6 Extensometer
T1-T4 Temperature sensors

K: core drilled
S: percussion drilled

	Datum	Name
geänd.	10.02	
geänd.	23.02.	
geänd.	06.03	
geänd.	26.04	

Projekt Felslabor Grimsel

Teilvorhaben Heatertest WT

Instrumentation

Datum	Bearbeiter	Maßstab:	Fig
08.02.84	J.S./T.M.	M 1:100	5

Gesellschaft für Strahlen- und Umweltforschung mbH München
Institut für Tieflagerung

3.1 INSTRUMENTATION OF HEATER TEST

Layout and instrumentation of the heater test are shown as schematic diagram in Fig. 5. In the testing phase I two electrical heater systems with a maximum power of 24 kW each will be used. The entire length of the actual heating element measures 6 m, its diameter app. 0,30 m. The heaters will be installed in the lamprophyre vein as well as south of it in solid granite leaving a tectonical discontinuity between them. The operating temperature after gradual heating up will be about 100°C. Initially only the heater placed in granite will be operated.

The generated temperature field will be determined in its extension and controlled by app. 120 temperature gauges (PT 100) which are installed in special boreholes on their own or together with other measuring equipment. For determination of borehole deformations rod extensometers, inclinometer probe and incremental extensometer will be used. Some of these measuring systems are temperature compensatet. Other instruments require special calibration by using master curves for various temperature gradients.

About 50 m apart from the inner test field 3 geophysical tiltmeters are installed at 3 different locations. The tiltmeters placed in boreholes app. 20 m below ground floor of the laboratory have an extreme high accuracy. They are capable for the detection of rock mass movements i.e. tilts in the order of magnitude of 10^{-4} seconds (angle). Together with 5 high-frequency acceleration transducers they shall identify and measure dynamic effects offside the actual heated area.

For the determination of the secondary stress field induced by the temperature gradient 3 boreholes are fitted with flat cells (Type Gloetzl). At least 6 cells in each borehole orientated in 6 different directions shall supply sufficient data for the computation of the actual stress tensor for the various heating and cooling stages. These hydraulic flat cells are designed for direct stress measurements. For this purpose the cells are casted in a concrete cylinder which has got mechanical properties almost aquivalent to those of granite. The cylinder diameter is only little smaller than the borehole diameter. Before installation the cells become calibrated in a pressure vessel. After sinking the cylinders into the boreholes the remaining gap between cylinder and borehole wall as well as the entire borehole will be backfilled with swelling mortar.

3.2 DATA RECORDING

All measuring data will be sampled periodically and

Time Schedule of GSF-programme in the underground laboratory grimsel

		1983			1984				1985				1986
		2.Q.	3.Q.	4.Q.	1.Q.	2.Q.	3.Q.	4.Q.	1.Q.	2.Q.	3.Q.	4.Q.	1.Q.
Planning Acquisition of instruments	VE	██	██	██	██	██							
	WT	██	██	██	██								
Development of Holes, Galleries,Cavernes	VE	██	██	██	██								
	WT	██	██	██	██	██							
Instrumentation	VE						██	██					
	WT						██						
Test Runs	VE						██						
	WT						██						
Start of Experiments	VE							██	██	██	██	██	██
	WT							██	██	██	██	██	██

transmitted via data line to a central computer station where they are stored on magnetic tape. The computerized data sets are being compared with each other in order to defect as soon as they occur. In case of any faults a warning signal calls the supervisors attention. Data processing will be performed by the Institut für Tieflagerung with a seperate computer system.

4. SCHEDULE OF GSF-PROGRAMME IN THE UNDERGROUND LABORATORY GRIMSEL

The development of testing tunnels and the construction of the central section for infrastructural installations will be completed in the near future. In May 1984 the GSF will commence the electrical installations at 3 testing sites i.e. ventilation test, heater test and tiltmeter boreholes. In the central section preparatory work will be done for the connection of the data recording computer. Beginning in June and lasting presumably till October installation work will be performed at the various testing sites. Hopefully at the end of this year all GSF tests will be in operation. It is being expected that by mid 1985 first interim results will be in hand.

REFERENCES

[1] Witherspoon, P.A. et al: "Large Scale Permeability Measurements in Fractured Crystalline Rock", Presented at the International Geologic Congress, Paris (1979).

[2] Wilson, C.R. et al: "Geohydrological Data from the Macropermeability Experiment at Stripa, Sweden", LBL-report 12520, (1981).

[3]   Brewitz, W. et al: "In situ Experiments for Determina-
        tion of Macropermeability and Relevant RN-Migration in
        Faulted Rock Formations at the Konrad Mine", CEC/NEA-
        Workshop on "Design and Instrumentation Experiments in
        Underground Laboratories", Brussels, (1984).

[4]   GSF-Jahresbericht 1982: "Aktuelle Ergebnisse der laufen-
        den F+E-Arbeiten in der Schachtanlage Konrad zur Endlage-
        rung radioaktiver Abfälle", Hrsg. Gesellschaft für
        Strahlen- und Umweltforschung mbH München, GSF-T 165.

[5]   Nationale Genossenschaft für die Lagerung radioaktiver
        Abfälle (NAGRA): "Sondierbohrungen Juchlistock-Grimsel",
        Technischer Bericht 81-07, Baden/Schweiz (1981).

[6]   Nationale Genossenschaft für die Lagerung radioaktiver
        Abfälle (NAGRA): "Sondiergesuch NSG 14 Felslabor
        Grimsel", 21.12.1981.

[7]   Swedisch-American Cooperative Program on Radioaktive
        Waste Storage in Mined Caverns in Cristalline Rock:
        Nelson, P.H.; Radvile, R.; Remer, J.S.; Carlsson, H.:
        "Water inflow into boreholes during the STRIPA heater
        Experiments", Technical Report, 35 (1981).

[8]   Gesellschaft für Strahlen- und Umweltforschung mbH
        München, Institut für Tieflagerung, ABT 12/83:
        Rothfuchs, T.; Schwarzianeck, P.; Feddersen, H.:
        "Simulationsversuch im Älteren Steinsalz Na2ß im
        Salzbergwerk Asse", Temperaturversuchsfeld 4 (TVF 4),
        Abschlußbericht August 1983, FE-Vorhaben 77201.

[9]   Carlsson, H.: "A Pilot Heater Test in the Stripa
        Granite", LBL-7086, SAC-06.

[10]   Cook, N.G.W.; Hood, M.: "Full-Scale and Time-Scale
        Heating Experiments at Stripa", LBL-7072, SAC-11.

[11]   Burleigh, R.H.; Binnall, E.P.; DuBois, A.O.; Norgren,
        D.O. Ortiz, A.R.: "Electrical Heaters for Thermo-
        Mechanical Tests at the Stripa Mine", LBL-7063, SAC-13.

Planned construction phase geomechanics experiments at the Underground Research Laboratory

P.M.THOMPSON, P.BAUMGARTNER & P.A.LANG
Atomic Energy of Canada Ltd, Pinawa, Manitoba

ABSTRACT

The program for geomechanics experiments during the construction phase of the Atomic Energy of Canada Limited (AECL) Underground Research Laboratory near Lac du Bonnet, Manitoba is described. The rock mechanics tests are discussed with particular reference to the test methods and instrumentation selected. Modifications to certain instruments are described and the measurement of thermal conditions around the shaft is emphasized.

The preliminary results of experimental activities in the shaft collar are outlined and the authors indicate how the early experience in the shaft collar has helped to refine the main construction phase rock mechanics experimental program.

1. INTRODUCTION

As part of the Canadian Nuclear Fuel Waste Management Program, an Underground Research Laboratory (URL) is being constructed by Atomic Energy of Canada Limited (AECL). The URL is located in a granitic pluton of the Canadian Shield near Lac du Bonnet, Manitoba. The facility will be used exclusively for in situ experiments designed to provide data required to assess the concept of the safe disposal of commercial nuclear fuel wastes deep in stable plutonic rock. No nuclear fuel wastes will be used in any of the experiments, and the site, which is being leased from the province of Manitoba, will be returned to the province in 2000. The URL concept and technical objectives were described earlier by Simmons and Soonawala (1982).

Construction at the site began in 1982 October. As of 1984 May, the shaft collar had been excavated to a depth of 15 m, and all the surface facilities, including headframe, mineworks, shops and offices, had been constructed.

2. CONSTRUCTION PHASE GEOMECHANICS EXPERIMENTS

2.1 General

Experiments have been scheduled for the underground construction phase of the URL project, either because excavation is an integral part of the experiment or because a required activity can most efficiently be completed during construction. The plan for the construction phase geomechanics experiments is described here. An overview of the overall construction phase experimental program is provided by Lang et al (1984).

The objectives of the URL construction phase geomechanics experiments are:

(1)   to determine the rock mass unloading deformation modulus by measuring the deformations and changes in rock mass stresses due to excavation,
(2)   to determine in situ stress field,
(3)   to determine the changes in hydrogeological conditions near the boundaries of an excavation,
(4)   to apply engineered blasting methodology to design blast rounds, predict the damage in the surrounding rock mass from blast rounds, validate these predictions by subsequent field measurement of excavation damage and compare the excavation damage due to blasting with that due to boring,
(5)   to measure and sample all groundwaters flowing from fractures into the excavation, and measure the changes in the groundwater systems caused by excavation,
(6)   to map the stress field in the rock mass around the shaft, and
(7)   to sample the rock excavated, and fractures exposed during excavation, to characterize the altered zones and fracture infillings and determine their geochemical and mechanical properties.

The underground construction of the URL has been divided into three phases:

(1)   Shaft sinking: 1984 May – 1985 March
(2)   Shaft Geotechnical Characterization: 1985 March – 1985 November
(3)   Horizontal Development: 1985 November – 1986 May

2.2 Shaft Sinking Experiments

2.2.1 General

During shaft sinking the excavation contractor will work 16 h daily in two shifts, with the day shift from 0800 to 1600 h being reserved for experimental work. There will also be three excavation shutdown periods of about 12 days each, during which the shaft will be continuously available to experimenters to install instrument arrays and to conduct tests. The selected depths for these instrument arrays are 63 m, 185 m, and 218 m (225 m, 105 m and 72 m Above Mean Sea Level (AMSL), respectively).

These locations were chosen from the core logs of surface boreholes to provide the best known rock conditions. Because the instrumentation arrays will be used to determine rock mass deformation moduli, overcoring stress measurements will be conducted at these locations during the geotechnical characterization phase. The interpretation of results is facilitated by good rock conditions and a minimum number of discontinuities. The variation of the rock mass deformation modulus with depth will be determined by monitoring several simple total convergence arrays to be installed at regular depth intervals throughout the shaft. These are described in Section 2.2.3.

Because the 12 day experimental periods disrupt the shaft sinking

229

contractor, only experiments that cannot be done during geotechnical characterization will be done during shaft sinking.

2.2.2 Shaft Convergence Measurements

During shaft sinking, the following parameters will be monitored via geomechanics experiments and tests:

(1) excavation-induced displacements
(2) excavation-induced stress changes
(3) the thermal regime in the rock adjacent to the shaft.

At each of the three instrumentation arrays, geomechanical instruments will be installed to monitor the shaft convergence and stress changes that will occur as the shaft excavation progresses. The general layout of each instrumentation array is illustrated in Figure 1. Each array will include the following instrumentation:

(1) Four 15-m long sonic-probe five-anchor borehole extensometers, manufactured by IRAD GAGE of Lebanon, New Hampshire, U.S.A., installed in horizontal BQ-3 (60-mm diameter) diamond drill holes. The holes will be collared within 0.5 m of the shaft bottom. Two extensometers will be installed in one shaft wall, with one in each of two other walls. The fourth wall is reserved for hydrogeological testing.

The IRAD GAGE extensometer system has been modified to include six thermistors along each extensometer, and a stainless steel flange has been added to the measurement head to prevent anchor slippage during blasting. The measurement heads will be anchored with epoxy and C-clips at the base of 0.15-m deep H-sized (100-mm diameter) collars. The thermistors permit detailed monitoring of thermal changes in the extensometer connecting rods, allowing corrections to be made for thermal expansion or contraction. Thermistors have been used for the temperature measurements rather than thermocouples, or other devices, based on recommendations contained in Koopmans (1982).

The extensometers will continue to be monitored over the long term to determine the magnitude of time-dependent deformations.

(2) Six convergence pins installed within 0.5 m of the shaft bottom. These will be monitored using a proving ring tape extensometer manufactured by the Slope Indicator Company (SINCO) of Seattle, Washington, U.S.A.. These measurements will give the total convergence across the shaft and provide a rough check on the borehole extensometers.

(3) Six Hollow Inclusion (HI) triaxial strain cells, developed by the Commonwealth Scientific and Industrial Research Organization (CSIRO) of Melbourne, Australia, installed in diamond drill holes angled down from the shaft walls at 75° and 83° from horizontal in the same three walls as the extensometers (north, east, and south). The HI cells will be installed approximately 8.5 m below the shaft bottom, and 1.3 m and 2.6 m outside the projected excavation lines. Finite-element analysis shows that at 8.5 m below the shaft bottom the HI cells will be beyond the zone of disturbed stresses when they are installed. At lateral distances of 1.3 m and 2.6 m from the projected excavation wall the HI

230

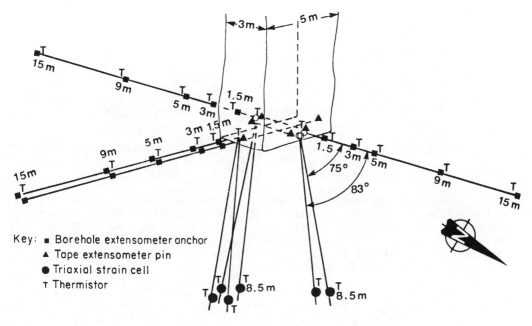

Key: ■ Borehole extensometer anchor
 ▲ Tape extensometer pin
 ● Triaxial strain cell
 ᴛ Thermistor

FIGURE I: URL SHAFT CONVERGENCE ARRAY

cells are beyond the zone that will likely be damaged by blasting, but
are within the projected area of excavation-induced stress changes. Each
HI cell will measure a different part of the stress field.

The HI cells are being used to determine the percentage of excavation-
induced displacement that occurs before the borehole extensometers are installed. Both the borehole extensometers and convergence pins record the
displacements that occur after extensometer installation. By the time the
extensometers and convergence pins are installed 0.5 m above the shaft bottom, a
large percentage of convergence has already taken place. By monitoring the
total strain change with the HI cells, the displacements that have occurred
prior to the time the extensometers were installed can be estimated.

All CSIRO HI cells to be installed at the URL will be modified by the
addition of a thermistor inside the cell body during fabrication. These thermistors, in addition to providing extra data on the thermal regime in the rock
mass, will be used to compensate for temperature changes to the HI cells during
the stress monitoring period, and also to monitor HI cell temperatures during
eventual overcoring, to take place during the geotechnical characterization
phase.

2.2.3 Total Convergence Arrays

In addition to the three main instrumentation arrays, a number of simple
total convergence arrays will be installed at approximately 15-m intervals
throughout the URL shaft. Each array will consist of six convergence pins, to
be monitored with the SINCO tape extensometer. The objective is to obtain
information on the variation of the rock-mass deformation modulus with depth and

231

varying rock conditions, including two fracture zones that have been identified from core logs obtained from surface boreholes.

2.2.4   The Excavation Damage Program

The objectives of the Excavation Damage Program are to develop

- a method of predicting excavation damage around an underground opening

- the equipment, procedures and interpretation methods necessary to determine excavation damage in the rock mass around an underground opening.

The program attempts to determine the nature and extent of excavation damage in the rock mass immediately around an underground opening. From this information, the appropriate access tunnel and shaft sealing methodologies can be developed. The excavation damage is expected to be due to the combined effects of both blasting, or machine-boring, and stress relaxation. It is believed that the two effects are inseparable.

The program is composed of two main components: the Engineered Blast Excavation Program and the Excavation Damage Assessment Program. These programs are being developed for general application in the Nuclear Fuel Waste Management Program and will be tested site-specifically at the URL.

The Engineered Blast Excavation Program attempts to develop an underground blast design methodology including the ability to predict blast performance (e.g. fragmentation) and perimeter wall damage, i.e. the first objective stated above. The second component, the Excavation Damage Assessment Program (see Section 2.3.3), is developing the equipment, procedures and interpretation methods to determine the nature and extent of the damage zone and will therefore assess the validity of the predictive model.

The methodology to be employed utilizes a proprietary blasting model, the ICRAX computer program, developed by Imperial Chemical Industries (ICI), Australia. This model will be used to predict the wall damage. The ICRAX model, supplied by the blasting consultant, CIL Inc. of Willowdale, Ontario, Canada, combines explosive properties such as composition, thermal-chemical data and reaction equilibrium constraints with the rock properties, such as dynamic intact rock modulus and tensile strength. The blasting calculations are based on the effects that dynamic and quasistatic gas pressures have around the blast holes. The result of these calculations is a "fracture map", which details a hypothetical crack pattern in and around the blast area. Inside the blast round itself, the density of "cracks" is related to fragmentation based on the Rosin-Rammler size distribution. The crack distribution outside the blast is similarly calculated and can be related to perimeter wall damage. The predicted wall damage for any particular blast round will be expressed as the frequency of hypothetical cracks versus depth into the rock.

In addition, the blasting consultant may design specific "careful" blast rounds for a 20-m section of the shaft. This section is planned to coincide with the central instrumentation array and will span depths from 90 to 110 m (200 to 180 m AMSL). The purpose is to minimize blasting damage for future shaft-sealing experiments, predict the damage, and compare the blast damage prediction with the Excavation Damage Assessment (see Section 2.3.3). If the ICRAX model is valid, then the correlation of Excavation Damage Assessment parameters to the ICRAX model would provide a modelling input for defining the excavation damage zone in a commercial waste disposal vault.

2.3 Geotechnical Characterization of the Shaft

2.3.1 General

During geotechnical characterization of the shaft, no construction activity is scheduled. The experimenters will have continuous access to the shaft, and upper and lower shaft stations. Drilling and experimental work will commence at the lower shaft station and progress up the shaft. Staging will be set up at five separate locations for drilling and testing. Three of these locations will be at the three instrument arrays installed during the excavation shutdown periods during shaft sinking; one will be near the shaft bottom and the other will be at 205 m depth (elevation 85 m AMSL). Some testing will also be conducted at the upper shaft station.

The geomechanical experiments planned for the geotechnical characterization phase are those that do not have to be done during shaft sinking, as explained in Section 2.2.1 and require long periods of uninterrupted access to the shaft and shaft stations. They include

(1)   stress field mapping
(2)   excavation damage assessment
(3)   stiffness/permeability/temperature relationships of natural fractures
(4)   testing of drill cores for intact rock properties
(5)   probe holes
(6)   horizontal level convergence array

2.3.2 Stress Field Mapping

Stress determinations by the overcoring method will be done at two locations in the lower shaft station, one location at the upper shaft station and four locations in the shaft. At each location two perpendicular drill holes will be tested.

At 218 m, 185 m and 63 m depths (72 m, 105 m, and 225 m elevations AMSL) in the shaft, both the far-field undisturbed stresses and the near-field excavation-induced stresses will be measured. The excavation-induced stresses will be measured by conducting overcore tests from near the shaft wall to 5 m from the shaft wall. Far-field stresses will be obtained by conducting over-core tests at distances of 11 m to 15 m from the excavation boundary. These test areas have been defined based on elastic continuum analyses of the shaft excavation, and from preliminary analyses of results of shaft collar tests.

Throughout the testing, the USBM biaxial cell will be alternated with one of the two triaxial cells (CSIRO HI cell or CSIR cell). At the first overcore location at the shaft bottom, the two triaxial cells will be compared. The better one will be used for the remainder of the overcoring.

The HI cells installed at the three instrument arrays during shaft sinking will also be overcored. The rock annulus containing the HI cell will be tested in a biaxial chamber so that Young's modulus and Poisson's ratio can be

233

determined. The strains recorded during shaft sinking and overcoring can then be converted to stress changes.

2.3.3   Excavation Damage Assessment

Excavation damage assessment testing will be done at the three instrument arrays and also at 250 m and 205 m depth (39 m and 85 m elevation AMSL). At each location, six EWG (38-mm diameter) diamond drill holes will be drilled horizontally, two into each of three walls; two holes will be 15 m long and the remainder 5 m long.

The following techniques may be used to assess excavation damage:

(1)   Vapour permeability measurements.
(2)   High-pressure borehole dilatometer measurements.
(3)   Cross-hole ultrasonic surveys.
(4)   Bulk rock permeability and porosity by tracer migration measurements through core samples.

The ability to measure fracture density and permeability and intact rock permeability by vapour permeability measurements will be assessed by using the equipment and methodology being developed by Jakubick (1983) and de Korompay (1980).

When the vapour permeability equipment has been removed from the shaft, it will be replaced with a high-pressure borehole dilatometer, a modified Colorado school of Mines (CSM) cell developed by Hustrulid and Hustrulid (1975). The CSM cell will be used to attempt to measure the difference in Young's modulus between the intact rock substance away from the shaft and that of the excavation damage region close to the shaft. A series of Young's modulus measurements will be taken at 250-mm increments along the length of each borehole.

The EWG holes will also be used in the cross-hole ultrasonic surveys. It is expected that the excavation damage zone will be delineated either by changes in signal arrival time (i.e. P- and S-wave velocities) or by signal amplitude and frequency alteration effects.

As a precaution against the potential ineffectiveness of these in situ damage assessment methods, a laboratory test is also being developed. This test will measure the change in bulk rock permeability and porosity by the measurement of tracer (KI) migration through rock samples (drill core from a shaft wall). The rock matrix permeability can be calculated as a function of the tracer concentration versus time data. Petrographic analysis of the core will also be done as a component of this test. H-sized core samples (100-mm diameter) will be required and may come either from untested portions of the overcore stress measurement holes or specially drilled short holes.

2.3.4   Stiffness/Permeability/Temperature Relationships of Natural Fractures

In order to begin developing a database on the mechanical, thermal, and hydraulic response of natural fractures, 200-mm diameter core samples will be drilled axially along natural joints at the 63 m, 185 m, 218 m and 250 m levels (225 m, 105 m, 72 m, 40 m elevations AMSL) in the shaft. Two joints will be sampled at each location. The undisturbed joint samples will then be tested under controlled laboratory conditions to measure normal and shear stiffness and the joint permeability and aperture under varying normal and shear stresses, at both ambient and elevated temperatures.

2.3.5 Testing of Drill Cores for Intact Rock Properties

The physical, mechanical, and thermal properties of selected drill cores will be measured in the laboratory. Samples from mutually perpendicular holes will be tested to determine any anisotropy in these properties.

2.3.6 Probe Holes

At least three sub-horizontal 150-m long NQ-3 (76-mm diameter) diamond drill holes will be advanced from the lower shaft station. The boreholes will be used to assess the geological conditions into which the URL experimental level is to be excavated. When a suitable volume of rock has been identified, a pilot hole will be drilled along the proposed centre line of the main access drift, and will be carefully logged. These data will be used to predict the rock conditions to be encountered along the drift. When the drift is excavated, the actual conditions will be mapped and compared with predictions. This will provide an idea of the distance over which NQ-3 core data can be extrapolated in a granitic pluton.

2.3.7 Hydrogeological Monitoring Boreholes

Three NQ-3 diamond drill holes will be advanced from the upper shaft station in an array over the planned layout of the experimental level. These boreholes will be tested and monitored to assess the contribution that subvertical fracturing makes to the hydraulic drawdown caused by excavation of the URL experimental level.

2.3.8 Horizontal Level Convergence Array

An instrumentation array to measure tunnel convergence will be installed at the face of the main access drift in the lower shaft station. It will be similar in design to those used during shaft sinking and will be monitored during horizontal development.

2.4 Horizontal Development

2.4.1 General

Plans for geomechanical experiments during horizontal development have not yet been finalized. Final layout of the testing rooms will depend on the results of stress-field measurements at the lower shaft station and further along the main access drift, and on the data gathered on geological and hydro-geological conditions in the probe holes. The items discussed below will be part of that plan.

2.4.2 Convergence Measurements

The convergence instrumentation array installed during geotechnical characterization will be monitored, and another array will be installed and monitored further along the drift. In addition, a number of simple total convergence arrays will be installed for tape extensometer measurements.

2.4.3 Stress Field Mapping

Overcoring stress measurements will be conducted at a minimum of two locations.

2.4.4 Excavation Damage Program

The Engineered Blast Excavation program will be continued.

3. WORK COMPLETED TO DATE

3.1 General

A program of geomechanical experimental work has been conducted to date at the 8-m and 15-m levels of the URL shaft collar. Convergence measurements have been taken on only the 8-m level instruments. The 15-m instrumentation array, similar in design to those described in Section 2.2.2, will be monitored when shaft sinking resumes. In addition, overcoring stress measurements have been made in two mutually perpendicular holes at the 15-m level.

The work done to date has been primarily aimed at

(1)   personnel training,
(2)   obtaining realistic estimates of the times required for drilling, installing, and monitoring instruments, and
(3)   debugging equipment and procedures.

3.2 Preliminary Results

The shaft collar work has been very successful in fulfilling the three primary objectives described above. The personnel are now well trained, the procedures have been refined, and most equipment problems appear to have been overcome. Time schedules proposed for shaft sinking instrumentation appear to be realistic based on our experience in the shaft collar.

The numerous instrument problems encountered can mostly be attributed to the fact that the rock at the URL site is of far better quality than normally experienced in instrumentation programs in civil and mining engineering. The instruments being used are standard geotechnical instruments that are generally used in soft rock, or high-stress conditions. For example, the sonic probe extensometer was designed for use in coal mines where displacements of many centimetres are common. In the shaft collar, we have been using these same instruments (modified with thermistors) to monitor displacements of less than 0.1 mm over 15 m during daily temperature changes of up to 20 C. Similarly, the HI cells have been required to provide strain data with a repeatability of $\pm$ 5 microstrains during a period of epoxy curing and changing rock temperatures.

Most solutions to problems in instrumental accuracy have involved refining the monitoring procedures and accurately calibrating the instruments for temperature effects. Some examples follow:

(1)   One of the main reasons that the IRAD GAGE sonic probe extensometer system was originally chosen was because the sonic probe transducer

236

could be removed between readings, thus eliminating the danger of blast damage. In actual field use it was determined that errors were introduced by removing and replacing the transducer. As a result, we now know that system accuracy can be maximized by leaving the transducer in the measurement head between readings.

(2)   When the CSIRO HI cells were used for stress monitoring in sub-vertical holes collared at the 8-m level, severe fluctuations occurred from one reading to the next. It was determined that these changes were proportional to ambient air temperature changes. The strain indicator readout box is now kept at a relatively constant temperature between readings and the strain indicator readout box is calibrated with a highly accurate stable resistor to negate temperature effects.

(3)   During stress monitoring, it was determined that the temperature of the rock around the CSIRO HI cells had a large effect on the strain readings. Because of this we now incorporate a thermistor inside each HI cell. During stress overcore measurements we carefully monitor and control the temperature of the drill water to eliminate any adverse temperature effects.

(4)   Abnormally high values of Poisson's ratio were obtained during biaxial pressure chamber testing of CSIRO Hl triaxial strain cell rock overcore cylinders. This same phenomenon was reported by Stillborg and Leijon (1982). It was determined analytically that with the standard biaxial pressure chamber (125 mm loaded length), inward "bending" of the rock cylinder occurs near the centre of the chamber, causing the triaxial strain cell to record erroneously high values for axial strains, and resulting in an incorrect value for Poisson's ratio. The standard biaxial pressure chamber had been originally designed for use with the U.S.B.M. biaxial strain cell which does not measure axial strains. In order to eliminate the bending effect, the loaded length of the biaxial pressure chamber was increased to 340 mm. Poisson's ratio values are now within the normal range.

3.3   Technical Results of Work Completed to Date

Preliminary findings based on the work done to date are as follows:

(1)   The total convergence at the 8-m bench was about 1.8 mm, indicating a rock mass modulus at that depth of about 10 GPa.
(2)   The measured shear displacement on a horizontal joint at the 8-m bench was approximately 1 mm.
(3)   Based on hydro-fracture tests performed before URL construction, the maximum principal stress is between of 5 and 10 MPa in the horizontal plane. Overcoring stress measurements conducted at the 15-m bench have not yet been analysed.
(4)   Biaxial testing of overcored rock annuli indicates that the Young's modulus of the intact rock is anisotropic. In most tests the modulus varied from about 35 to 50 GPa, with the higher value tending to be in the vertical direction.
(5)   Temperature monitoring at the 15-m level indicated that the maximum rock temperature at that depth was 8.8 C attained in January, which is 6 months out of phase with the surface temperature. This agrees with earlier work by Hooker and Duvall (1971).

(6)   None of the excavation damage assessment methods used to date has been
       able to detect an excavation damaged zone around the shaft. The
       compliance of the CSM cell has subsequently been reduced and it is
       being tried again at the 15-m bench.

4.   CONCLUSIONS

As a result of the URL underground construction phase geomechanics experimental program, when the URL operating phase commences in 1986 we should
already have a sound knowledge of the basic geomechanical rock parameters that
exist at the URL. This knowledge can be used to design operating phase experiments that will expand our understanding of the behaviour of plutonic rock under
a variety of conditions. These data will be used to help assess the concept of
safe disposal of commerial nuclear fuel wastes deep in stable plutonic rock.

REFERENCES

(1)   Hooker, V.E. and W.I. Duvall "In Situ Rock Temperatures, Stress
       Investigations In Rock Quarries", U.S. Bureau of Mines Report of
       Investigations Number RI 7589 (1971).

(2)   Hustrulid, W. and A. Hustrulid "The CSM Cell - A Borehole Device for
       Determining the Modulus of Rigidity of Rock", Proceedings of the
       Fifteenth Symposium on Rock Mechanics, ASCE (1975).

(3)   Jakubick, A.T. "Permeability Measurement of the Near-Field of
       Excavations by Transient Pressure Testing", Atomic Energy of Canada
       Limited Technical Record (in preparation).

(4)   Koopmans, R. "A State-of-the-Art Report on the Determination of Rock
       Properties in Boreholes", Ontario Hydro Research Division Report No.
       82-180-K pp. 25-30 (1982).

(5)   de Korompay, V. "A Vacuum Hole Monitoring System to Detect Fractures in
       Mine Roof." Div. Rep. Can. Met., MRP/MRL 80-87(TR) (1980).

(6)   Lang, P.A., P. Baumgartner, P.M. Thompson "Overview of Proposed
       Construction Phase Experiments at the URL, and Preliminary Results of
       Rock Mechanics Work Completed to Date", Atomic Energy of Canada Limited
       Technical Record (in preparation).

(7)   Simmons, G.R. and N.M. Soonawala (editors) "Underground Research
       Laboratory Experimental Program", Atomic Energy of Canada Limited
       Technical Record Number TR-153 (1982).

(8)   Stillborg, B. and B. Leijon "A Comparative Study of Rock Stress
       Measurements at Luossavaara Mine", Swedish Mining Research Foundation,
       Kiruna, Sweden, Report Number FB-8217(1982).

In-situ experiments in JAERI

M.KUMATA
Japan Atomic Energy Research Institute, Tokai, Ibaraki

ABSTRACT

For the purpose of research and development of the measurement technique concerning in-situ experiments for geological
disposal of high level waste, several field tests were carried
out in Japan Atomic Energy Research Institute (JAERI). This
paper is a short introduction of JAERI work on in-situ
experiments.

INTRODUCTION

The Japan Atomic Energy Commission announced guidelines for
radioactive waste management in 1976. The time schedule for
these guidelines was established in 1980 for high-level waste
(HLW) management.

For the safety assessment on deep geological disposal for
HLW, it is necessary to know the geological, geochemical, geo-
phsical and hydrological conditions of a underground repository.
In order to predict the behaviour of hard rocks and nuclides migration in circumstances that would exist in a radioactive waste
repository, it is necessary to perform in-situ experiments and
measurements at similar depth to that of the repository.

Japan Atomic Energy Research Institute (JAERI) performed to
develop the measurement technique, instruments and analytical
methods concerning in-situ experiments. The drift at the depth
of 380m under the earth surface in the Akenobe mine in southwest
Japan was used for the in-situ experiments. The rock type was
shalstein of Permian age.

HEATER TEST

For the research of the heat transfer in the rock mass and
the mechanical behaviour of rock under thermal stress, field
tests have been conducted since the end of 1977 [1]. The temperature and thermal stress distributions have been investigated
using two electrical heaters to simulate the heat of vitrified
HLW. Also, the behaviour of water which are contained in rock
mass around the heater was investigated (Figure 1-1).

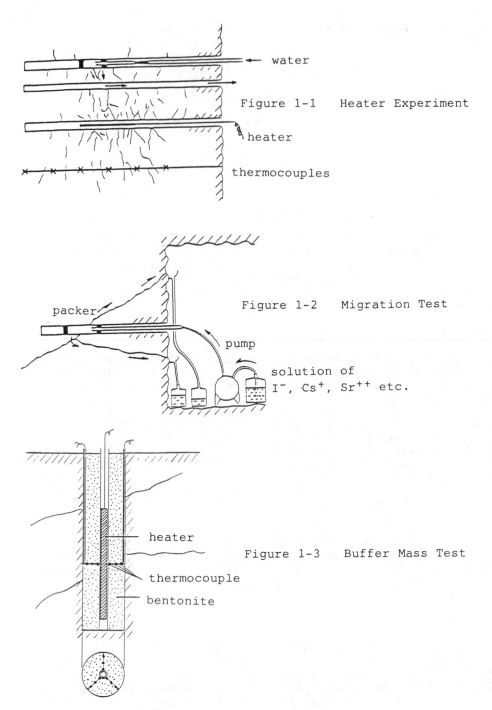

Figure 1-1   Heater Experiment

Figure 1-2   Migration Test

Figure 1-3   Buffer Mass Test

Figure 1   Conceptual design of in-situ experiments
at Akenobe mine

240

The heaters were composed of nichrom wire surrounded by
magnesia insulation and sheathed with stainless steel tube.
The dimension was 2.2 m in length and 25.4 mm in outer diameter.
Two 29 mm diameter boreholes were drilled for the heaters and
seven boreholes (the same diameter) were for thermocouples.
The boreholes were horizontal and approximately 7.5 m deep from
the wall-rock surface. The heaters were separated from each
other in such away that each heater simulates a single canister
emplacement. The electric power of the heaters were kept constant at around 2kW for first 47 days, and after 26 days cooling
the power was raised to 4kW and kept for further 48 days. Temperatures were measured by the twenty chromel-alumel thermocouples which were placed in each boreholes. The accuracy of
temperature measurment was within 0.1°C. Comparing the observed
temperature with calculated one, effects of vaporization of water
in the rock mass was clearly observed. Partly, the observed
temperature agreed with calculated one using the core sample's
thermal conductivity, 2.9 W/m°C.

The thermal stress of rock mass was measured by strain
gauges and strain meters. The thermally induced stresses has
been measured by using six rosette gauses with dummy gauges on
the wall-rock surface and two mold gauges in a horizontal
borehole.

For the research of thermal effect on permeability of rock,
the double packer method was used to measure permeability of a
rock mass around the heater. Measurements were performed for
three times: before heaters on, at elevated temperature and after
cooling. The distance between two rubber packers was set to be
45 cm and water was supplied with a pressure of 3, 6, 10 and 18
kg/cm^2.

MIGRATION TEST

Retardation of the migration of radioactive nuclides is due
to the sorption of ions on minerals, not only in rock but also in
fractures. Importance of in-situ test is that natural and real
structure of a fracture is only available. The experiment was
carried out at the same site used for heat conductivity test.
Double packers were inserted in a hole at about 18cm and 68cm in
depth from the wall-rock surface (Figure 1-2). The solution,
including nonradioactive elements as tracers, was charged into
the borehole at 4 kg/cm^2 pressure by compressor. LiI, $CsNO_3$,
$Ba(NO_3)_2$, $Sr(NO_3)_2$ and $ZrO(NO_3)_2$ were dissolved in water to
prepare the solutions of 50 - 100 μg/ml of metal elements.

Typical break through curves are shown in Figure 2.

BUFFER MASS TEST

Buffer material is expected to play an important role as an
engineered barrier in a multi-barrier system of geological
disposal of HLW. Thermal conductivity and water content
change of bentonite were measured. The experimental vertical

241

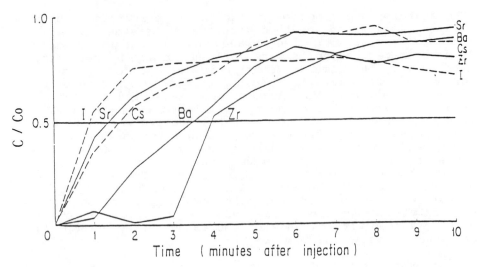

Figure 2   Concentration curve for the fracture flow path

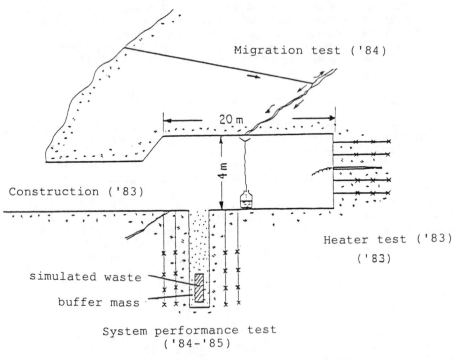

Figure 3   Conceptual design at a new underground laboratory

hole was drilled to 330 mm in radius and 3.3 m in depth.
After drilling, several fissures were observed on the inner wall
of the hole and the water seepage was found at three parts of
these fissures.   In this hole KUNIGEL VA Bentonite, almost Na-
montmorillonite, was solidified by use of 20 ton oil pressure
jack with a pressure of 20 kg/cm^2.   An electric heater and
thermocouples were emplaced in such a geometry as shown in Figure
1-3.

Measurments of the water content change of the buffer mass
was carried out used neutron scatter method with ^{252}Cf, before
the heater power on and after power off.

These experiments finished in May 1984 and a part of the
results was already published [2].

FUTURE EXPERIMENTS

More systematical experiments of in-situ test is designed at
a new site of granite rock mass.   A new underground laboratory
has been constructed at the near surface for an exclusive use on
safety research for geological disposal of HLW.   Excavation of a
gallery and test room were already finished.   A microcomputer
system would be used for data logging.   A borehole television
set would be also used for direct observation of fractures in a
borehole.   Heater test, migration test and system performance
test are planned at the labolatory (Figure 3).

OTHER STUDY CONNECTED WITH IN-SITU EXPERIMENTS

Considering deep underground state, for geological disposal
of HLW, some geological informations were collected from several
areas in constructing a underground facility.   Especially, for
permiability data from these area, statistical treatment was
planned.   Initial stress was measured using a acoustic emission
(AE) instrumetation system for core samples gathered from
granitic rock areas.   In this results, granitic rock mass was
stressed corresponding to lithostatic pressure at that point in
the vertical direction.

REFERENCES

[1]   Araki, K. et al. : "Preliminary Research on Geological
         Isolation of High-level Radioactive Waste at the Japan
         Atomic Energy Research Institute" Proc. Symp. Underground
         Disposea of Radioactive Waste, Finland (1979).

[2]   Shimooka, K. et al. : "Pilot Research Projects for
         Underground Disposal of Radioactive Waste in Japan" Proc.
         Conf. Radioactive Waste Management, Seattle (1983).

Summary of discussion

The discussion, chaired by Mr. Robinson, can be divided into three themes:

(a) Rock-mechanics tests

Miss Cooling noted other in-situ stress measurements undertaken in the locality of the shallow mine in Cornwall in which BRE was researching. These were at South Crofty Mine (at a depth of 790 m) and the Geothermal Project at Rosemanowas (to depths of 2000 m). The results of the stress measurements in all three programmes had been compared and the directions of the principal stresses agreed well.

A series of tests using four different systems of measuring in-situ stress had been undertaken in the shallow mine in order that a comparative assessment of the techniques could be made.

Mr. Patrick pointed out that hydrofracturing measurements have to be considered with caution; at any test location, both pore pressure and rock tensile strength are not always well known.

Mr. Dietz wondered whether flat-jacks are accurate enough to detect stress changes near heater tests. Mr. Schneefuss specified that the devices used for the Grimsel lab will be calibrated in a temperature vessel up to 40 MPa pressure, the accuracy being 0.1 MPa.

(b) Hydrogeological and tracer tests

Mr. Saari wondered whether the investigators have given thought to the possibility of turbulent flow taking place in some hydraulic tests because it seems that in some tests the pressures used are quite high.

In the Stripa experiments, injections are performed with the smallest overpressure possible (about 10% in excess of the natural flow). Therefore, pertubations are minimal and the flows remain laminar (Mr. Abelin). Tracer tests include (i) non-sorbed tracers such as dyes, iodide, Cr-EDTA, and (ii) sorbed tracers such as non-active Cs, Sr, Eu, Nd, Th, U.

As regards the disturbance of the hydrogeological regime brought about by driving the test galleries themselves (Mr. Brewitz), it should be low compared to the influence of the mine itself. Anyway, the purpose of the tests are to compare the behaviour of sorbed and non-sorbed tracers in similar conditions, to be found at present.

Mr. Black raised the important point that an underground cavern in hard rock may not be best suited for hydrogeological and tracer measurements because of the stress modification at the cavern wall which (i) tends to close fractures or cracks, hence giving lower values for fracture numbers, and (ii) reduces permeability, hence reducing the recovery of tracers injected away from the gallery.

Mr. Brewitz stressed that this situation is mainly the result of the excavation technique. Mr. Carlsson mentioned that no stress concentration have been measured at Stripa.

According to Mr. Lake, the shape of the gallery must also be considered to minimize stress pertubations (a rectangular cross section, inducing compressions in the corners and tensions in the walls, is probably the least suitable one).

Turning back to hydrogeological aspects, Mr. de Marsily mentioned that "radial" flow towards a gallery can be used to estimate dispersion coefficients, although the preferable arrangement would make use of "parallel" flow.

To conclude, Mr. Black proposed to reserve underground rock laboratories for rock-mechanical tests, and to do migration experiments from surface boreholes. These would be the most appropriate conditions to obtain reliable measurements.

(c) Transferability of the test results

Mr. Kühn observed that the present hard rock underground laboratories are in the depth range of 250-400m, which is "shallow" compared to planned repository depths (800 - 1,000 m), and feared problems of test extrapolations to these depths. Mr. Simmons pointed out that models can be validated, for instance, at the URL site, then transferred and applied to other sites. Therefore, validated models, and not results, can be transferred.

Session 4/Séance 4
Design and instrumentation for in situ experiments in clay
Conception et instrumentation d'expériences in situ dans l'argile

Chairman/Président:
L.BAETSLE
CEN/SCK, Mol, Belgium

Expérience acquise à l'occasion de la réalisation
d'une campagne géotechnique dans une argile profonde*

P.MANFROY, B.NEERDAEL & M.BUYENS
Section Géo-Technologie, CEN/SCK, Mol, Belgium

RESUME

La Belgique a retenu l'option de l'argile pour le rejet en formation géologique des déchets radioactifs conditionnés.
Dans ce cadre, le CEN/SCK de Mol a entrepris un important programme de Recherche et de Développement sur l'argile de Boom, formation présente sous le site nucléaire de Mol-Dessel.
Une installation expérimentale souterraine a été construite à 220 mètres de profondeur dans le but de procéder à de nombreuses expériences (géomecanique, corrosion, migration, transfert de chaleur).
Sa construction a été mise à profit pour placer un grand nombre d'appareils de mesure à caractère géotechnique sur le revêtement ou au sein du massif argileux.
Les conditions de chantier en souterrain sont de nature à rendre la réalisation de telles campagnes de mesures très difficile.
Cette communication énonce les premiers enseignements qui peuvent à ce jour en être tirés.

ABSTRACT

Belgium has selected clay as a possible disposal medium for conditioned radioactive waste.
CEN/SCK has launched an important research and development programme to evaluate the disposal potential of the Boom clay formation present under the nuclear site Mol-Dessel.
An underground facility has been built at 220 m. depth in order to proceed to geomechanical, corrosion, migration and heat transfer experiments.
During its construction numerous geotechnical measuring instruments were emplaced on the lining and in the clay medium.
Successful realization of such measurement campaigns was hampered by the very difficult underground working conditions.
This paper describes what can be learned from the experience gained so far.

1. INTRODUCTION

L'argile a été retenue par la Belgique comme roche potentielle pour le rejet des déchets radioactifs conditionnés en formation géologique.

(*) This work was performed under contract with the European Atomic Energy Community in the framework of its R & D-programme on Management and Storage of Radioactive Waste.

Afin d'améliorer nos connaissances sur les propriétés générales de l'argile à grande profondeur(géotechniques,de corrosion en milieu argileux, de transfert thermique, etc.) il fut décidé de creuser une installation expérimentale dans l'argile de Boom (âge Rupelien-oligocène) en dessous du site nucléaire de Mol-Dessel.

Cette installation expérimentale comportant principalement un puits d'accès et une galerie d'essai établie à la profondeur de -223 m, a été creusée par la technique de congélation eu égard aux formations sableuses aquifères et boulantes qu'il fallait traverser.

La construction de cette installation a été mise à profit pour réaliser une campagne géotechnique intensive d'auscultation de l'argile. La saisie des données se poursuivra pendant encore longtemps après l'achèvement des travaux et donnera lieu à un énorme travail d'interprétation. Différentes mesures portant sur les contraintes et les déformations tant dans l'argile que dans les revêtements furent effectuées à l'aide d'appareils de types et de principes divers.

Un nombre considérable de données furent ainsi récoltées dans des conditions d'autant plus difficiles qu'elles étaient liées souvent à l'avancement des travaux de type minier, sous des variations importantes de températures, et dans des atmosphères susceptibles d'endommager les appareils de mesure ou de nuire à leur bon fonctionnement.

Malgré ces conditions extrêmes, l'auscultation systématique du massif argileux a pu être menée à bien,permettant d'enrichir notablement nos connaissances sur le comportement géomécanique d'un massif argileux profond et d'acquérir une grande expérience sur la métrologie géotechnique en conditions difficiles.

2.   DESCRIPTION DE L'INSTALLATION EXPERIMENTALE (FIG. 1)

L'installation expérimentale constitue un laboratoire souterrain en soi et a été considérée comme tel dès le début des travaux. Elle se compose principalement de deux unités distinctes : un puits d'accès et une galerie expérimentale complétée par des installations d'essai complémentaires.

2.1.   Le puits d'accès

Le puits d'accès est de section circulaire (diamètre excavé : 4,20 m; diamètre utile :2,65m; hauteur : 215 m, il est terminé par un accrochage de section également circulaire (diamètre excavé :5,80 m, diamètre utile : 4,00 m, hauteur hors tout : 15 m).

L'ensemble puits-accrochage a été creusé à l'intérieur d'un cylindre de terrain préalablement congelé sur toute la hauteur à excaver. Cette congélation a été obtenue par circulation de saumure froide (≈ -25°C) de chlorure de calcium dans deux faisceaux concentriques de 16 tubes congélateurs chacun forés par injection et alimentés par une centrale cryogénique à l'ammoniac de 600.000 K frigorie/H installée en surface. Cette technique de creusement était la seule envisageable dès lors que le puits devait traverser des terrains sableux et aquifères avant d'atteindre l'argile dont le toit s'établit à ≈ -190 m.

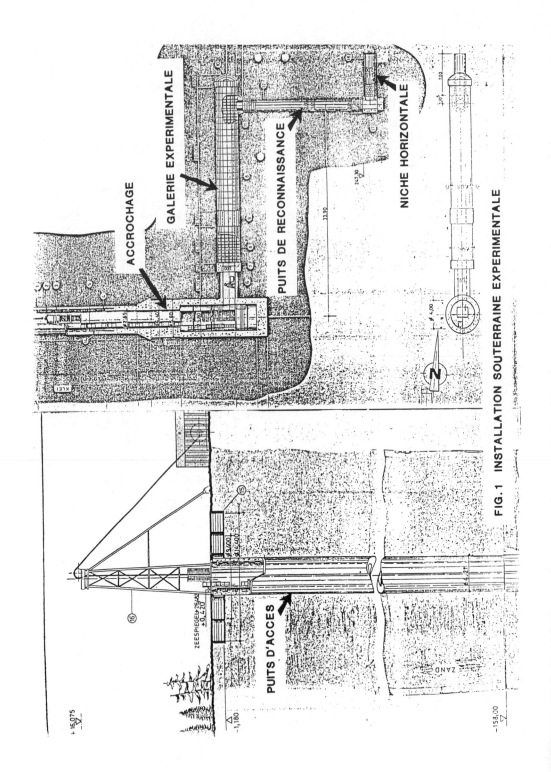

ACCROCHAGE

GALERIE EXPERIMENTALE

PUITS DE RECONNAISSANCE

NICHE HORIZONTALE

PUITS D'ACCES

FIG. 1 INSTALLATION SOUTERRAINE EXPERIMENTALE

Le revêtement du puits est constitué de deux parois concentriques distinctes en béton non armé à haute résistance. La première, externe et coulée en descendant au fur et à mesure de la progression des travaux de creusement, est prévue pour reprendre la pression lithostatique. La deuxième, interne et coulée en remontant après l'achèvement du creusement de la totalité de l'ouvrage, est prévue pour reprendre la pression hydrostatique. Une feuille d'étanchéité intermédiaire de polyéthylène, placée en remontant juste avant le bétonnage de la paroi interne assure par sa continuité l'imperméabilité de la portion du puits traversant les sables aquifères.

Cette feuille d'étanchéité est raccordée en dessous du niveau du toit de l'argile au moyen de joints toriques à une virole d'acier inoxydable noyée dans un béton monolithe qui constitue, à partir de cette profondeur, le seul revêtement de la portion restante du puits et de la totalité de l'accrochage.

2.2. La galerie expérimentale

Pour des raisons de prudence devant le risque d'un comportement rhéologique incontrôlable de l'argile en cours de creusement, il fut décidé pour la galerie d'utiliser également la technique de congélation préalable du massif. Celle-ci fut réalisée en deux phases successives à l'aide de tubes congélateurs disposés en faisceaux divergents, forés à l'air comprimé et alimentés par la saumure refroidie cette fois par des groupes cryogéniques au fréon permettant à la saumure d'atteindre des températures plus basses (-32°C). La première phase de congélation permit de creuser une première portion de galerie séparée de l'accrochage par un pertuis rectangulaire de 2 m x 1,5 m. Ce pertuis était destiné à s'éloigner suffisamment de l'accrochage afin d'inscrire à l'intérieur du faisceau de congélateurs l'excavation de 4 m de diamètre nécessaire à l'implantation de la galerie. La deuxième phase de congélation entreprise à partir de l'extrémité de la première portion de galerie permit d'achever le creusement de la partie restante.

Cette galerie, de section circulaire, de diamètre utile de 3,5 m présente donc une longueur totale utile de 26 m revêtue de voussoirs en fonte nodulaire, de 1 m de longueur boulonnés les uns aux autres et est terminée par un bouchon de béton armé de 2 m d'épaisseur. Chaque anneau de voussoirs est constitué de 8 pièces plus une clef. De nombreux voussoirs sont pourvus d'orifices de dimensions variées munis de couvercles de manière à disposer d'un accès aisé au massif argileux.

2.3. Installations pour essais complémentaires dans l'argile non gelée

Du fait que les travaux de creusement de la galerie et de la portion du puits située dans l'argile se sont effectués sous la protection de la congélation, ces travaux nous ont beaucoup appris sur la rhéologie de l'argile gelée mais peu sur celle de l'argile à l'état naturel. Comme d'autre part il n'est pas réaliste financièrement d'envisager le creusement par congélation d'une installation industrielle d'enfouissement avec les grandes longueurs de galerie que cela implique, il fallait absolument procéder à un creusement dans l'argile n'ayant jamais été congelée pour en observer la rhéologie tout en ne compromettant en aucune façon la stabilité de la galerie existante. C'est ce qui fut fait par le creusement d'un petit puits vertical de reconnaissance (diamètre excavé : 2,00 m, diamètre utile : 1,40m), réalisé en bout de la galerie expérimentale.

Ce petit puits de reconnaissance est revêtu de claveaux en béton séparés par des intercalaires de bois compressible et a été creusé sur une hauteur de 23 m dans une argile n'ayant jamais été gelée précédemment à l'exception de quelques mètres proches de la galerie expérimentale.

Le creusement du petit puits de reconnaissance est actuellement complété par celui d'une niche horizontale d'environ 7 m de longueur de même diamètre et pourvu d'un revêtement identique à celui du petit puits.

3. CONCEPTION ET INSTRUMENTATION DES TESTS IN SITU DANS L'INSTALLATION EXPERIMENTALE

L'installation expérimentale est destinée à l'exécution de tests et de mesures in situ de différentes natures sur l'argile. En dehors des mesures géotechniques qui constituent l'objet de cette présentation et dont une grande partie a déjà été réalisée, il sera procédé, dès le moment où l'installation deviendra opérationnelle, à des expériences de corrosion, de transfert de chaleur, de migration et de colmatage qui s'étendront sur plusieurs années. Les expériences de corrosion seront décrites au courant de la même session par Monsieur Casteels du Département de Métallurgie du CEN-SCK. Nous nous limiterons donc, dans la présente communication, à ne traiter que les tests et mesures géotechniques déjà réalisés ou en cours d'exécution.

Le principe de la campagne géotechnique menée à Mol dans l'installation expérimentale était l'auscultation systématique de l'argile en cours de creusement et après creusement, ce qui impliquait des mesures de contraintes, de déformations et de températures.

Des mesures de contraintes portaient sur les contraintes naturelles (pression lithostatique et pression hydrostatique) et les contraintes dans les revêtements. Les mesures de déformations portaient sur l'étude du comportement rhéologique du massif argileux et sur les déformations subies par les revêtements. Une série de mesures couplées contraintes-déformation ont en outre été exécutées à l'aide du pressiomètre Ménard tant dans l'argile gelée que non gelée. Enfin, les mesures de températures, intensives pendant la phase de congélation et moins fréquentes par la suite, constituèrent un complément indispensable aux mesures contraintes-déformations pour la mise en évidence de l'influence de l'état congelé sur celles-ci.

3.1. Placement des appareils de mesure

3.1.1. Puits d'accès

L'instrumentation du puits d'accès s'est déroulée en 2 phases distinctes: la première au cours du creusement, et la deuxième après achèvement du revêtement définitif du puits.

3.1.1.1. Instrumentation pendant la phase de creusement

Cette instrumentation concerne exclusivement les cellules de pression totale hydrauliques GLÖTZL et les sondes de température couplées au cellules. Deux types de cellules de pression totales ont été placées (circulaire 60 cm de diamètre et rectangulaire 20 cm x 30 cm) pour la mesure des pressions verticales, horizontales tangentielles et horizontales radiales le long du puits. Pour ces 51 cellules des niches ont été creusées à la paroi et colmatées à l'aide d'argile et/ou betonite avant la mise en place du revêtement bétonné externe.

3.1.1.2 Instrumentation sur l'intrados du revêtement après achèvement du puits d'accès et de l'accrochage

Un certain nombre d'équipements furent installés sur le périmètre intérieur pour les mesures de déformation à différents niveaux :
- Seize extensomètres à corde vibrante (type Telemac SC 5)
- Huit extensomètres à variation d'inductance (capteurs inductifs LVDT).
- Un grand nombre de plots repères, scellés dans la paroi, tant pour des mesures de déformations périmètrales que diamètrales à l'aide respectivement de déformètres mécaniques (DEMEC type Mayes & Son) et de cannes de convergence à capteur inductif.

3.1.1.3. Instrumentation dans le terrain après achèvement du puits d'accès

Durant le creusement du puits d'accès et la mise en place du revêtement bétonné, une vingtaine de fourreaux traversant la paroi ont été installés dans celle-ci afin de ménager des passages pour les conduits d'huiles alimentant des cellules de pression et de permettre un accès ultérieur direct au massif argileux. Après pose des équipements de mesure les fourreaux ont été parfaitement colmatés à l'aide d'une résine appropriée.
Les équipements de mesure placés dans le massif à partir de forages radiaux se composaient :
- de cannes à thermistances pour la mesure des températures à différentes distances durant la phase de congélation et de dégel,
- de cellules piézométriques hydrauliques GLÖTZL pour la mesure de la pression interstitielle dans l'argile
- de tubes fendus permettant l'introduction d'une sonde Menard pour l'exécution d'essais dilato-pressiométriques tant dans l'argile gelée que dans l'argile non gelée.

3.1.2. Galerie expérimentale

Tout comme pour le puits d'accès, l'instrumentation de la galerie expérimentale a été exécutée en deux phases distinctes : pendant le creusement et après la mise en place du revêtement.

3.1.2.1. Instrumentation pendant la phase de creusement

Pendant cette phase seules les cellules de pression totales GLÖTZL de forme rectangulaire (20 cm x 30 cm) et couplées avec des thermistances ont été installées, au fur et à mesure de l'avancement du creusement dans des niches colmatées avec de l'argile remaniée. Ces cellules, au nombre de vingt cinq, ont été positionnées en différents endroits et dans différentes orientations afin de mesurer les contraintes tangentielles et radiales dans le massif au niveau du radier, des piédroits et de la clef de la galerie.

3.1.2.2. Instrumentation après revêtement

Après la pose du revêtement de fonte nodulaire certains des éléments constitutifs de celui-ci ont été équipés de jauges de déformations (jauges à variations de résistance). Ces jauges collées en divers endroits des voussoirs de fonte (nervures, semelles et bords) permettent de surveiller l'état de contrainte du revêtement en fonction de l'établissement des pressions externes. Certaines de ces jauges de déformation ont également été placées sur les arma-

253

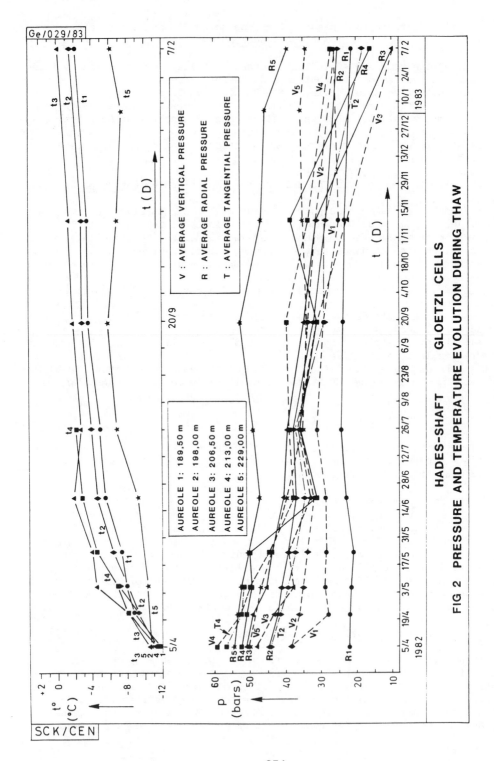

HADES-SHAFT GLOETZL CELLS

FIG 2 PRESSURE AND TEMPERATURE EVOLUTION DURING THAW

tures du bouchon intermédiaire en béton entre le sixième et le septième anneau de voussoirs de la galerie et sur certains cadres du pertuis rectangulaire. Après l'achèvement de la galerie, il a été procédé au placement de cellules piézométriques GLÖTZL ainsi qu'à des thermistances à différentes distances dans le massif d'argile autour du revêtement de la galerie.

3.1.3. Travaux complémentaires de creusement dans l'argile non gelée

La réalisation des sondages de congélation avant le creusement de la galerie expérimentale nous avait permis de déterminer certains paramètres de fluage mais seulement pour de petits diamètres et dans des conditions non représentatives d'un creusement réel. Le creusement du puits de reconnaissance vertical et de la niche horizontale de section circulaire a été l'occasion d'enrichir considérablement nos connaissances dans le domaine de la rhéologie de l'argile à grande profondeur.
Pour cela, il a été procédé à une mesure systématique des déformations et des contraintes tant dans le massif d'argile que dans le revêtement.

3.1.3.1. Mesures de déformation

Avant le fonçage du puits de reconnaissance vertical, plusieurs dispositifs de mesure ont été installés à partir de la galerie de manière à pouvoir suivre les déformations du terrain en cours de creusement, ces dispositifs se composent :
- de quatre inclinomètres dont deux verticaux dans le plan de l'axe de la galerie et deux obliques dans un plan perpendiculaire pour la mesure de la déformation radiale (torpille inclinométrique descendue dans un tube rainuré)
- de 3 extensomètres à tiges ancrés à différentes profondeurs dans l'axe du puits de reconnaissance pour la mesure de la déformation longitudinale relevée au fur et à mesure du fonçage.

Complémentairement à ces dispositifs et aux mesures systématiques de déformations en paroi faites par l'entrepreneur, 2 anneaux de revêtement de la petite niche horizontale seront équipés de plots repères pour la mesure des déformations diamètrales (DISTOMATIC).

Avant le creusement de la niche horizontale plusieurs dispositifs ont également été installés :
- un extensomètre TELEMAC DISTOFOR à bases multiples placé dans un sondage légèrement oblique foré à partir de l'extrémité de la galerie permettant de mesurer les déformations radiales du massif dans un plan passant par les axes de la galerie et de la niche, au fur et à mesure du creusement.
- un extensomètre à tige placé horizontalement dans l'axe de la niche avant creusement de celle-ci pour la mesure des déformations longitudinales

3.1.3.2. Mesures de contraintes

Durant le creusement du puits de reconnaissance huit cellules de pression totale GLÖTZL ont été installées dans l'argile à différents niveaux et suivant différentes orientations derrière le revêtement tandis que des cellules de charge étaient placées entre les claveaux à plusieurs endroits du revêtement.
La niche horizontale qui est actuellement en cours de creusement sera équipée également au fur et à mesure de l'avancement des mêmes dispositifs de mesure des contraintes.

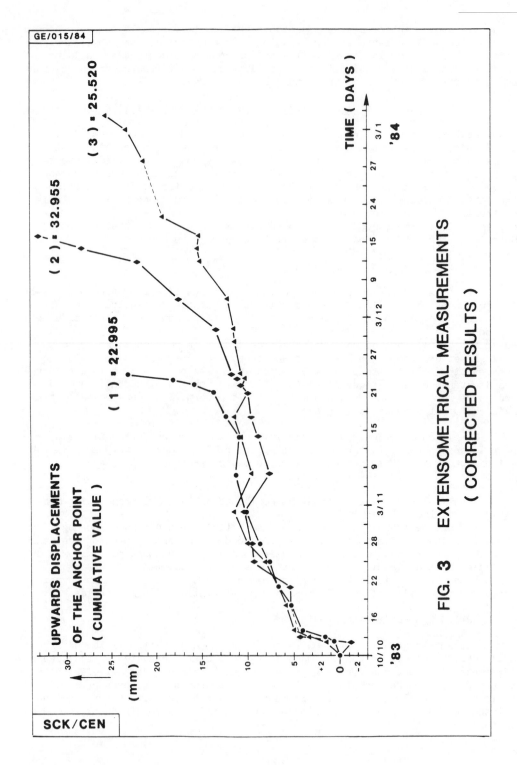

FIG. 3   EXTENSOMETRICAL MEASUREMENTS

(CORRECTED RESULTS)

3.1.3.3. Mesures dilato-pressiomètriques

Juste avant l'achèvement du puits de reconnaissance il a été procédé à des essais pressiométriques de type Menard à sonde nue dans des sondages horizontaux et verticaux forés à la tarrière.
Des essais au dilatomètre MAZIER sont prévus dans les prochaines semaines.

4. PREMIERS RESULTATS ET COMMENTAIRES

4.1. Puits d'accès/galerie

Cette première phase de travaux réalisée dans sa totalité par congélation doit néanmoins être subdivisée en 2 étapes bien distinctes, aussi bien en ce qui concerne la campagne d'essais que les travaux d'excavation.

La campagne de mesures le long du puits d'accès avait essentiellement 2 objectifs à savoir :
- acquérir les premières données (contraintes, déformations) au sein du massif et à la paroi
- tester les différents appareils de mesure, leur méthode de placement et leur fiabilité

La mise en pression très rapide du revêtement peu après bétonnage entre l'argile gelée et le fluage important du massif gelé autour de l'excavation ont amené des perturbations dans le déroulement de l'auscultation comme : la perte de certaines mesures "origine", la fissuration du revêtement à certains niveaux contrariant les mesures extensométriques correspondantes, la rupture des canalisations d'huiles alimentant les cellules hydrauliques placées dans le massif et le cisaillement de tubes de mesure (inclino, extenso) au contact béton/argile.
De plus, les conditions de fonctionnement et de travail (humidité, basse température, dommages causés à l'instrumentation, mesures manuelles et ponctuelles) ont été autant de contraintes supplémentaires au niveau du dépouillement et de l'interprétation. Les premiers enseignements qui peuvent être tirés à ce jour sont les suivants :
- les mesures de pressions totales (cellules GLÖTZL) et de température (thermistances) ont fonctionné correctement en suivant de façon fidèle les différentes phases de construction.
- le placement des piézomètres (cellules GLÖTZL) reste un problème très délicat (saturation). Les valeurs mesurées à 200 mètres de profondeur en argile non gelée sont bien de l'ordre de 20 kg/cm2 et semblent dans cet argile indépendantes du type de filtre utilisé (céramique ou métal fritté).
- l'interprétation des mesures de déformation du revêtement sont en cours.
  A ce stade et malgré les différents phénomènes qui les ont affectées, les premières informations disponibles semblent indiquer une relativement bonne corrélation au niveau des différents appareillages d'une part (zones de compression et de traction) et de la relation contraintes-déformations d'autre part.
  L'illustration nous paraissant le plus significative pour cette auscultation du puits d'accès est donnée par la figure 2 traduisant l'évolution, après arrêt de la congélation, des valeurs moyennes des températures et pressions (V = vertical, R = radial, T = circonférentiel) aux niveaux de mesures, n°s 1 à 4, (puits) et sous l'accrochage (niveau 5) pour la période s'étendant d'avril '82 à janvier '83.

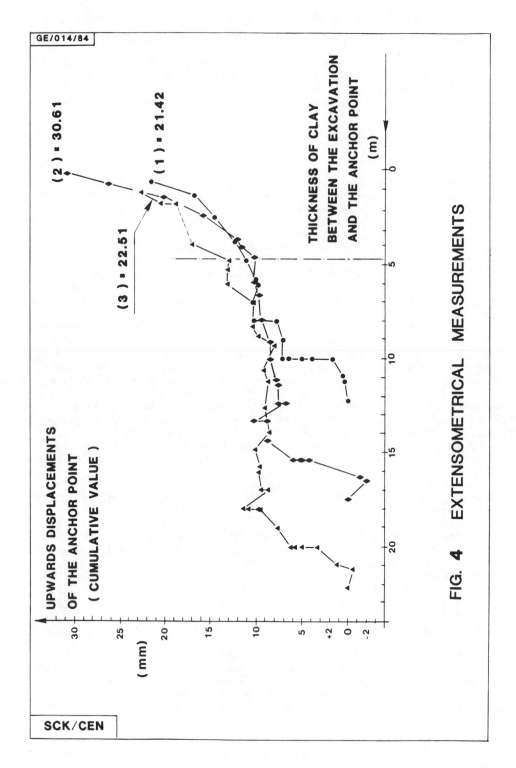

FIG. 4 EXTENSOMETRICAL MEASUREMENTS

Les considérations suivantes peuvent en être déduites :

* alors que la pression des terres à cette profondeur devrait être de l'ordre de 40 à 45 bars, l'évolution des pressions varient de 60 bars (supplément de pression dû à la congélation) à des valeurs bien inférieures à 30 ou même 20 bars (modification du contact cellule-terrain dûe au dégel).

* la corrélation pression-température est très nette

* seul le niveau 5(fond d'accrochage) a été affecté par la congélation de premier tronçon de galerie (Stabilisation de la pression correspondant à une première chute momentanée de température).

* le champ des contraintes est relativement hydrostatique excepté à l'auréole n° 1 caractérisée par la présence d'argile assez silteuse.

Une méthode de construction de la galerie différente de celle prévue initialement et les observations et expériences tirées de la construction et de l'auscultation du puits d'accès nous ont amené à revoir le programme de mesures. Cette campagne a finalement consisté en mesures de pressions et températures autour de la galerie et en mesures de contraintes dans le revêtement ("strain gauges") constitué en grande partie de claveaux en fonte nodulaire.

L'annulaire revêtement-paroi excavée a été remblayé par des blocs d'argile provenant du front . Comme on s'y attendait, cet anneau "compressible" a conduit à une mise en pression beaucoup plus lente de la galerie ce qu'ont très bien montré les cellules hydrauliques de mesure dont les indications ont par alleurs suivi aussi régulièrement que dans le puits, le dégel du massif. La pression maximale enregistrée a été de l'ordre de 35 bars. La technologie de placement et de mesure des jauges de déformation s'est révélée correcte et l'interprétation des résultats est en cours.

4.2. Travaux de reconnaissance complémentaires en argile non gelée

Le puits d'accès et la galerie expérimentale ne nous ont pas permis l'observation directe de l'argile non gelée en place, ni l'examen de son comportement lors de travaux miniers. C'est précisément dans cette optique que ces travaux complémentaires ont été instrumentés.

L'accent a été mis sur les mesures de déformations (inclinomètres-extensomètres) sans pour autant négliger les mesures de pression totales et interstitielles. Outre les cellules de mesure de pression (GLÖTZL) installées pour la première fois au sein du massif "vierge", l'utilisation d'un revêtement souple constitué de claveaux en béton avec intercalaires en bois compressible nous a permis d'y insérer des cellules de charge (modèle GLÖTZL) à lecture directe.

Bien que les déformations se soient révélées être beaucoup plus faibles que prévu dans les estimations les plus optimistes (de l'ordre du cm seulement pour les 2 mois de réalisation du petit puits) le matériel a bien rempli sa fonction par une sensibilité plus que satisfaisante aux différentes phases d'exécution.

Ces considérations sont bien illustrées aux figures 3,4 et 5 (petit puits). La figure 3 montre que les déplacements totaux des points d'ancrage sont de l'ordre de 2 à 3cm seulement pour la durée des travaux; les courbes expérimentales relevées s'apparentent très bien aux courbes classiques de fluage. La figure 4 illustre les même phénomènes tout en intégrant au diagramme un para-

259

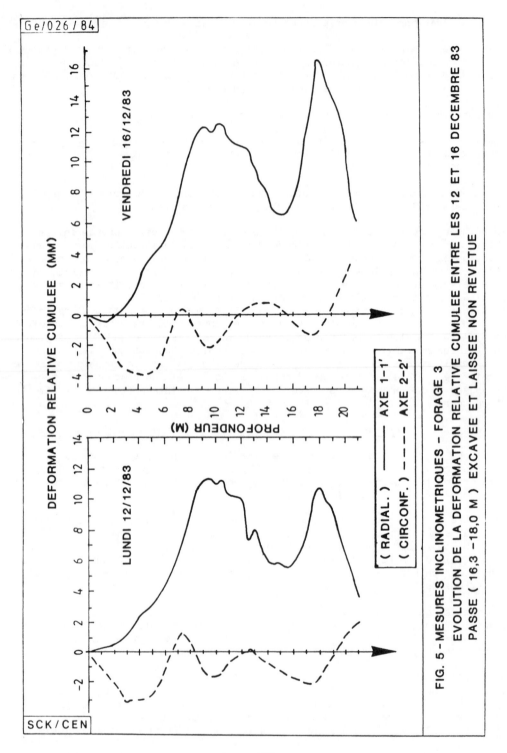

FIG. 5 – MESURES INCLINOMETRIQUES – FORAGE 3
EVOLUTION DE LA DEFORMATION RELATIVE CUMULEE ENTRE LES 12 ET 16 DECEMBRE 83
PASSE (16,3 –18,0 M) EXCAVEE ET LAISSEE NON REVETUE

mètre supplémentaire à savoir l'avancement des travaux dans le temps; le diagramme montre clairement l'accentuation des déplacements des ancrages lorsque l'excavation s'en rapproche de moins de 5 mètres.

La figure 5, relative aux mesures inclinomètriques, met en évidence l'évolution de la déformation relative cumulée entre le lundi 12 et le vendredi 16 décembre, alors qu'une passe (16,30 m - 18,05 m) avait été volontairement excavée et laissée non revêtue.

A noter la déformation maximale de l'ordre du 15 mm et la bonne correspondance entre la profondeur atteinte et la forme de la courbe; la déformée plus faible vers 15 mètres de profondeur s'explique par la présence d'un banc de septaria (nodules calcaires de grandes dimensions) qui confère à l'argile une raideur plus importante.

Quant aux pressions, aussi bien les cellules de charge (revêtement) que les cellules placées dans le massif n'indiquent toujours que des valeurs minimes, alors que 6 mois se sont écoulés depuis leur pose.

Pour clôturer ces informations relatives à l'auscultation géotechnique en argile non gelée, signalons que des essais pressiométriques ont montré que la pression limite était bien supérieure aux 75 bars appliqués à la paroi lors de leur réalisation. Des essais dilatométriques vont être entrepris dans les prochaines semaines en collaboation avec le BRGM.

CONCLUSIONS

- la méthode de congélation utilisée a entraîné des modifications de température, de volume et de caractéristiques mécaniques non négligeables aux 2 milieux auscultés (argile - béton), ce qui était bien mis en évidence à la fois par les essais de laboratoire et les essais in situ.

- après cette auscultation détaillée en terrain gelé, les travaux de reconnaissance en argile non gelée nous permettent d'acquérir les données essentielles sur le creusement de l'argile à 240 mètres de profondeur. Les données, convenablement interprétées, permettront grâce à l'utilisation d'une loi rhéologique adéquate, d'être extrapolées au creusement de galeries horizontales de plus grand diamètre et serviront alors de base pour le dimensionnement optimal du revêtement et la conception du bouclier et de l'outil d'une machine du type tunnelier apte à creuser et revêtir de grandes longueurs de galerie.

- Si au fur et à mesure du développement de l'auscultation dans les excavations successives, il a été possible de tirer profit des expériences précédentes et de remédier aux problèmes rencontrés, le choix des plages de variation et de la sensibilité des appareils de mesure a été pour nous le problème fondamental vue la différence de comportement du massif entre la théorie et la pratique et cela aussi bien en argile gelée que non gelée.

REMERCIEMENTS

L'ensemble de l'installation expérimentale ainsi que les travaux complémentaires ont été réalisés par la firme FORAKY de Bruxelles.
Les travaux préparatoires pour forages de congélation ont été exécutés par la firme SMET de Dessel.
L'Ingénieur Conseil pour la construction de l'ouvrage ainsi que pour les missions d'assistance technique diverses y afférentes était l'association momentanée des deux firmes TRACTIONEL-COURTOY de Bruxelles.

Le CEN/SCK, Maître d'Oeuvre de l'ouvrage et du projet, a reçu sous forme de programmes d'actions indirectes, une aide substantielle de la Commission des Communautés Européennes et de l'Organisme National des Déchets Radioactifs (ONDRAF-NIRAS),pour divers programmes de recherches liés au projet.

Différentes institutions publiques et universitaires ont participé activement aux tâches techniques et expérimentales en cours de travaux. Parmi celles-ci, il convient de citer :
- l'Université de Louvain-la-Neuve, Laboratoire du Génie Civil (L.G.C.) -
  Louvain-la-Neuve
- l'Institut Géotechnique de l'Etat, (IGE-RIGM) Gent
- l'Institut National Interuniversitaire des Silicates et Matériaux (Inisma),
  Mons
- le Bureau de Recherches Géologiques et Minières (BRGM) - Orléans
- l'agence Nationale pour la Gestion des Déchets Radioactifs (ANDRA). Paris
- l'organisme de contrôle qui a procédé au suivi de l'ensemble des travaux
  était l'Association des Industriels de Belgique (A.I.B.)

Les auteurs remercient ces firmes ou institutions ainsi que leurs agents pour leur travail qui a rendu cette publication possible.

BIBLIOGRAPHIE SUCCINCTE

- FIELD MEASUREMENTS DURING THE CONSTRUCTION OF AN UNDERGROUND
  LABORATORY IN A DEEP CLAY FORMATION
  *B. NEERDAEL, M. BUYENS, M. LEJEUNE, J.F. THIMUS, R. FUNCKEN,
  B. DETHY*
  International Symposium on Field measurements in Geomechanics,
  September 05.08.1983

- PROJET DE CREUSEMENT D'UN PUITS D'ESSAI DANS L'ARGILE PLASTIQUE
  A GRANDE PROFONDEUR ET MESURES DE CONVERGENCE
  P. MANFROY, P. GONZE, E. LOUSBERG
  Symposium International sur la Géologie de l'Ingénieur et la
  Construction en Souterrain - Lisbonne 12-15 septembre 1983

- CONSTRUCTION OF AN EXPERIMENTAL LABORATORY IN A DEEP CLAY FORMATION
  R. FUNCKEN, P. GONZE, P. VRANCKEN, P. MANFROY, B. NEERDAEL
  Symposium "Eurotunnel '83" Basel, June 22-24, 1983

- REJET DES DECHETS RADIOACTIFS EN FORMATION ARGILEUSE PROFONDE,
  CONSTRUCTION D'UN LABORATOIRE SOUTERRAIN POUR UN PROGRAMME EXPE-
  RIMENTAL APPROFONDI   - IAEA-CN 43/54 -
  P. MANFROY - CEN/SCK - Mol

  POSSIBILITES D'EVACUATION DE DECHETS RADIOACTIFS DANS UNE FORMA-
  TION PROFONDE D'ARGILE PLASTIQUE- IAEA-CN 43/56
  R. HEREMANS - CEN/SCK - Mol

  Conférence internationale sur la gestion des déchets radioactifs
  Seattle, Washington (Etats-Unis d'Amérique), 16-20 mai 1983.IAEA.

Geotechnical instrumentation for in-situ measurements in deep clays

D.BRUZZI, F.GERA & S.PICCOLI
ISMES, Bergamo and Rome, Italy

E.TASSONI
ENEA, Casaccia, Italy

RIASSUNTO

Al fine di determinare le proprieta' geotecniche di argille Plioceniche in un laboratorio sotterraneo a Pasquasia nei pressi di Enna in Sicilia, sono stati sviluppati strumenti per la misura della pressione dei fluidi interstiziali, pressione totale e deformazioni radiali. Tutti gli strumenti sono stati controllati per determinare la loro risposta ai fattori ambientali e per determinare tecniche che generino accurate correzioni sulle misure. Il controllo in sito e in laboratorio degli strumenti prevede la registrazione delle misure per mezzo di un sistema automatico di acquisizione.

ABSTRACT

Borehole instrumentation has been designed and developed to carry out pore pressure, total pressure and strain measurements, in order to assess the geotechnical properties of stiff Pliocene clays in an underground laboratory at Pasquasia near Enna in Sicily. All the instruments have been tested to evaluate their response to environmental factors and to determine a suitable technique for generating accurate corrections. In situ and laboratory calibration and monitoring of the instruments include computer logging of the transducer outputs.

1.0 INTRODUCTION

ENEA and ISMES, with the economic assistance of the CEC, cooperate on the study of different aspects of disposal of heat-generating, long-lived radioactive wastes in deep clay formations.

Many different clay formations exist in Italy. They vary greatly in age, mineralogical composition, dimensions, consolida-

263

tion history and physico-chemical properties. The so-called "blue clays" of Plio-Pleistocene age are among the most abundant and are considered particularly promising from the viewpoint of radioactive waste disposal. They are consequently the object of the most detailed studies at present.

No specific area has yet been selected as a candidate for a repository site and no choice of this kind is expected to take place in the near future. Consequently no particular variant of the blue clays, in terms of geotechnical properties, is currently considered more suitable than any other. The search for a site suitable for construction of an underground laboratory has thus been controlled more by the availability of access to a suitable formation, by the depth of the clay, and by the willingness of the site owners to host the facility.

The Pasquasia mine in central Sicily was eventually selected as being suitable in all these respects. It is an operating potash mine located within a large body of rock salt. A recently excavated adit provides convenient access to a layer of marly blue clays about 100 m thick contained in the cover of the salt formation. The adit crosses the blue clay layer at a depth ranging between approximately 100 and 200 m.

The underground laboratory will be located about 170 m below the surface.

2.0   ROCK PROPERTIES AND FIRST PHASE TESTS

Available information on the response of deep clay formations to both excavation and emplacement of heat-generating waste is inadequate. Data on the geotechnical behaviour of deep clays, on heating effects and on techniques for backfilling and plugging repositories and boreholes are essential in order to bring basic knowledge on disposal in argillaceous sediments to the same level as for other geologic disposal options.

Owing to the great variability in clay properties data obtained at a particular site, in a particular formation, cannot be applied elsewhere with confidence. Nevertheless underground laboratories are justified since a generic understanding of the behaviour of deep clays is a basic requirement for a logical search of potential repository sites.

It is undoubtedly true that most investigations carried out in underground laboratories will have to be repeated for the characterization of potential repository sites. This is itself a major justification for the present phase of construction of underground laboratories since they allow development of instrumentation, experimental techniques and operational procedures. Subsequent investigations and operations at potential repository sites will thus be carried out more efficiently.

During excavation of the Pasquasia mine adit samples of clay have been obtained for initial laboratory rock mechanics observations. These preliminary results show the rock to be a stiff clay, gray-blue in color, with latent fissures. The age of the clay is lower Pliocene (Piacenziano). Mineralogical analyses have shown that clay minerals constitute about 50% of the rock, with carbonates comprising 25 to 30%, and silt/sand particles between 20 and 25%. The most abundant clay minerals are illite and kaolinite with smectite and chlorite both below 5%; traces of vermiculite and interlayered clay minerals are also present.

Geotechnical testing has confirmed the hardness of the rock with values of undrained shear strength greater than 10 kg/cm2. Cohesion in terms of effective stresses is between 1.0 and 1.5 kg/cm2 and the friction angle is between 30 and 34°.

At the depth of the underground laboratory the overall vertical stress is about 36 kg/cm2. Since the water table is estimated to lie at the contact between the sandy clays and the overlying calcarenites the resulting hydrostatic pressure can be calculated to be about 12 kg/cm2. The effective stress at the level of the underground laboratory should then be 24 kg/cm2.

It is anticipated that a number of in-situ investigations will be performed at the Pasquasia underground laboratory during a time period of several years. Individual investigations will be initiated at different times over this period and some expansion of the experimental gallery is likely to be required in the future. In the present phase planned investigations comprise:

- rock mechanics measurements (to study the effects of excavation on the rocks);

- a standard heater experiment.

The experimental gallery will be excavated from the side of the mine adit and will be 50 m long. The experiments will be carried out in the last 20 m in order to avoid the influence zone of the adit.

The effects of the excavation will be investigated by means of the following measurements:

- total pressure variations in proximity of the opening;

265

- variations of pore fluid pressure;

- rock deformation by means of inclinometers;

- rock deformation by means of extensometers;

- tunnel convergence.

In the thermal field of the heater experiment the planned measurements are:

- temperature;

- pore fluid pressure;

- total pressure;

- rock deformation by means of inclinometers;

- borehole closure.

Some of these measurements are known to be difficult in clays. In particular the measurement of pore fluid pressure and total pressure are occasionally troublesome due to the very low permeability and plastic behaviour of the rock. In the heated zone the difficulties could be compounded by the high temperatures.

The other types of measurements are reasonably reliable and are not expected to cause unusual problems in the underground laboratory.

3.0 GENERAL PRINCIPLES FOR THE DESIGN OF IN-SITU INSTRUMENTATION

Generally speaking the most cost-effective approach to instrumentation design is to adapt general purpose transducers to the particular conditions envisaged. Essential requirements of geotechnical instrumentation are good stability with time combined with high resolution and precision. Different approaches can be adopted to achieve this end.

Electrical sensors are a possible solution. The vibrating-wire and force-balance types are among the most used sensors.

Vibrating-wire sensors have various advantages such as low cost, good repeatability, low hysteresis and high resolution and are easily connectable even in the case of long distance data transmission. The disadvantages of these sensors are their sensitivity to temperature change and to shocks and vibrations, large non-linearity, and a low frequency response.

Another potentially useful electrical measuring system is represented by force-balance sensors, such as the servo inclinometer. The servo system technique provides high accuracy, good stability and wide pressure range selection but it also has disadvantages such as low frequency response and a requirement for considerable electronics, which are sensitive to shocks and vibrations.

The use of electrical sensors gives good results in many applications, but it requires a careful study of transducers and the development of particular construction techniques. The high stability required for long-term measurements is not easy to reach.

These difficulties do not exist if non-electrical instruments are used. In this case the measurement is generally performed by balancing the force or pressure of the soil. On the other hand if an automatic data acquisition system is needed then the use of electrical sensors is practically unavoidable. During recent years new technologies have been developed in order to verify the long term stability of transducers. At present the best choice is to use electrical sensors provided with devices for monitoring long-term stability and checking zero drift.

For the Pasquasia underground laboratory all instruments have been designed to operate at relatively high temperature. As far as possible the measurement sensors that have been chosen are based on proven technologies so as to obtain a good performance-cost ratio.

On the basis of the preceding considerations specially designed instruments have been developed for the measurement of pore pressure, total stress, and strain.

The sensor choice has been based on the evaluation of transducer accuracy according to two basic methods:

- the error band envelope;

- parameter specification.

The former has been used to evaluate transducers under stat-

ic conditions and to obtain an immediate real-time accuracy
tolerance. The latter has been used to define transducer behaviour under expected conditions during the tests.

4.0   INSTRUMENTATION FOR THE MEASUREMENT OF PORE PRESSURE, TOTAL
       PRESSURE AND STRAIN

The instrumentation design effort has been limited to those
types of instruments which we believed to require development and
adaptation to satisfy the anticipated testing conditions. Piezometers, total pressure cells and calipers for the measurement of
borehole closure are the only non-standard instruments that will
be utilized in the initial testing phase.

4.1   Pore pressure measurement

Having considered the advantages and disadvantages of presently available sensor types it was decided to use extensometer
sensors. Two options exist: either to use commercial sensors,
which are easily available, well tested and inexpensive, or to
develop a custom made sensor.

In either case the critical problem is to check the transducer zero drift. To this end two approaches have been developed:

- unloading the transducer by inducing a pressure by means of an
   electrical system;

- pneumatic balancing of the pressure acting on the transducer
   membrane.

Both approaches have been tested extensively. The result has
been the development of three different piezometer designs. Two
types make use of commercial transducers, while the third one
uses a sensor design especially developed for this project.

One of the designs comprises an electric zero drift check;
the other two types use the pneumatic balancing approach. The
three piezometer designs are described briefly below.

Figure 1 shows the scheme of an electrical piezometer which
makes use of a commercial pressure gauge and a magnetic valve for
checking zero drift.

The magnetic three-way valve is placed between the transducer and the porous filter. In the measurement position the valve

268

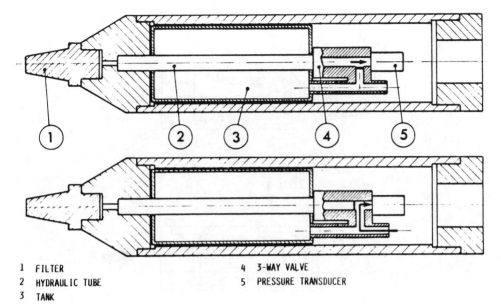

1 FILTER
2 HYDRAULIC TUBE
3 TANK

4 3-WAY VALVE
5 PRESSURE TRANSDUCER

Fig. 1 - Scheme of piezometer n. 1

permits the pore fluid pressure to load the transducer. The valve
can be switched by an electrical impulse. In the check position
the transducer is not affected by pore pressure; instead the
transducer is connected with the atmosphere through a nylon hydraulic tube included in the electrical cable. This arrangement
allows checking of zero drift.

This piezometer is particularly suitable for automatic data acquisition systems. The only disadvantage is that during switching
of the magnetic valve a small amount of water is ejected. This
water must be collected in a tank, which limits the number of
checks. However, it is possible to carry out more than 1200
checks during the piezometer life. The accuracy and sensitivity
characteristics are respectively 0.5% of full scale (f.s.) and 2
cm H2O with f.s. = 10 bars.

Figure 2 shows the scheme of an electrical piezometer with
pressure balancing and a commercial differential transducer. The
use of a differential transducer permits loading of both faces of
the sensor with the same pressure. When the balance point of the
transducer is reached the zero drift can be read. A thin membrane
of teflon separates the pressure sensing diaphragm from the porous filter. During measurement the membrane adheres to the sensing diaphragm.

Air or an inert gas is pumped inside the instrument to check zero
drift. The pressure must be high enough to cause the detatchment
of the membrane and consequent unloading of the transducer. The

269

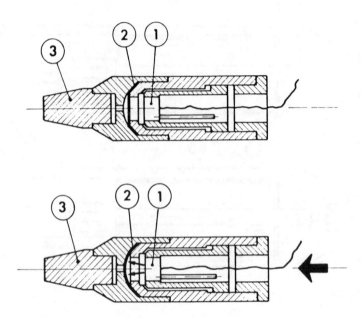

1 DIFFERENTIAL PRESSURE TRANSDUCER
2 DIAPHRAGM
3 FILTER

Fig. 2 – Scheme of piezometer n. 2

air or gas is pumped through a tube contained in the electrical cable.

This piezometer has good accuracy and sensitivity; respectively 0.3% of full scale and 2 cm H2O with f.s. = 10 bars.

Figure 3 shows a piezometer which has been designed and developed by ISMES. This instrument operates along the same principles as the previous one, but it makes use of a pressure balancing transducer and a purpose built differential pressure transducer. This solution has been chosen because few small size differential transducers are available on the market and are very expensive.

The ISMES piezometer has good characteristics, with accuracy and sensitivity respectively 0.5% of full scale and 3 cm H2O with f.s. = 10 bars, and costs about half of commercially available ones.

All three piezometer types allow checking of zero drift more than one thousand times.

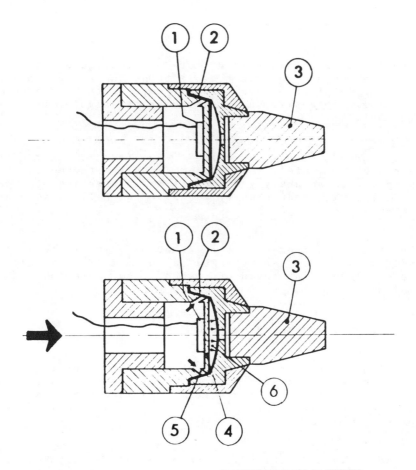

1 STRAIN GAUGE   4 PRESSURE SENSING DIAPHRAGM
2 DIAPHRAGM   5 GAS ENTRY PORT
3 FILTER   6 COMPENSATION CHAMBER

Fig. 3 - Scheme of piezometer n. 3

In the piezometer with the magnetic valve, no variation of measurement accuracy has been observed, while in the other two types an accuracy change amounting to 0.1% of full scale was noted.

Considering that the piezometers will be used in deep clays, special attention has been given to specific points. In particular the three following requirements are considered important:

- to minimize the volume of the measurement chamber;

- to check the sensor temperature;

- to prevent the loss of filter saturation.

The first requirement is dictated by the need to minimize time-lag in order to reach equilibrium quickly and be able to measure shortly after both instrument emplacement and zero drift checks. This objective was achieved by redesigning the measurement chamber.

The second requirement has been met by introducing a thermo-resistor inside the piezometer. The thermo-resistor, a PT-100 type with accuracy 0.5°C, allows correction of measurements during data reduction by means of the calibration curve of each sensor.

The third requirement has been met by using filters characterized by an "air entry value" as high as 5 to 6 kg/cm2. The filters are made of sintered or baked clay material (particle size $\sim$ 2 μm).

4.2   Total pressure measurement

Total pressures are measured by means of electrical hydraulic pressure cells in which the external pressure is transmitted to oil contained in an element (cell) and measured by an electrical transducer. The scheme of a total pressure cell is shown in Figure 4.

These cells have the advantage of having a very low elastic modulus. Therefore the effect of cell emplacement on the deformation behaviour of the rock is negligible. A very good cell-rock contact can be assured by the repressurizing system.

Potential problems associated with pressure cells are due to the deformable parts of the probe which must match the elastic modulus of the rock. Other critical areas are the welds (which must not cause local stiffening), oil saturation, and pressure transducer zero drift.

The problem of zero drift checking has been solved by using pressure transducers with pneumatic zero drift checks similar to those used for the piezometers.

Laboratory testing has allowed cheking of the properties of the structural components of the cell and verification of their conformity with requirements.

These particular cells, with parallel expanding faces, will be used within boreholes. Consequently the cells will be completed with semicylindrical elements in order to fill the borehole and transmit the rock pressure to the cell. For the fabrication of these semicylindrical elements the following criteria have been followed:

- the elastic modulus of the material must match the modulus of the rock;

- the filling elements must make good contact with the borehole walls;

- the material must be non-conducting and non shrinking.

The laboratory tests indicate that these requirements can be met.

An additional problem specific to those cells that will be placed in proximity to the heater is the effect of temperature

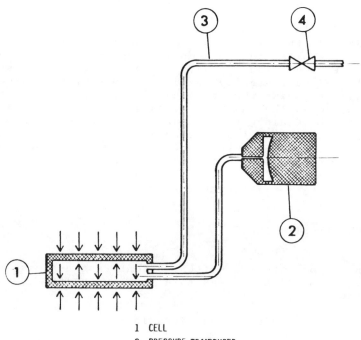

1 CELL
2 PRESSURE TRANSDUCER
3 REPRESSURIZING TUBE
4 VALVE

Fig. 4 – Scheme of hydraulic pressure cell

changes. Oil volume changes as a function of temperature and consequently the value of measured total pressure is affected. Due to the relatively large volume of oil present in the cell this effect is not negligible. The solution is again to place a thermo-resistor in the cell and to correct measured pressure values on the basis of a calibration curve predetermined for each cell.

The performance of the cells in laboratory tests has been satisfactory. The accuracy has been calculated to be 0.5% of full scale with f.s. = 40 bars. Field testing has confirmed that measured values can be affected by the installation procedure, therefore cell emplacement will have to be performed with the greatest care.

4.3   Strain measurement

In-situ rock deformation can be measured with a variety of different instruments. In the Pasquasia underground laboratory inclinometers and extensometers will be employed. These instruments will be of commercial design and no difficulties are anticipated in their use.

In addition the closure of the tunnel and of boreholes will be measured. While the measurement of tunnel wall convergence will be carried out with standard methods, i.e. by fixing markers to the wall and periodically measuring the distance between them, a new caliper type instrument is under development for measuring borehole closure. The instrument measures the diametral deformation of a borehole along two orthogonal diameters. Movement of the borehole wall is transmitted to a magnetic servo system by 0-ring sealed pistons mounted through the wall of the gauge body. The initial zero point is detected by a proximity sensor. The instrument is suitable for automatic data acquisition. No testing of the instrument has yet been carried out.

5.0   FUTURE IN-SITU TESTS

Planning for future in-situ tests at Pasquasia is in the initial stage. Three main areas have been indentified as deserving investigation:

- water flow and radionuclide migration;

- waste emplacement procedures;

- plugging and backfilling.

Due to the very low permeability of the material, in-situ permeability measurements in clays are usually difficult, and their interpretation in terms of potential groundwater advection is problematic. The Pasquasia clays are stiff and contain latent fissures. However no water has been observed in the clay section of the mine adit. An initial conventional injection hydraulic test will be carried out in wells 10 to 15 m below the laboratory floor. One objective will be to determine whether any component of fracture flow is to be expected under normal hydraulic conditions.

A more sophisticated permeability test is planned to be similar to the large scale permeability measurement carried out at Stripa. Experience with tunnels in clay and, in particular with the mine adit, shows that no water enters the excavation from the clay formation. On the contrary exposed clay dries-up, indicating that, at least initially, evaporation rates exceed the flow of water towards the cavity. The experiment could consist of closing a small drift, checking air moisture at the entrance and at the exit and measuring water content in the clay away from the excavation wall. The saturation profile and water flow at equilibrium should allow calculation of the permeability.

As far as radionuclide migration is concerned no plans for in-situ testing exist yet. The principal migration envisaged is by diffusion, and consequently in-situ tests are likely to require very long times in order to show any movement. Their value compared to laboratory experiments is thus very debatable. A worthwhile migration experiment is thus only likely to be envisaged if the hydraulic tests indicate that substantial advection could occur.

The conceptual design study for repository disposal currently carried out by ISMES for ENEA has highlighted two further engineering issues that require eventual in-situ testing.

First, in relation to waste emplacement procedures, the objective is to assess the feasibility of emplacing waste packages directly in contact with the clay. The stability of unlined disposal boreholes will thus be assessed by measuring closure of boreholes of differing depth and diameter. This is essential in order to assess the possible depths of disposal holes and their stand-up time between drilling and waste emplacement.

Second, for the study of disposal hole plugging and tunnel backfilling the main questions are seen to be : bonding, permeability along the plug-rock interfaces and long-term physico-chemical stability of the system. Emplacement procedures for plugging and backfilling materials will also need to be demonstrated. Initial emphasis will be placed on investigation of injectable "setting" clay seals, reconstituted mechanically compacted clay backfills, and various cement-clay mixtures. Eventually it is hoped to link these studies to tests of partially removable tunnel liners and bulk tunnel backfilling in a large scale experiment.

In situ testing and corrosion monitoring in a geological clay formation*

F.CASTEELS , J.DRESSELAERS & J.KELCHTERMANS
Metallurgy Department, SCK/CEN, Mol, Belgium

R.DE BATIST* & W.TIMMERMANS
Materials Science Department, SCK/CEN, Mol, Belgium
Also Rijksuniversitair Centrum Antwerpen, Belgium

ABSTRACT

The corrosion rate and mechanism of a number of candidate canister materials, structural materials of the repository and simulated vitrified waste forms will be tested in two environments representative during the storage of radioactive waste in a geological clay formation. These environments are: direct contact with clay and a humid atmosphere loaded with clay extracts. The corrosion experiments will be carried out at $\approx 15°C$, $50°C$, $90°C$ and $170°C$ for total exposure times of 50.000 h.

RESUME

Le mécanisme et la vitesse de corrosion des matériaux possibles pour les "containers", des matériaux de construction et des déchets vitrifiés simulés seront testés dans deux environnements représentatifs de ceux existants lors du stockage des déchets radioactifs dans les couches d'argile. Les environnements sont: le contact direct avec l'argile et une atmosphère humide chargée d'extraits d'argile. Les expériences de corrosion seront réalisées à des températures de $\approx 15, 50$, 90 et $170°C$ et pour des temps totaux d'exposition de 50.000 h.

INTRODUCTION

Over the past several years, S.C.K./C.E.N. has been pursuing an extensive research programme aimed at evaluating the suitability of a clay formation, situated under the nuclear site at Mol, as a possible repository for storing conditioned nuclear waste. As part of this programme, an underground laboratory has been constructed allowing gathering of essential geotechnical and geomechanical information required for the feasibility study of the concept (see the contribution of R. Heremans, this workshop [1]). This underground laboratory will further allow performance of a wide variety of in-situ experiments, as described by P. Manfroy and B. Neerdael (this workshop [2]). In this contribution, the present authors describe in particular two types of experiments designed for investigating the corrosion behaviour of various materials in a clay environment. The experimental "loops" are conceived in such a way that they can be used for corrosion tests on metallic container materials and structural materials and on vitrified simulated waste forms. One

---

*Supported in part by the EC and by NIRAS/ONDRAF.

Figure 1.   Cast iron segment

Figure 2.   Pressure tube for in situ corrosion tests
in direct contact with clay.

278

type of loop is used for investigating corrosion of the materials in contact with the geological clay formation. The other type of loop is aimed at studying the corrosiveness of the atmosphere emanating from the clay formation. To simulate the thermal effect of the waste canisters, the experiments are to be carried out not only at the temperature of the clay formation ($\approx$ 15°C), but also at two or three higher temperatures.

The experimental loops and the control panels are being prepared and will be inserted in the underground laboratory in the fall of 1984.

1. CORROSION IN DIRECT CONTACT WITH CLAY

The corrosion loop systems to carry out corrosion experiments in direct contact with clay will be mounted on the 76 cm flanges mounted on bolted cast iron segments (see Figure 1). The length of the loops (5,33 m) is sufficient to reach into the undisturbed zone of the clay formation. (A layer of about 3 m has been frozen for construction of the gallery). The samples of candidate canister materials including stressed specimens, structural materials (coated with different anti-corrosion layers) and different simulated α-waste forms and high active waste forms to be tested in direct contact with clay are located at the outer side of a pressure tube at a distance of about 5 m of the linings of the underground experimental room. The corrosion tests will be carried out at $\approx$ 15°C, 90°C and 170°C.

The metallic specimens will be tested as welded rings. The specimens will be either tested in the as-received condition or after a heat treatment. The temperature-time dependence of the heat treatment has been derived from temperature measurements carried out on the outer side of a canister during the conditioning of high level waste using the French AVM* process.

The materials to be tested are given in Table I.

The specimens will be heated up by a retractable furnace situated at the inner side of the pressure tube (see Figure 2). The furnace is constructed from heating wires spirally wound around a stainless steel support tube. The temperature and redox-potential will be measured at the clay-candidate container interface. The diffusion and migration of corrosion products and radionuclides will be evaluated on clay plugs removed together with the corrosion loops at the end of each individual corrosion experiment with exposure times up to 50.000 h.

2. CORROSION EXPERIMENTS IN CLAY ATMOSPHERE

The configuration of the experimental loops designed for the clay atmosphere tests is shown in Figure 3. Each of the four loops is inserted vertically downwards in the clay formation and connected to the experimental gallery (cfr. Figure 1). The loops terminate in a porous, stainless steel plug. The loops contain heating elements and a series of test specimens, which can be removed and exchanged during the experimental period. Argon is used as a carrier gas to collect the corrosive products perspiring from the clay into the experimental loop and to circulate these over the test samples. The resistive heaters are used to maintain the sample chamber at temperatures of 170°C,

*Ateliers de Vitrification de Marcoule.

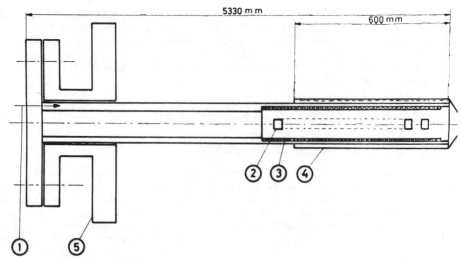

Figure 3.   Schematic view of the experimental loop for in situ corrosion tests in contact with clay derived atmosphere.
1 gas inlet; 2 container and waste form samples;
3 heating elements; 4 porous filter; 5 gallery.

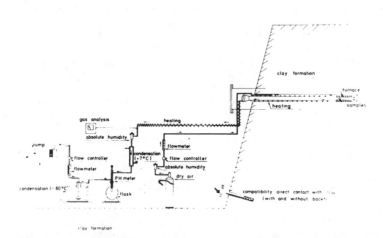

Figure 4.   Block diagram of the instrumentation for corrosion experiments in clay atmosphere.

TABLE I. Test materials for clay corrosion

I. Structural and container materials [3]:
   - EC reference materials: carbon steel; Ti 0.2 Pd; hastelloy C-4.
   - Others: IMI 115 (technically pure Ti); inconel 625; UHB 904 L;
       AISI 316; 1803 T.

II. Conditioned waste forms:
   - Simulated HLW forms [4]:
       UK209, SON58, SON64, SON68, C31, SM58, SAN60.
   - Simulated αW forms [5]:
       UWG119, UWG122, UWG123, UWG124, FLK (HP).

N.B. - The HLW forms are borosilicate glasses except for C31, which
        is a glass ceramic; the αW forms are laboratory made silicate
        glasses or hot-pressed incinerator granulate.
     - Some of the glasses will also be corroded in the devitrified
        state.
     - Some of the glasses will be tracered with ^{239}Pu, ^{90}Sr and
        ^{134}Cs, for the experiments in direct contact with clay.

90°C or 50°C; one loop is to be operated at clay ambient temperature ($\approx$ 15°C). Tests are expected to last for 1500 h for a first run and up to 50.000 h for a second run.

Samples are disposed in teflon boats in the sample compartment of the loops. They are in the shape of square plates of lateral dimensions 30 x 30 mm and 3 mm thickness. In each loop, there is room for 36 specimens. The same types of samples will be used as for the direct contact with clay.

Figure 4 gives a block diagram of the instrumentation. In addition to thermometry and temperature control, there are provisions for the characterization of the corrosiveness of the atmosphere aspired into the specimen compartment. This characterization is performed sequentially for each of the four loops. It consists of on-line gas analysis (by IR absorption for Cl_2, SO_2, NH_3, F_2, CO, CO_2, NO_x, H_2S) and of the determination of the dew point of the loaded sweep gas. In addition, the corrosive products are condensed (at 8°C and also at - 80°C) and the condensates, after heating to room temperature, are analysed for pH and composition.

Following the predetermined corrosion times, the specimens are to be withdrawn from the loops and characterized by means of weight change, surface analysis (SEM, EMPA and AUGER) and RX diffraction.

3.   CHARACTERIZATION OF THE ENVIRONMENT

The corrosion of underground metallic structure is primarily a result of corrosive factors of the soil itself, and of stray currents. Of the various factors which have been associated with soil aggressivity or corrosivity, several have been qualified as being significant. These include low soil resistivity, low redox potential in association with anaerobic sulphate reducing bacteria, low pH, high concentration of soluble iron, and structural properties of the soil which could contribute to oxygen concentration cells. A number of these soil characteristics will be monitored as a function of clay temperature and time.

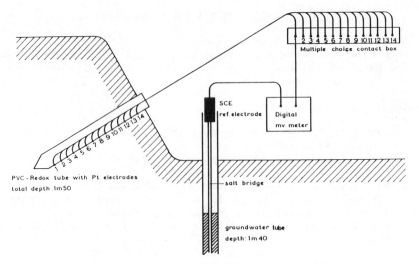

Figure 5.   Redox potential measurements in clay at different depths.

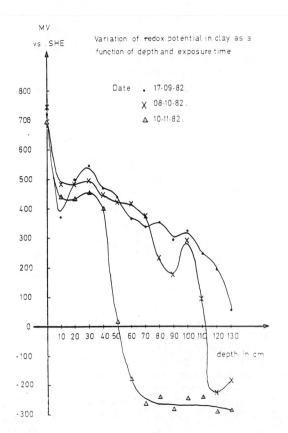

Figure 6.   Variation of redox potential in clay as a function of depth and pressure time.

3.1 Redox potential measurements

The potential of a platinum electrode measured against an external reference electrode (temperature and pressure reducing type) has been found to be an adequate solution for measuring the redox potential. The experience acquired in the past with electrochemical corrosion work at high temperature and pressure will be used to adapt the prototype, used at the moment at the near surface testing site of Terhaegen, for use in the underground experimental room [6].

In order to measure redox potentials under field conditions, a redox potential measurement device was constructed and tested at the Terhaegen site. Into a PVC tube of 2.5 cm outer diameter, horizontal openings are sawn every 10 cm, in which a Pt wire is fixed (L = 30 mm, Ø = 0.7 mm). These Pt-electrodes are connected to multiple contact plugs at the top of the tube through isolated copper wires. Finally the tube is filled with polyester (araldite) in order to improve the durability. This set stays permanently in the soil. The redox potential measurements are carried out using a digital mV-meter placed between the Pt-electrodes and a reference electrode (saturated calomel electrode). The circuit is closed by a KCl-agar-agar salt bridge dipped in the ground water tube (Figure 5).

Results of redox potential measurements as a function of exposure time are given in Figure 6. After two months the clay is already in good contact with the Pt wires at more than half the depth.

Systems for measuring "in situ" the pH and electrical resistivity of the soil are under development.

4. AUTOMATED MEASUREMENT OF INSTANTANEOUS AND INTERGRATED CORROSION RATES

Four different automated systems for continuous measurements of corrosion rates have been adapted for underground experiments.

4.1 Corrater measurements

The corrater apparatus measures the corrosion rate of a metal or alloy in a soil or aqueous solution based on the linear polarization resistance (LPR) method. This method is derived from the fact that approximately a linear relationship exists between the electrical polarization of a metal in contact with a corrodent and the interface resistance between the metal surface and the fluid. The basic relationship derived by Stern and Geary [7]:

$$\frac{\Delta E}{\Delta I} = \frac{B_a B_c}{(2.3 I_c)(B_a + B_c)}$$

states that if a metal in contact with a corrodent is polarized from its free corrosion potential E_c by an amount ΔE, a current ΔI will flow between the metal and corrodent which is related to the free corrosion current I_c and to the slopes of the logarithmic local anodic and cathodic polarization curves B_a and B_c.

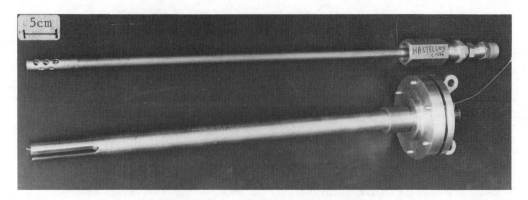

Figure 7.   Retracting device to monitor the corrosion.

Figure 8.   Automatic corrosion rate measurements.

The instrument circuitry measures the interface resistance and converts it directly to corrosion rate in 10^{-6} m.y^{-1}. In addition to general or uniform corrosion, localized corrosion may occur in a system. A vigorous pit on a metal surface tends to cathodically protect nearby metal surfaces connected to it, as pits are very strong anodes. The corrater takes advantage of the distinction between the current induced by an external potential and the current resulting from pitting activity. The sensing probes are designed for use at temperatures up to 150°C and pressures of 34 N/mm^2. The sensing probes employ replaceable electrodes mounted on and electrically isolated from a metal pressure tube which can be introduced in the clay from the underground experimental gallery. Eight of these probes (see Figure 7) will be used equipped with a mechanical retracting device to monitor the corrosion rate of Ti 0.2 Pd, Hastelloy C-4, carbon steel, and of cast iron (grade 60), the construction material of the linings of the underground experimental room, protected or not with different anti-corrosion layers.

The electrical conductivity of consolidated clay is 2.55 mS. This conductivity is compatible with the selected corrater device.

4.2   The corrosometer

The corrosometer operates on the fundamental principle that electrical resistance increases as the cross-sectional area of a metallic conductor decreases. The heart of a corrosometer system is a probe which is introduced into the corrosive environment and with which the corrosion measurements are made. A corrosometer probe functions as an in situ sensor which accumulates the corrosion history of the environment and displays or records the corrosion information on an instrument. The resistance of an element which is exposed to corrosion is compared with the resistance of an element being protected against corrosion. The circuitry of the corrosometer instrument is designed to determine the resistance ratio of the measuring element to the reference element and ensure that corrosion readings are directly proportional to the amount of metal loss. The sensing probes are installed so that the sensing element (made of the alloy of interest) available in a number of forms and sensitivities is permanently exposed to the corrosive environment. A probe consists of any sensing element mounted on one end of a metal body, hermetically sealed and electrically isolated.

The probes can be inserted in the clay formation using fixed and retractable designs (see Figure 7).

A CBM computer, supported by the proprietary soft ware provides graphic display and print-out of long-term corrosion rate and metal loss data. From the plotted data, general trends, corrosion history and present corrosion status can be obtained (see Figure 8).

Specially designed probes, the so-called marine probe, are equipped with a grounding strap to insure that the probe measuring element is at the same electrical potential as the linings of the underground experimental room. These special probes made from cast iron grade 60 will be tested with different anti-corrosion coatings (galvanization treatment, duplex system based on a galvanization treatment and painting system) (see Figure 9).

285

Figure 9.   Marine probe.

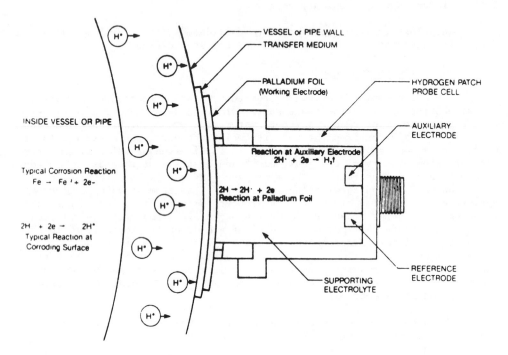

Figure 10.   Hydrogen patch probe.

4.3 PAIR technique

Another corrosion rate monitoring instrument based on the "PAIR" (polarization admittance instantaneous rate) is also based on the formula, developed by Stern and Geary [7] and is a completely computer-compatible corrosion rate monitor. Instantaneous corrosion rates as low as $0.25 \cdot 10^{-6} \, \text{my}^{-1}$ and as high as $25.000 \cdot 10^{-6} \, \text{my}^{-1}$ can be measured. The PAIR method of determination of corrosion rate is similar to the LPR method. The PAIR technique requires three electrodes which establish electrical contact with the corrosive solution, an adjustable current source, an ammeter and a voltmeter.

The sensing probes can be inserted in the clay from flanges available on the linings of the experimental gallery.

4.4 Hydrogen rate monitor

The corrosion of metals is an electrochemical process consisting of simultaneously occurring oxidation and reduction reactions. The iron is oxidized to ferrous iron with two positive charges and the hydrogen ions each gain an electron and are reduced to hydrogen gas. Some of the hydrogen gas formed during the corrosion process is not reduced to hydrogen gas but remains as dissociated hydrogen atoms. These atoms may enter the lattice structure or framework of process component steel. Factors affecting the diffusion of atomic hydrogen through steel are: corrosion rate, temperature, thickness of steel structure. The partial pressure of hydrogen gas in a system does not result in steel entry by atomic hydrogen unless temperatures are sufficiently high to cause dissociation of hydrogen gas to hydrogen atoms.

The hydrogen patch probe (HPP) features a sensing cell and does not require a hole to be cut in the pipe or vessel of interest. The HPP is mounted directly to the outside of a pipe wall by simple mechanical straps or is easily mounted on a large vessel (Figure 10). The pipe or vessel to be monitored is first surface cleaned. Then a transfer medium (wax) and a small piece of 0.10" thick palladium foil are placed on the cleaned metal surface. The HPP is then mounted over the foil. A pair of gaskets and an insert, shaped to the general contours of the pipe or vessel, provide a leak-tight seal against the foil. The probe is then filled with a suitable electrolyte (96 % sulphuric acid). When the palladium foil is polarized by the instrument, the foil acts as a working electrode, quantitatively oxidizing the hydrogen as it emerges from the metal surface. After the probe is allowed to come to equilibrium, the voltage indicated by a customer-supplied voltmeter or recorder attached to the instrument is directly equivalent to the real-time hydrogen penetration rate and monitors the corrosion rate.

CONCLUSIONS

The underground experimental room in Mol will be equipped with corrosion loops to carry out measurements, using the coupon method, on candidate canister materials, structural materials and simulated waste glasses as a function of clay temperature either in direct contact with clay or in a humid clay atmosphere. The loops are equipped with the necessary measuring devices to characterize either the corrosiveness of the clay or the nature of corrosion products present in the humid clay atmosphere.

"In situ" monitoring techniques will be used for measuring the instantaneous and integrated corrosion rates of metallic materials in direct contact with clay and in a humid clay atmosphere. These systems monitoring the corrosion rate of metallic materials without removal of the specimens are based on the following principles: change of electrical resistivity, linear polarization resistance, diffusion of hydrogen and polarization admittance instantaneous rate.

REFERENCES

[1]   Heremans, R.H.:"Expériences en place en terrains argileux", this work-shop.

[2]   Manfroy, P. and Neerdael, B.: "Design and instrumentation of in-situ experiments in underground laboratories in clay", this workshop.

[3]   Casteels, F. et al.: "R & D programme on radioactive waste disposal into geological clay formation", Semi-annual report n° 10 (1980).

[4]   De Batist, R. et al.: "Testing and evaluation of solidified high-level waste form", Joint annual progress report, EUR 8424-EN (1981).

[5]   Sambell, R.A.J. et al.: "Characterization of low and medium level radioactive waste forms", Joint annual progress report, EUR 8683-EN (1981).

[6]   Casteels, F. et al.: "R & D programme on radioactive waste disposal into geological clay formation", Semi-annual report n° 14 (1982).

[7]   Stern, M. and Geary, A.L.: J. Electrochem. Soc. 104 (1957) 645.

In situ experiments for the determination of macro-permeability and RN-migration in faulted rock formations such as the oolithic iron ore in the Konrad Mine[1]

W.BREWITZ, H.KULL & W.TEBBE
GSF, Institut für Tieflagerung, Braunschweig, FR Germany

ABSTRACT

Main subjects of the hydrogeological feasibility studies for radioactive waste disposal in the Konrad Mine (FRG) were the identification of water pathways - existing or in future possible ones - and the determination of size and extension of these pathways. Due to extreme dryness of the rock mass and the absence of any open water flow in the mine advanced investigation techniques had to be found for measuring the macro-permeability.

Rock mass properties especially to be considered were elastoplasticity and hydraulic anisotropy. So far good results were achieved by pressure pulse tests in boreholes and ventilation tests in a mine gallery. It was proved that the undisturbed rock mass is almost impermeable. Only in the near field of mine workings where the primary stress field is disturbed the rock can be regarded to a certain extent as permeable. For additional proof of the hydraulic parameters of a distinct dislocation zone a specific tracer experiment is in preparation.

INTRODUCTION

The disused iron ore mine Konrad is situated near the city of Salzgitter in the eastern part of Lower Saxony (FRG). At present the mine consists of 2 shafts, 1000 m and 1200 m deep and app. 23 km of mine workings such as travelling ways, ventilation raises, testing galleries, and exploration developments.

In April 1982 the feasibility studies for radioactive waste disposal carried out in the mine under the leadership of the Gesellschaft für Strahlen- und Umweltforschung mbH (GSF) were concluded with success [1]. As a result the Physikalisch-Technische Bundesanstalt (PTB) in accordance with the German

---

1) This paper was prepared under the research contract KWA 53078 with the Federal Minister for Research and Technology (BMFT) of the Federal Republic of Germany

Atomic Act applied for licensing with the appropriate authorities of the Federal State of Lower Saxony [2]. Today on behalf of the PTB various specific site investigations are under way in order to collect additional data for verification of the repository's long term safety i. e. the effective inclusion of the radioactive wastes by the geological barriers. The termination of the entire programme including data processing and data evaluation is scheduled for March 31, 1985 [3].

Despite of these activities the GSF carries out a research and development programme on behalf of the BMFT which lasts until end of 1985. This programme is aimed specifically at the development of new in situ investigation techniques for the determination of rock mass parameters which are most sensitive for safety assessments and calculations of radioactive waste repositories in hard rock formations. These investigations refer to neotectonical movements, hydraulic conductivity, and radionuclide retention behaviour.

HYDROGEOLOGICAL PROPERTIES OF THE IRON ORE FORMATION

As the geology of the Konrad mine has been described in several papers, e. g. [1], [4], only the main hydrogeological features are being summarized as far as they are important for the evaluation of the macro-permeability tests.

The iron ore beds measure up to 18 m in thickness. They are part of a synsedimentary jurassic trough which has been cut off by a Lower Cretaceous transgression. Thus the iron ore has no outcrops at surface and is restricted to depths between 600 m and 1400 m. The covering rock formations are mainly marlstones and claystones forming an impermeable barrier against meteoric water. Aquifers next to the iron ore formation are the "Hils"-sandstone (Lower Albium) 330 m to 700 m above and the cornbrash-sandstone (Dogger epsilon) app. 170 m below [5]. From both formations no water intrusions have been registered within 20 years of mine operation.

When the mine was developed about 39 faults cutting through the iron ore produced water flows of up to 50 l/min at the utmost. Most of them dried up after a short time. Only few of them remained flowing for a couple of months and more subsequently becoming smaller. Today some of them can be classified as seepage water forming halite crusts and stalagtites as visible products of the high salinity of the formation water [1], [6]. Only recently a similar water intrusion was experienced during underground development work. The initial flow rate was as much as 24 l/min decreasing rapidly measuring only 3,5 l/min after 11 days and app. 1 l/min after 3 months [7]. Unfortunately the chance was missed to performe conventional pumping-in tests in order to measure the actual hydraulic pressure at the water bearing local discontinuity.

From such tests rough figures can be derived for the rock mass permeability. On account of the measuring techniques being in use, rock mechanical side-effects such as hydro-fracturing, and the conversion of the hydraulic storage capacity into the coefficient of hydraulic conductivity K there remains a certain degree of faultiness [8].

Beside the system of tectonical faults and joints which is in no way saturated with groundwater the iron ore as such contains 3 to 5 weight per cent of saline water. The rock moisture occurs as filling of rock pores which occupy between 6 and 16 per cent of the total bulk volume. Only half of the porosity is related to interconnected interstices. At some locations underground tiny halite crusts or hair-like flowerings on rock walls resulting from evaporation of saline rock moisture indicate the low hydraulic conductivity of the rock.

Only a few horizons of claystone and marlstone in the near field above and and below the iron ore show – where opened up by mining – moist rock faces or produce non-measurable amounts of seepage water. In all cases the water flow is so little that it is being dried up already by the mine ventilation.

Altogether the hydraulic potential of the rock mass affected by mining is extremly small and obviously limited proving the aquifuges present at site as being perfect barriers against intrusions of near surface groundwater.

THEORETICAL CONSIDERATIONS OF WATER FLOW AND PERMEABIILITY MEASUREMENTS

The intrinsic permeability k measured in Darcy (m^2) or the hydraulic conductivity defined as coefficient K ($m \cdot s^{-1}$) are the most important hydrogeological parameters for the characterization of rock masses and underground water flows. In particular faulted hard rock formations show often anisotropic hydraulic properties because of interstices, pores, voids, or other openings. Therefore the movement of seepage water in unsaturated hard rock formations follows its own principles which are difficult to determine by conventional hydrogeological methods [9]. So far there is very little known from literature about permeability tests performed in situ from underground workings, since such tests play an important role in oil exploration where commonly boreholes from surface are being used.

The hydraulic conductivity of hard rock formations are mainly connected with the permeability of tectonical joints and faults. In most cases the water flow in open interstices can be regarded as laminar provided their aperatures are not too wide. Such openings are commonly experienced only in shallower depths (< 200 m) or at tear faults and in the vicinity of mine workings [9]. Other important parameters are mineralization, dip, strike, extension and interconnection of faults (or system of faults) and the evenness of fault planes.

Numerical modelling of in situ tests aimed at the determination of the macro-permeability of hard rock formations is possible under certain boundary conditions on the basis of the Darcy Law.

$$K = \frac{Q}{A} \cdot \frac{dl}{dh} \qquad\qquad v = \frac{Q}{A} = K \cdot \frac{dh}{dl}$$

Q = volumetric water flow rate
A = area normal to flow direction
v = Darcy velocity

$\frac{dh}{dl}$ = hydraulic gradient

Developed for porous media the Darcy Law describes the hydraulic transport as a function of flow potential and transport coefficient, which is defined as a quotient of rock permeability and viscosity of the flowing medium. For faulted hard rock formations it is only valid if the flow is homogeneous, i. e. one phase, laminar and the flow medium is non-compressible. For extreme small porosities or turbulent flow correction factors are needed.

For calculation of the laminar flow in a two dimensional system with parallel and even planes WITTKE, W. & JÜNGLING, H. (1979) [10] state an equation which has been derived from the Darcy Law. For the determination of the Darcy velocity the following parameters have to be known: hydraulic gradient, acceleration of gravity, kinematic viscosity, aperatures and distance of fractures.

Regarding the performance of in situ permeability tests in the Konrad mine the following aspects have to be considered:

- On account to elastoplasticity and deformation of the rock mass the original permeability can be measured only outside the mechanical stress field caused by mining.

- In the near field the rock mass has been affected considerably resulting in secondary fractures which in sections increase the permeability substancially. The disturbed zone extends 5 to 10 m into the side walls and up to 20 m and more into the hanging and foot wall of galleries [1].

- Effects on the rock mass permeability caused by mining are directly dependent on the proportion of mine working areas and pillar areas in defined mine sections.

- Greatest secondary rock mass permeability occurs at locations where rock mass deformation i. e. cavity convergence

292

is highest. Without being anchored by rock bolts the foot-
wall shows significant permeability and considerable water
storage capacity.

As experienced by 20 years of mine operation the rock
mass permeability caused by mining seems to predominate. Not
only in this respect but also in general it is the question
whether the Darcy Law is totally suited for the numerical modelling of such a hydrogeological medium. Possibly there are
other transport equations which permit better modelling of water flows in unsaturated rock masses with low hydraulic conductivity.

PRESSURE PULSE TESTS IN BOREHOLES

In the Konrad mine pressure pulse tests were performed
first by F. RUMMEL [11] in order to determine the rock mass
permeability in the near field of underground workings. These
tests were carried out in six boreholes at three different locations 860 to 1215 m below surface. The boreholes measuring
80 mm in diameter and 30 m in length were drilled horizontally
as well as vertically. The borehole probe used consists of two
packer units sealing off an injection interval of 1 m. The injection pressure was generated by pumps capable of maximum
pressures up to 650 bars.

The measurements were performed in various depths of
the boreholes. After an initial pressure pulse of 10 to
100 bars depending on the rock quality the hydraulic pressure
decrease was registered over a period of about 300 s measuring
the diffusion rate of the injected water into the rock mass.

For computation of the rock mass permeability k and the
hydraulic transmissivity T following equations (JUNG, R. 1978)
[12] derived from the Darcy Law have been used:

$$k = \frac{dv}{dp} \cdot \frac{\eta}{\Pi \, 1} \cdot \frac{1}{t_D} \qquad\qquad T = \frac{r_s^2}{t_D}$$

$\frac{dv}{dp}$ = stiffness of hydraulic system

1 = length of injection interval
η = viscosity of injection fluid
t_D = time difference; to be derived from comparison of
    measured and master pressure curves
r_s = radius of injection interval

For the interpretation of the results the chosen length
of the injection interval is of considerable importance. Initially there were good chances to measure the macro-permeability of undisturbed rock masses as well as of fractured

zones by using a 1 m interval. Later even better results were achieved with an 0,75 m injection interval.

The results from 98 spot measurements produced k-values for intact marlstone, claystone, and iron ore in the range of 1 to 10 μD. Higher k-values were determined for fractured rock in the near field (14 to 34 μD) and for faults (100 μD to 1 mD). The scatter in permeability values of 1 μD to 1,5 mD expresses clearly the influence of joints and fractures to the direction of high values [11].

For a more accurate determination of the near field effects on the rock mass permeability caused by mechanical stresses and strains around mine workings the GSF performed two more testing campaigns in 1982/83 and 1983/84. For this purpose the borehole probe was substancially altered in order to fit into smaller boreholes and to use packer systems which can stand the specific testing conditions in fractured rock.

In the first campaign measurements were carried out in an undeveloped mine section with only small rock deformations. The results show a distinct interrelation between permeability values and distance from mine cavity. Up to 5 m the permeability may reach up to 80 μD. Beyond that distance the permeability of the rock mass measures less than 20 μD (Fig. 1) [6].

The second campaign was performed in a developed mine section with larger cavity area. Since the data computation has not been finished yet no figures can be presented in this case. But it is being assumed that the permeability is substancially higher due to intensive rock deformation.

From such permeability measurements also the storage coefficient S can be estimated if compressibility, porosity, and the thickness of the storage medium are known [13]. As the parameters compressibility and porosity have to be determined by laboratory measurements, these calculations have only limited validity in particular in respect of larger rock masses.

VENTILATION TEST IN A MINE GALLERY

The ventilation test was firstly developed by P. A. WITHERSPOON in the Stripa-Mine (Sweden) in order to measure the macro-permeability i. e. the hydraulic conductivity of a larger volume of fractured granite [14]. The local hydraulic gradient was determined by pressure-in tests in boreholes. The water flow into the testing gallery was measured either directly where possible or as water vapour increase in the ventilation air.

As in the Konrad mine no underground water flows exist a similar test had to be chosen in order to measure the water release from the iron ore caused by evaporation. Ventilation

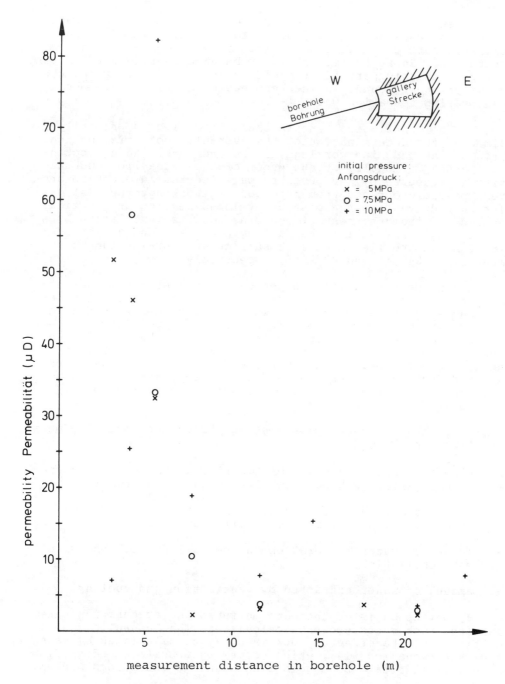

Fig. 1: Rock mass permeability K as a function of distance from underground galleries. Results of pressure pulse tests in oolithic iron ore in a low developed mine section 1215 m below surface.

air measurements in the mine had proved that there is a water
vapour increase underground of 20 to 50 l/min depending on
climatic surface conditions as well as on specific mining
operations. In a preliminary water balance carried out for the
entire mine only part of it could be accounted for by the use
of water underground. An indeterminable rest is assumed being
evaporating formation water [1].

The ventilation test in the Konrad mine is primarily
aimed at the determination of the hydraulic potential of the
iron ore around mine workings, i. e. the water storage potential and additionally at the proof of the hydraulic conductivity as measured by the pressure pulse tests. In order to prevent possible effects from neighbouring workings the test gallery was driven in an almost undeveloped section of the mine.
Sealed off from the rest of the mine and its ventilation by a
1 m strong brick wall it has a length of 74 m and a size of 35
to 40 m^2 in profile. Free rock surface and total volume are
measuring 1649 m^2 and 2379 m^2 respectively.

Inside the gallery temperature and humidity measuring
gauges were installed at different points. At ventilation inlet and exhaust ducts another two sets of gauges were fitted
in order to measure air temperature and air humidity differences when the ventilation is in operation. The ventilation
system connected consists of a fan (max. capacity 800 m^3/h),
an air cooler (app. 13 kW power), and a heater. The water extracted from the cooler can be measured separately. In addition the circulating air volume is being controlled continuously by an anemometer.

In several successive testing phases the following data
are to be determined:

-   time span reaching state of equilibrium in the atmosphere
    of the gallery without ventilation and after ventilation,

-   relative air humidity and air temperature level as func-
    tion of time or volumetric ventilation rate,

-   amount of water evaporation in the gallery with and with-
    out ventilation,

-   amount of water extracted by ventilation and cooling.

On the basis of the data measured the evaporation rate
per free rock surface can be calculated. The hydraulic conductivity of the effected rock mass may be calculated by using
the THIEM equation [14], which describes an uniform radial flow
into a cylindrical opening in a homogeneous and porous medium.

$$K = \frac{Q}{2 \Pi b} \cdot \frac{\ln{r_1/r_2}}{h_2 - h_1}$$

For the calculation of the hydraulic coefficient K the hydraulic gradient ln r_1/r_2 / h_2-h_1 has to be known as well as the volume of water flow Q (l/min) and the length of the cylinder b. The volumetric flow rate Q is being measured directly, but so far the effective hydraulic gradient is still unknown since water head measurements (h_1, h_2) at the radial distances (r_1, r_2) were impossible to be performed.

Tab. I: Measuring data from ventilation test phase I

time span (days)	water vapour content (kg)	evaporation rate (g/min)
0 - 238	100,63	0,29
0 - 523	110,5	0,15
0 - 752	118,82	0,11
0 - 1018	123,0	0,084
0 - 1073	120,65	0,078

Phase I lasted from December 1980 to December 1983. No ventilation was used in order to reach state of equilibrium in the locked atmosphere and as far as possible water saturation in the permeable rock mass. The curve computed from individual measurements shows a sharp rise of water vapour content at the beginning of the test gradually levelling off towards the end (Tab. I)[6].

At the end of phase I after 1073 days 80 % of relative air humidity at 43 °C were measured. The vapour saturation at this level is mainly a result of the high salinity (app. 210 g/l NaCl) of the evaporating water (Raoult Law). The difference between the vapour content measured and the possible maximum vapour content calculated for pure water as a function of temperature remained stable during the second half of the test. In succession to water vapour saturation no open water flow was detected in the testing gallery. Schlumberger-measurements proved only a slight increase in the electrical conductivity of the neighbouring rock mass, possibly resulting from water accumulation in the fracture zone around the gallery caused by rock mechanical strain.

Phase II started in January 1984 after the installation of the ventilation and water extraction unit (Fig. 2) and after connecting ventilation ducts for air inlet and exhaust. Due to technical alterations ventilation was interrupted at several occasions i. e. up to now the testing conditions were not yet that stable as they were supposed to be. But in spite of these difficulties the results gained so far indicate the order of magnitude of water accumulation inside and in the near field of the testing gallery.

Between the 1074th and the 1153rd day of the test the relative air humidity was reduced from 80 % to app. 64 % by combined air ventilation (app. 7 m^3/min) and cooling. Accordingly the water vapour content in the atmosphere dropped from

Fig. 2: Front side of brick wall sealing off the ventilation
test gallery. The air tight cover is a synthetic
mixture with glassfibre and gypsum. Right top the
exhaust duct. Bottom left ventilation and cooling unit.

app. 120 kg to 95 kg. After two days without ventilation
the relative air humidity increased to app. 70 % before ventilation got it down again to app. 64 %. Since the 1159th day
the ventilation is not in operation awaiting automatic switch
on at about 70 to 72 % (Fig. 3).

Over the entire ventilation period of 85 days a total
of 3150 l of water was recovered from the cooling trap. As
such an amount of water was impossibly contained in the atmosphere at one time there is the only explanation that intense evaporation took place during ventilation. This required
a good water storage in the rock mass adjoining the ventilation test gallery.

The latest measurements show that after the water extraction so far accomplished the evaporation rate is slowing
down. This fact indicates that either the storage capacity or
the water content is limited.

Although it is too early for any more interpretations a
simple calculation produces a volumetric evaporation rate of
$0,016$ cm^3 $\cdot$ min^{-1} per m^2 of the free rock surface.
By comparison: In the Stripa Mine a volumetric water flow of
about 0.076 cm^3 $\cdot$ min^{-1} per m^2 was computed corresponding with a calculated hydraulic conductivity K of
$1 \cdot 10^{-10}$ m $\cdot$ s^{-1} [14].

TRACER TEST AND RN-MIGRATION EXPERIMENT

Both the pressure pulse tests as well as the ventilation test give a chance for measuring the macro-permeability
of the rock mass. But the hydraulic conductivity of a distinct
fault zone can be determined more precisely by specific tracer
tests. In addition the application of radionuclides as tracers
makes it possible to gain information on the retention behaviour of faulted rock under defined in situ conditions. The
theory of such tracer tests has been published e. g. by
WITTKE, W. & JÜNGLING, H. (1979) [10] und SCHNEIDER, H. (1982)
[15].

In the Konrad mine one of the major faults has been selected for a tracer test including a RN-migration experiment
which will follow later. At the location of the experiment the
fault is app. 1 m wide and strikes north-south. It dips 50°
east and there is a vertical downthrow of about 15 m shifting
the hanging marlstone formation against the iron ore.

From a parallel mine working at a distance of about 7
to 14 m there have been drilled 7 boreholes into the fault
zone in a fan-like order. Two of these boreholes will be used
as injection boreholes while the others will serve as discharge holes. For the experiment the boreholes measuring 86 mm
in diameter have been fitted with packers sealing off
injection and pressure intervals where the faulted zone is intersected (Fig. 4).

After the installation of the pumping and control unit
the tracer fluid will be pumped into the injection boreholes
applying a pressure of up to 10 bars. Hydraulic pressure, volumetric flow rate, and temperature of the fluid will be measured in the injection and/or in the discharge holes.

After pressing saline water into the dry fault zone
first of all the main flow directions will be determined just
by measuring the volumetric flow receiving in the various discharge holes. This testing phase will be finished when the entire system is saturated and the laminar flow is steady. In
the second phase suitable tracers will be used in order to
measure flow time and Darcy-velocity between the various boreholes under steady state conditions. As the hydraulic gradient will be measured by pressing-in tests and the aperatures
as well as the distances of the fractures are known from the
drilling cores it will be possible to compute the Darcy-velocity after WITTKE, W. & JÜNGLING, H. (1979) [10].

In a later stage radioactive tracers might be used in
order to determine specific travelling times as well as the
sorption capacity of the fractured rock mass adjoining the
fault zone. In addition it is being intended to investigate
the sorption mechanism on rock samples gained by additional
core drillings.

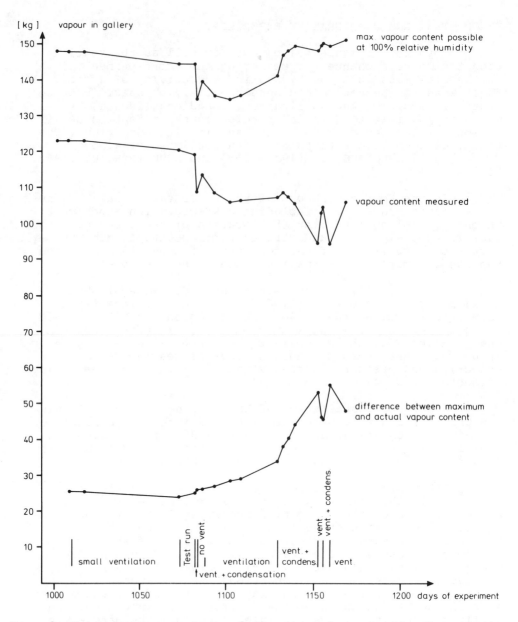

Fig. 3: Water vapour content of the atmosphere inside the ventilation test gallery and its ventilation during the first
ventilation campaign with and without water extraction.

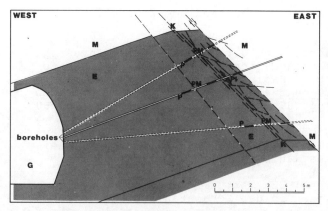

Fig. 4: East-west section of the migration test location (G)
in the Konrad Mine. Direction and length of three boreholes with injection/observation intervals (PI/PM),
fitted packers (P) and fault zone (K).

CONCLUSION

Although the in situ experiments are not finished yet
and the information gained so far is not fully converted by
computation into hydraulic parameters there is a strong belief
that these tests will help to fill the gap between the hydrogeological data already known and the data requirements set by
an overall safety analysis. In the specific case of the Konrad
Mine the absence of effective water pathways and the very
small order of magnitude of in future possible ones cause
problems not only for the experimental techniques but also for
the computation and the interpretation of the results. What is
being considered as most favourable for the long term safety
of a radioactive waste repository is on the other hand a major
disadvantage for the proof of the site's feasibility. A problem that cannot be solved completely by these experiments is
the determination of the actual hydraulic gradient in the area
opened up by mining. For that matter we have to wait for a
real water flow which can get investigated by conventional methods such as slug tests.

REFERENCES

[1]   GSF-Abschlußbericht: "F+E-Programm zur Eignungsprüfung
       der Schachtanlage Konrad für die Endlagerung radioakti-
       ver Abfälle", Ed. Gesellschaft für Strahlen- und Um-
       weltforschung mbH München, GSF-T 136 (1982).

[2]   Brewitz, W.: "Disused Konrad Iron Ore Mine - a Future
       Low-Level Waste Repository in the FRG", Nuclear Europe
       9, 21 - 23 (1983).

[3]   PTB: "Rahmenterminplan Schachtanlage Konrad",
       Info-Blatt 4/83 (1983).

[4]   Brewitz, W. & Löschhorn, U.: "Geoscientific Investiga-
       tions in the Abandoned Iron Ore Mine Konrad for Safe
       Disposal of Certain Radioactive Waste Categories",
       Proc. Int. Sypm. on the Underground Disposal of Radio-
       active Wastes, pp. 89 - 102, IAEA/NEA, Otaniemi, 1980.

[5]   Kolbe, H. & & Simon, P.: "Die Eisenerze im Mittleren
       und Oberen Korallenoolith des Gifhorner Troges", Bei-
       heft Geologisches Jahrbuch 79, 256 - 338 (1969).

[6]   GSF-Jahresbericht 1982: "Aktuelle Ergebnisse der lau-
       fenden F+E-Arbeiten in der Schachtanlage Konrad zur
       Endlagerung radioaktiver Abfälle", Ed. Gesellschaft für
       Strahlen- und Umweltforschung mbH München, GSF-T 165
       (1983).

[7]   Unpublished report, Konrad Mine, January 1984.

[8]   Heitfeld, K.-H.: "Hydro- und baugeologische Untersu-
       chungen über die Durchlässigkeit des Untergrundes an
       Talsperren des Sauerlandes", Geol. Mitt. 5, 1 - 210
       (1965).

[9]   Hähne, R. & Franke, V.: "Bestimmung anisotroper Ge-
       birgsdurchlässigkeiten in situ im grundwasserfreien
       Festgestein", Z. angew. Geol., 29, 219 - 226 (1983).

[10]   Wittke, W. & Jüngling, H.: "Sickerströmungen in klüfti-
       gem Fels", Mitt. Ing.- u. Hydrogeol., 9, 219 - 263
       (1979).

[11]   Heuser, U.: "Das Druckstoß-Test-Verfahren zur Bestim-
       mung der Gebirgspermeabilität in Bohrungen dargestellt
       am Beispiel von Permeabilitätsmessungen im Eisenerz-
       bergwerk Konrad/Salzgitter und in der Forschungsbohrung
       IRDP in Ostisland", Ruhr-Universität Bochum, Institut
       für Geophysik, Arbeitsgruppe Gesteinsphysik (1982).

[12]   Jung, R.: "Bericht über Permeabilitätsmessungen im Fal-
       kenberger Granit-Massiv/Nordbayern", BGR Archiv-Nr.
       80 443 (1978).

[13]   Ferris, J.G. et al.: "Theory of Aquifer Test", U.S.
       Geological Survey Water Supply Paper 1536-E (1962).

[14]   Lawrence Berkeley Laboratory: "Geohydrological Data
       from the Macropermeability Experiment at Stripa,
       Sweden", LBL-12520 (1981).

[15]   Schneider, H.J.: "Der in-situ Durchlässigkeitsversuch
       in geklüftetem Fels - Probleme der Interpretation",
       Proc. Symposium Rock Mechanics: Caverns and Pressure
       Shafts, pp. 135 - 144, ISRM, Aachen, 1982.

The Stripa Buffer Mass Test instrumentation for temperature, moisture, and pressure measurements

R.PUSCH
Swedish Geological, University of Lulea, Uppsala

ABSTRACT

The Stripa Buffer Mass Test involves recording of temperature, water uptake, swelling pressure, and water pressure. The temperature recording is made by use of more than 1200 copper-constantan thermocouples. Swelling, or rather total pressures, are measured by means of about 130 Gloetzl pressure cells, and this system is also applied for recording water pressures in heater holes, backfill and rock. BAT-piezometers are used as a back-up of the Gloetzl system.

The water uptake of the highly compacted bentonite in heater holes and the sand/bentonite backfill that covers the holes is made by 560 electric capacitance gauges. Displacements in the rock that associates the swelling of highly compacted bentonite is measured by applying Kovari's technique.

1. INTRODUCTION

The scope of the Buffer Mass Test (BMT) is to investigate the suitability and predicted functions of certain smectite-based buffer materials under real conditions with respect to the rock environment and the hydrology. The main features of the test arrangement is a tunnel with an array of six 0.76 m diameter and about 3 m deep heater holes, drilled from the tunnel floor (Fig. 1). The inner 12 m long part of the tunnel is separated from the rest of the space by a heavy bulwark. This inner part, which hosts two heater holes, is backfilled with clay/bentonite mixtures while the holes contain electrical heaters surrounded by highly compacted bentonite [1]. The outer four heater holes are covered by the same backfill material as in the tunnel but these volumes are covered by removable concrete lids.

Recording is required of temperatures, water uptake, and swelling and water pressures in the highly compacted bentonite in the heater holes as well as in the tunnel backfill. Also, internal displacements in the clay materials and possible changes in joint apertures in the rock are being measured.

2.   TEMPERATURE

2.1 Gauges

The choice of a suitable instrumentation system for temperature
measurement was based on the criteria that the gauges have to be
mechanically and chemically stable, that the system must offer a
simple way of recording, and that the net accuracy of the readings must be better than $\pm 0.5^\circ$C. A further requirement was that
the gauges must yield stable signals for at least four years at
temperatures up to 90°C. Steel-sheathed copper/constantan thermocouples (Pentronic ISA type T) with MgO insulation were found to
be suitable, the steel quality being SS 2343 to resist corrosion
[2].

The elements located in the blocks of highly compacted bentonite
were inserted in drilled holes with a close fitting of the
gauges (Fig. 2). The manufacturer´s specified minimum radius of
curvature of the wires was 2 cm and this was achieved by using a
special device for the bending operations. The same type of elements is also used to measure rock temperatures around heater
hole no 5. These gauges were embedded in sand in 30 mm diameter
percussion-drilled boreholes.

The wires from each heater hole and instrumented section in the
rock and backfill are collected in bundles that extend to switch
boxes in the outer part of the tunnel, which are in turn connected to the data acquisition system in the mine by insulated
multicables.

2.2 Positions

The superimposed temperature fields of the heater holes called
for locating the gauges in the longitudinal plane through the
centers of all the heater holes, as well as in the perpendicular
plane through each hole. In each of these planes, the gauges are
located in radial arrays on five levels so that the detailed
temperature variation with the distance from the heater surface
can be identified. The arrangement is illustrated in Fig. 3 which
also illustrates a computer-plotted temperature diagram. The
tunnel backfill has three instrumented sections, two intersecting
heater holes no 1 and 2 and one at half distance between these
two holes.

3.   SWELLING PRESSURE

3.1 Gauges

The expected swelling pressures in the highly compacted bentonite
range from a few hundred kPa to about 10 MPa, while the corresponding pressures in the sand/bentonite backfills are assumed to
be lower than 500 kPa, except for the immediate vicinity of the
heater holes where pressures are transferred from the heater

304

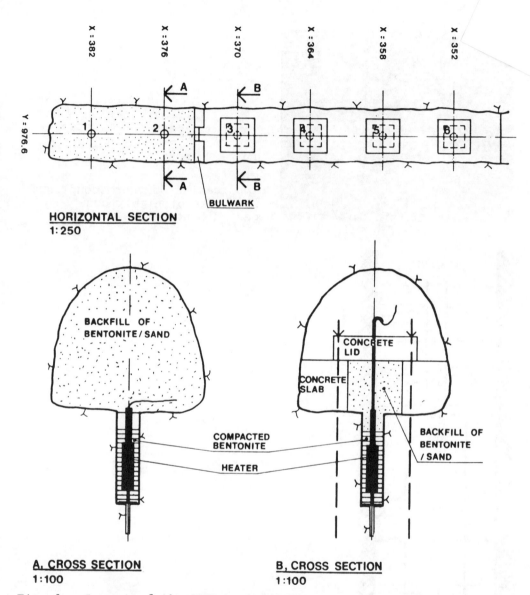

HORIZONTAL SECTION
1:250

A. CROSS SECTION
1:100

B. CROSS SECTION
1:100

Fig. 1.   Layout of the BMT test site.

holes. This large span and the long duration of the BMT experiment called for pressure gauges with proven durability and
mechanical strength and Gloetzl pressure cells, type B 15/25 QM
100 F, with mercury-filled pads were chosen for this experiment
(Fig. 4). The read-out unit is equipped with manometers and
transducers, digital reading and printing facilities. The time
for measuring and recording of each individual circuit is 1-10
minutes with the automatic device. The total number of cells is
128 so a complete recording cycle takes about five hours [2].

305

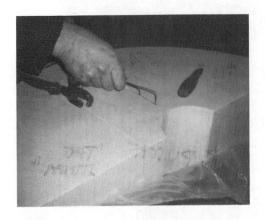

Fig. 2.
Application of thermocouples in blocks of highly compacted bentonite.

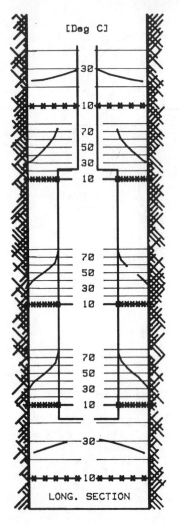

LONG. SECTION

Fig. 3.
Cross section through heater hole no. 3 showing the positions of the thermocouples (x) and the temperature situation 15 months after test start.

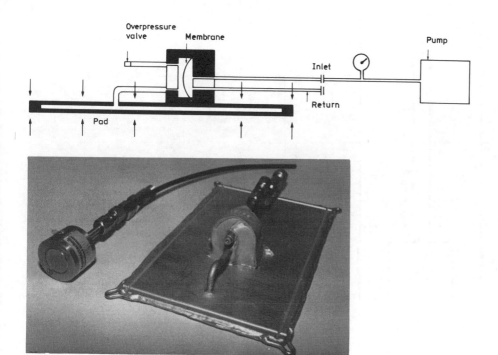

Fig. 4. The Gloetzl pressure cell. The photo shows a piezometer
gauge to the left.

The accuracy of the readings, which give the sum of swelling and
water pressures, i.e. the total pressure in soil mechanical
terminology, is ± 4 % when the system is operated at room temperature. The influence of temperature is significant and needs to be
taken into consideration. Thus, current investigations indicate
that the readings overestimate the pressure by about 0.03 MPa per
degree centigrade.

3.2 Positions

The large dimensions of the pressure cells limited the number
that could be installed in the heater holes without interfering
too much with the rock. Each heater hole holds ten cells; three
on each side of the hole, one on the top of the heater and one
immediately below it, and two more as illustrated in Fig. 5,
which also shows a computer-plotted pressure diagram. Three sections of the tunnel backfill were also monitored and additional
pressure cells were applied on the bulwark as well.

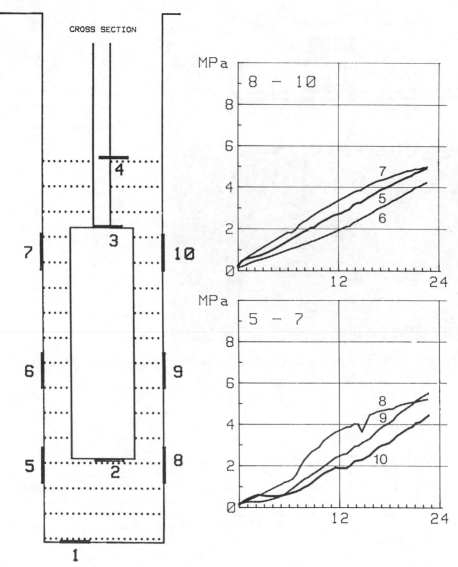

Fig. 5.   Cross section through heater hole no. 1 showing the
          positions of the Gloetzl soil pressure cells and the
          pressure state of cells 5-10 in the first two years.

4.   WATER PRESSURE

      4.1 Systems

Three different systems for the measurement of water pressures
are used: precision manometers, Gloetzl piezometers, and BAT
piezometers.

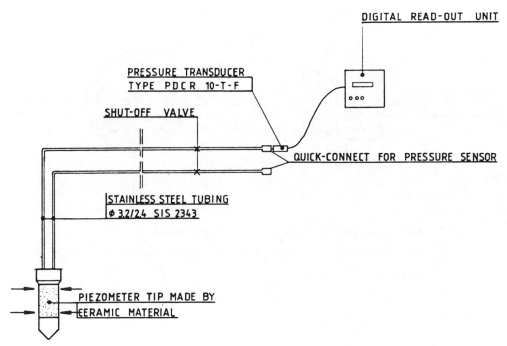

DIGITAL READ-OUT UNIT

PRESSURE TRANSDUCER
TYPE PDCR 10-T-F

SHUT-OFF VALVE

QUICK-CONNECT FOR PRESSURE SENSOR

STAINLESS STEEL TUBING
⌀ 3.2/2.4 SIS 2343

PIEZOMETER TIP MADE BY
CERAMIC MATERIAL

Fig. 6.   The BAT piezometer operating principle.

4.2   Manometers

Manometers are used for measuring the water pressure in 15 long ⌀ 76 mm boreholes which extend radially from the tunnel periphery or from the tunnel front [2, 3]. The recordings are indispensable for the interpretation of the hydrological interaction of the rock/backfill. The holes are grouted to within 5 m distance from the tunnel periphery so the recorded pressures represent averages over the distance 5-20 m from the periphery of each hole. The pressures are in the interval 1-1.5 MPa.

4.3   Piezometers

In addition to Gloetzl piezometers, which have treshold opening pressures of 80 kPa, BAT piezometers are used for the recording of water pressures as low as about 1 kPa. The operation principle of the BAT gauge is illustrated in Fig. 6. From the ceramic filter tip, two separate stainless steel pipes with shut-off valves lead to easily accessible "quick connects" to which a pressure transducer associated with a digital read-out unit is adapted when readings are taken.

Piezometers are installed at the base of the heater holes, and the tunnel, and in shallow boreholes from the tunnel periphery.

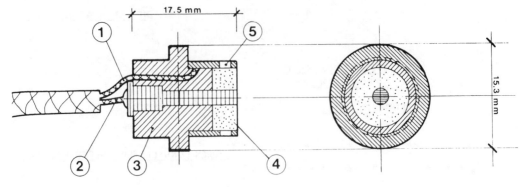

LEGEND:

1　WIRE TO OUTER ELECTRODE OF GOLD-PLATED BRASS

2　WIRE TO CENTRAL ELECTRODE OF BRASS PIN WITH GOLD-PLATED END

3　PLEXIGLASS INSULATOR

4　COMPACTED SOIL (BENTONITE OR BENTONITE/SAND MIXTURE)

5　HOLE FOR ENTRANCE OF MOISTURE

Fig. 7.　The moisture sensor.

5.　WATER UPTAKE

5.1　Gauges

Moisture measurement by utilizing the electrical resistivity of the soil is frequently applied in agriculture and various industries. Commercially available sensors are usually large and expensive, however, and not well suited for the BMT test.

A new sensor was therefore developed at the University of Luleå, the main features being an outer cylindrical electrode and a central pin electrode between which the same clay material is applied as the sensor is inserted in (Fig. 7). For use in the highly compacted bentonite in the heater holes, small cylinders of precompacted bentonite with central holes to fit the pin electrode were applied in the electrode gap before inserting the sensor in its prebored hole in the bentonite block. The sensors in the bentonite/sand backfill were filled with this sort of mixture which was compacted to approximately the same density as that of the backfill into which they were pushed [2].

The brass electrodes were gold-coated (3 micrometers) in order to prevent corrosion. The cables, which consisted of an outer plastic shielding surrounding the wires, were tightly connected to the sensor body by a polyurethan jacket, confined by a heat-shrinkable tubing. This arrangement was chosen so as to prevent water from entering the cables.

At the measuring, an AC voltage with a frequency of 200 Hz passes through the circuit consisting of the sensor and its content of clay material. The recorded current through the circuit, which is a measure of the soil resistivity or rather its capacitance, is a complex function of the water content, the water chemistry, and the temperature.

The laboratory calibration (cf. Fig. 8) gave a fairly definite relationship between the recorded electric potential and the water content of samples embedded by clay material under completely confined conditions, i.e. with a constant dry density. On site, however, the water uptake is associated with a change in dry density due to the swelling, and the recordings are therefore used merely as a check of whether a change in water content has taken place in a certain period of time.

5.2 Positions

A considerable amount of moisture sensors are applied in the three instrumented sections of the backfill, but most of the 560 sensors are located in the heater holes to get a detailed pic-

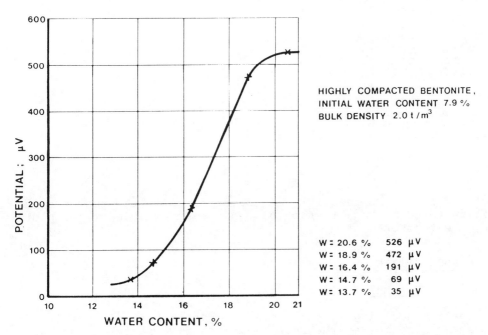

HIGHLY COMPACTED BENTONITE,
INITIAL WATER CONTENT 7.9 %
BULK DENSITY 2.0 t /m^3

W = 20.6 %   526 μV
W = 18.9 %   472 μV
W = 16.4 %   191 μV
W = 14.7 %   69 μV
W = 13.7 %   35 μV

Fig. 8.   Relationship between water content and recorded electric potential of highly compacted bentonite.

311

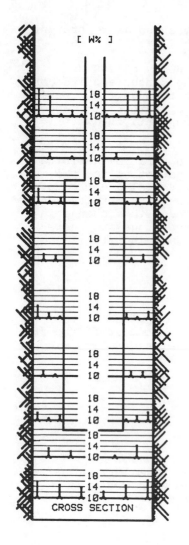

Fig. 9.
Location of moisture sensors in heater
hole no 3. Recorded water contents
15 months after test start.

ture of the water uptake. The instrumentation in the heater holes
is illustrated by Fig. 9 which also gives an example of the plotting of water contents.

6.   ROCK STRAIN

    6.1 General

The expected swelling of the highly compacted bentonite will produce shear stresses at the rock/clay interface in the upper part
of the heater holes and this may lead to a widening of horizontal
or subhorizontal rock joints. Also, the development of swelling

pressures acting radially on the rock surface in the heater holes
may affect the aperture of certain joints, i.e. those which are
steeply oriented and intersect the holes. An attempt is made to
measure possible effects of this sort and this requires precision instruments, such as the Kovari "sliding" micrometer - ISETH",
which is being used in the present investigation [2].

6.2 Principle of measurement

In the BMT, the ISETH micrometer is used for measuring axial displacements in boreholes. For this purpose a steel casing is
grouted in the hole, the casing being subdivided into 0.5-1 m
long elements separated by measuring marks. These marks consist
of cone-shaped seats of stainless steel into which the sphere-
shaped marks ("heads") of the instrument fit. The instrument is
simply an installing rod at which end an about 1 m long invar
steel measuring rod with two heads is attached. When measurements are made, the instrument is inserted in the hole and twis-
ted so that the measuring rod can enter the respective casing
section. A defined pulling force is then applied so that complete
contact is established between the cone and sphere marks, the
actual distance between the cast-in marks then being recorded by
the transducer in the measuring rod.

The instrument is temperature-compensated, the net accuracy of
measurements being better than ± 10 micrometers. An important
requirement is that the seats are absolutely clean, which calls
for flushing the hole before measuring.

6.3 Positions

Four vertical $\emptyset$ 86 mm boreholes were drilled parallel and close
to heater hole no 5, each hole being equipped with 13 measuring
sections, located so that the possible changes of the width of a
well defined, steeply oriented joint can be identified. Also the
expected expansion of several, less well defined, subhorizontal
joints and fractures near the tunnel floor can be recorded. No
aperture changes due to the swelling of the clay have been recorded so far, but the initial temperature change had a measurable
influence.

7. COMMENTS

The BMT has been running for almost three years now and considerable experience has been gained with respect to the suitability and durability of the instrumentation. No major failure
has occurred and very few gauges are not in operation. The piezometers and moisture sensors all seem to be intact and less than
one percent of the thermocouples have failed. The least durable,
but still reasonably reliable gauge, is the Gloetzl pressure cell.
Half a dozen of these cells are malfunctioning due to membrane
failure which has led to mercury outflow.

Also, the recording and plotting units, including manometers, the Kovari equipment and the BAT system, as well as the programmed data acquisition system have operated quite satisfactorily. The latter comprises of data logging units at the 360 m level in the mine, and Hewlett Packard computer units on the ground [4].

8.   REFERENCES

1.   Pusch, R., Nilsson, J. "Buffer Mass Test - Rock drilling and Civil Engineering". Stripa Project Internal Report 82-07.

2.   Pusch, R., Nilsson, J. & Ramqvist, G. "Buffer Mass Test - Instrumentation". Stripa Project Internal Report (in press).

3.   Wilson, C.R., Long, J.S., Galbraith, R.M., Karasaki, K., Endo, H.K., Dubois, A.O., Mc Pherson, M.J., and Ramqvist, G. "Geohydrological data from the macropermeability Experiment at Stripa, Sweden". Swedish-American Cooperative Program on Radioactive Waste Storage in Mined Caverns in Crystalline Rock. LBL-12520, SAC-37, US-70, 1981.

4.   Hagvall, B. "Buffer Mass Test - Data Acquisition and Data Processing Systems". Stripa Project Internal Report 82-02, 1982.

Summary of discussion

In this discussion chaired by Mr. Baetslé, the following subjects were examined:
    (a)   clay behaviour stricto sensu;
    (b)   interactions between clay and other test components;
    (c)   instrumentation.

Clay behaviour

Replying to Mr. Skytte Jensen, Mr. Pusch mentioned that compacted bentonite with densities as high as 2.1 would act as a total barrier against anion diffusion (e.g. chlorine, iodine).

It is well known that in situ measurements of clay permeability are very difficult. Therefore, an indirect method could be used, based on diffusion of radon 226, a decay product of uranium contained in clay (Mr. de Marsily). This radon could be collected in galleries. If this gallery acts as a drain for groundwater, higher amounts of radon would be detected. Therefore, the knowledge of U concentration in clay, together with measurements of radon content in the ventilation air and of radon diffusion coefficient in clay could lead to an estimate of clay permeability to water.

The variability of clay properties was also mentioned (Mr. de Marsily). There are in fact two scales of variability (Mr. Gera): a "larger-scale" one (from clay basin to clay basin) and a "smaller scale" one (inside the basin). In the case of Mol, the data gained from the existing laboratory could also be valid for a future pilot facility excavated from the existing shaft, as well as for the entire nuclear site of Mol where reconnaissance works have shown the same lithology (Mr. Neerdael)

Clay/component interactions

No intense radiation field will be used for in situ corrosion testing at Mol (Mr. de Batist). However, effects of radiolysis will be considered, such as production of hydrogen (Mr. Baetslé).

Instrumentation

Mr. Gera recalled that inserting a sensor in clay disturbs the property which is to be measured. Therefore, the large number of measuring devices in the Buffer Mass Test at Stripa could be detrimental. Mr. Pusch replied that decreasing the number of sensors could lead to unresolved questions due to lack of data in critical periods (changing rates, transients, etc). Presently, 10 to 20 millions of measurements are taken and loaded in the computers.

It was also observed that, in general, thermocouples and Glötzl cells work well; moisture gages are satisfactory; piezometric measurements are questionable.

Session 5/Séance 5
Design and instrumentation of in situ experiments in salt
Conception et instrumentation d'expériences in situ dans le sel

Chairman/Président:
H.RÖTHEMEYER
Physikalisch-Technische Bundesanstalt, Braunschweig, FR Germany

In situ test requirements for repository in salt

HEMENDRA N.KALIA
Nuclear Waste Isolation, Battelle Project Management Division, Columbus, OH, USA

ABSTRACT

For the characterization and evaluation of salt as suitable geologic media, a comprehensive investigative program is being implemented in the U.S.A. A key element of the program is in situ testing. In situ tests have been performed in the past at Lyons, Kansas, and are currently under way at Avery Island, Louisiana and at the Asse Mine in the Federal Republic of Germany (FRG). This paper presents the rationale and requirements for in situ testing in salt, status of current testing programs, and planning for in situ tests from an exploratory shaft.

RESUME

On est en train de mettre en oeuvre aux E.U. un programme de recherche comprehensif pour l'etude detailee et pour l'evaluation de sel entant que milieu geologique convenable. Un element clef du programme est l'essais sur place. Les essais sur place ont ete realises dans le passe a Lyons, Kansas, et sont actuellement en cours de realisation a Avery Island, Louisiana et a Asse Mine dans la Republique Federale d'Allemagne (RFA). Cette etude presente l'analyse raisonnee et les conditions pour les essais de sel en place, l'etat des programmes d'essais actuels, et la planification pour les essais sur place des puits d'exploration.

This paper addresses the in situ testing requirements for the purposes of the site characterization, developing test rationale and methodology, and U.S. experience with in situ testing in salt and current thoughts on developing the in situ test plan to produce data that can be used for the validation of numerical models, input to design engineering, performance evaluation input and for the purposes of characterizing the salt formations. The site characterization process is defined as the activities, whether in laboratory or in the field, undertaken to establish the geologic condition and the range of parameters of a candidate site relevant to the location of a repository.

In general, the methodology to be applied and the tests to be implemented in situ for the purposes of site characterization are dictated by the requirements to predict the long-term response of the salt. In the past, investigations have been conducted to determine the behavior of salt in situ such as at Lyons, Kansas (Project Salt Vault). At present, tests are being conducted at Avery Island Mine, Louisiana (Avery), Waste Isolation Pilot Plant at Carls-

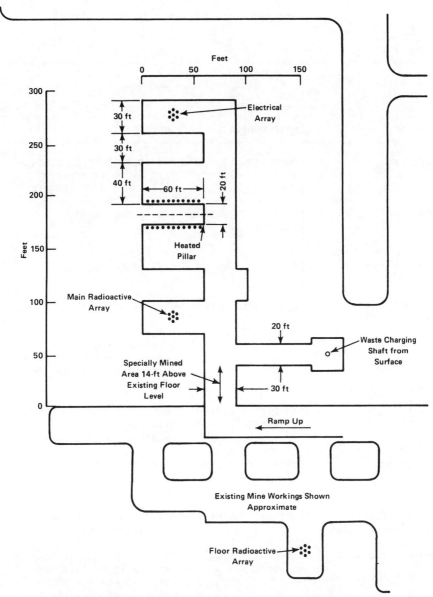

FIGURE 1. SCHEMATIC SHOWING HEATER AND CANISTER TEST LAYOUT

bad, New Mexico (WIPP), and at the Asse Mine in the Federal Republic of Germany (Asse).

Project Salt Vault. In September of 1955, a committee established by the National Academy of Science-National Research Council considered the disposal of HLW in geologic structures within the continental United States. They proposed storage in natural salt formations[1] and

TABLE I. RESPONSES MONITORED: PROJECT SALT VAULT[1]

Rock Type: Bedded Salt	Location: Lyons, Kansas	
Process Monitored	**Instrumentation**	**Comments**
Temperature	Thermocouples	393 thermocouples used; read manually and by multipoint data loggers
Strain	Multiposition Wire Extensometers	Dial gauge readout
Displacement: Vertical Room Closure	Pipes anchored in roof and floor	Dial gauge readout
Horizontal Room Closure	Tape Extensometer	Dial gauge readout
Floor Heave	Leveling Survey	¼" diameter by 1½" long roundhead stainless steel rivets installed in the floor
Roof Sag	Sag wire anchored above sagging bed	Read by potentiometric displacement transducer anchored to sagging bed
Stress Change	Potts Stressmeter	Selected on basis of availability; 17 installed
Integrated Radiological Dose to Salt	Chemical Dosimeter	Radiation effects on salt are not detrimental
Chlorine Production	Off-Gas Chemical Analysis	Chlorine production due to radiolysis of salt is possible but was not measured at elevated temperatures

[1]Extracted from Reference [3]

TABLE II. RESPONSES MONITORED: AVERY ISLAND[1]

Rock Type: Domal Salt	Location: Avery Island	References: 16, 17
Process Monitored	Instrumentation	Comments
Temperature	Thermocouples	129 at Site A, 38 at Site B, and 44 at Site C.
Strain	Multiple-Point Extensometers anchored at depth of 8, 10, 14, 18, 20 and 36 feet	4 at Site A, and 3 at Sites B and C. Dial gauge readout
Displacement: Floor Heave	Optical Level Survey	50 roundhead bolts (at each site) glued into the floor
Stress Change	Vibrating Wire Stressmeter	28 at Site A, 28 at Site B, and 20 at Site C
Brine Migration	Moisture Collection	Natural and synthetic brine migration measure in the heated experiments

[1]Extracted from Reference [3]

recommended Project Salt Vault. The engineering and scientific objectives of the Project Salt Vault were [2] to: demonstrate waste handling equipment and techniques; determine the possible gross effects of (up to 10 rads) on hole closure, floor uplift, salt shattering temperature, etc., in an area where salt temperature is in the range of 100°C; determine the radiolytic production of chlorine; and collect information on creep and plastic flow of salt at elevated temperatures which could be used in later design of an actual disposal facility.

The demonstration work was carried out in a portion of the Lyons Mine. The test site was at a depth of approximately 312m from the surface. For this site 14 irradiated Engineering Test Reactor (ETR) fuel assemblies,

contained in seven cans, were used. The cans were placed in circular array of holes in the floor with one can in the center and the remaining six cans located peripherally on 1.52m centers, Figure 1.

An identical array consisting of electrical heaters was operated as a control to determine the combined effect of radiation and heat on the salt. In addition, one rib pillar was heated with a number of electrical heaters placed in the floor around its base to obtain creep and plastic flow related data for salt at elevated temperatures. The test site layout and the instrumentation that was installed is shown in Figure 2. The response that was monitored at Salt Vault is presented in Table I [3].

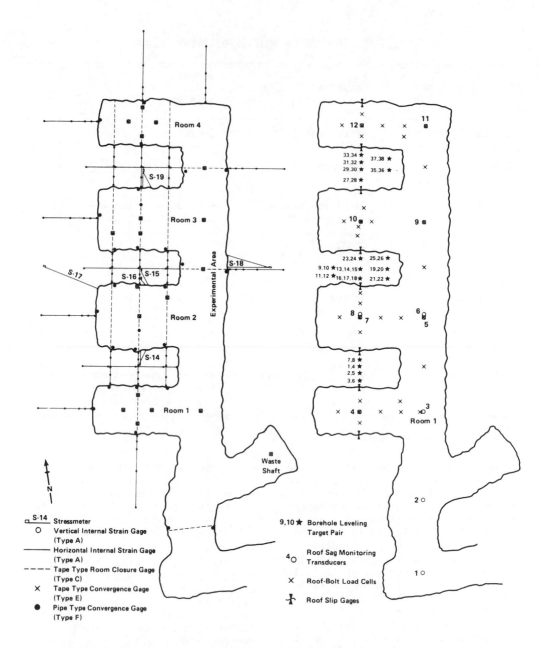

FIGURE 2. PLAN OF EXPERIMENTAL AREA SHOWING ROCK MECHANICS INSTRUMENTATION

At the termination of the Salt Vault experiment, it was concluded that a facility could be safely constructed for the long-term storage of solid high-level and long-lived low-level radioactive waste in salt. However, no facility was constructed because of the presence of vertical exploratory boreholes at the site.

The Salt Vault Project was then terminated and research was initiated at Avery to evaluate domal salt as a host media.

Avery Island Tests. Access by ONWI, for test purposes at Avery, was gained in 1977. By June 1978 three heater tests were set up and started. Avery provided an additional data base for the purposes of developing and validating numerical models, evaluating the design (as it existed) of enhancing retrievability through protective sleeves. The Avery tests compliment the work that was done at Lyons, but differ in that they are being performed in domal salt and that radiation is not used. The objectives of the initial test plan are presented in Table I. The test location and configuration is shown in Figure 3 [4].

At Avery, tests were also performed to obtain brine migration related data [4]. The experimental configuration used is shown on Figure 4. The brine migration tests were designated as AB, NB, and SB. The tests objectives were as follows: Test AB - Evaluate the movement of natural brine under ambient conditions; Test NB - Evaluate natural brine movement at elevated temperatures, with a 1000 watt electrical heater used to heat the salt; Test SB evaluate the migration of synthetic brine under elevated temperatures. A 1000 watt electrical heater was used to heat the surrounding salt. The synthetic brine was tagged.

During the initial 315 days of heating [4], the experiments exhib-

ited steady moisture collection rates, ranging from 0.026 to 0.04 grams/day. At present, DOE is also working with the government of the Federal Republic of Germany at Asse.

Tests Being Performed at Asse. The Asse test program is under the US/FRG Cooperative Radioactive Waste Management Agreement. The cooperative experiments, in progress, are designed to simulate a nuclear waste repository; to determine the effect of gamma radiation on brine migration; and to monitor salt decrepitation and disassociation of brine, the thermomechanical behavior of salt, salt creep and room closure, thermal gradient at the test site and the associated stress in the salt. The data obtained is being used for the purposes of model validation, to evaluate candidate waste package material, and to evaluate candidate test equipment and procedures under repository conditions with a view toward future at depth testing of potential repository tests. The Asse tests thus add to the understanding of the response of salt under simulated repository environment.

Waste Isolation Pilot Plant (WIPP). Another major in situ experimental program that is currently under way in the United States is WIPP. Since 1974, the Department of Energy has been conducting studies, experiments, and surface and subsurface investigations to characterize the Los Medanos site in southeastern New Mexico as a possible site for the isolation of defense radioactive waste.

The WIPP program is implemented in three general phases: technical development; in situ tests without radioactivity; and in situ tests with radioactivity. The WIPP research and development program is organized into three program areas that address the technical issues for waste disposal in bedded salt. These program areas are site characterization and evalu-

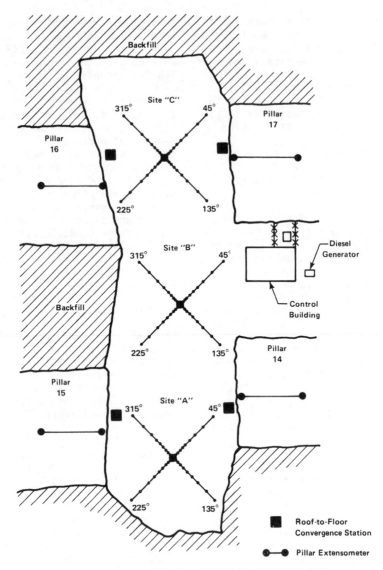

FIGURE 3. INSTRUMENTATION LAYOUT AT AVERY ISLAND

ation, repository development, and waste package interaction. The purposes of the WIPP program [5] are to present the technical issues that apply to the problem of isolating defense radioactive wastes in bedded salt; summarize the status of the art of isolating wastes; describe the role of in situ experiments in validating analytical procedures (including models and computer codes) and in confirming system designs; indicate that in situ testing programs are derived from model development, laboratory testing, and field testing to establish confidence in waste isolation; describe the in situ experiments needed to provide adequate

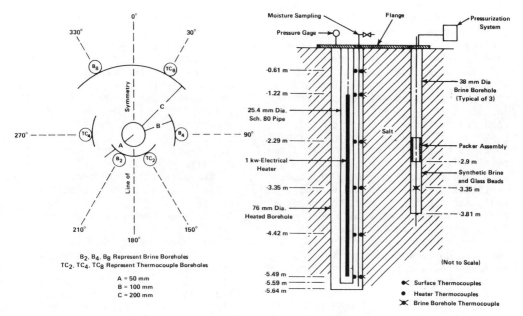

FIGURE 4a. ORIENTATION AND LAYOUT OF BOREHOLES FOR
EACH BRINE MIGRATION SITE (AB, NB, AND SB)

FIGURE 4b. CROSS-SECTIONAL VIEW OF HEATED BOREHOLES
AND BRINE BOREHOLE FOR SITE SB

understanding of various phenomena; and describe demonstration tests for retrieving transuranic (TRU) and defense high level waste (DHLW). The test facility is at a depth of 655.49m and field testing is in progress.

The in situ test program discussed and laboratory work performed to date has generated a great deal of data that are being used to develop and validate numerical models and in situ testing concepts, instrumentation, and test methodologies for salt. It is considered necessary to perform in situ tests at any site that is nominated as a potential repository site. Lessons learned to date suggest that the site specific investigations are necessary.

ONWI has developed test plans [6] to be implemented at a facility to be excavated from an exploratory shaft (ES). The in situ testing can be performed in three inter-related phases: Site Suitability Testing; At-

Depth Testing; and Test and Evaluation Facility. These three phases overlap in time. The purpose of the in situ testing is to develop information that can be used to characterize the nominated site.

The key issues to be resolved by site characterizations are: whether the mined geologic disposal system, of which the site is a part, will meet all applicable standards to contain and isolate radioactive wastes to the extent necessary to ensure that releases of radionuclides to the accessible environment do not result in unacceptable increases in doses to individuals and to the general population; and whether the changes in rock properties caused by construction of the repository and by waste emplacement will unacceptably affect system performance and if the occupational safety of the repository personnel will be consistent with applicable regulations.

Much of the data needed to re-

325

solve the site characterization issue can be adequately addressed by the surface based characterization processes. However, a great deal of information such as spatial response of salt in repository environment cannot be assessed without performing in situ tests. In situ testing is required to verify the geological, geotechnical, and lithological data obtained by surface methods. In situ testing is also required to obtain site specific data on creep characteristics of salt with and without radiation and heat. For the purposes of orienting the repository workings, it is necessary to know the in situ stress and orientation of the fractures to determine what is the most suitable horizon for the development of the repository.

The role played by in situ testing in characterizing the site is illustrated in Figure 5. The in situ testing that is planned is to support the following:

1. To develop data that can be used for license application
2. To evaluate the suitability of salt sites to isolate the radioactive waste for tens of thousands of years
3. To evaluate the capabilities of the numerical models to predict the repository performance
4. To evaluate the response of the salt formations when subjected to thermomechanical hydrological environment similar to the repository environment
5. To evaluate the geological, geohydrological, and geochemical environment of the repository horizon
6. To evaluate the anisotropy of the repository horizon
7. To evaluate the background radiological status of the repository horizon
8. To evaluate the seismic environment of the site
9. To evaluate the geochemical and mineralogical environment of the repository horizon
10. To evaluate the layout of excavations as manifested by changes in stress, strain, and closure of the openings for the purpose of predicting long-term behavior of the repository rocks
11. To evaluate the corrosiveness of the fluids at the repository level to determine their impact on waste containers
12. To evaluate the in situ state of stress, stress gradient, orientation of the in situ stress and evaluate its impact on the construction of repository and its

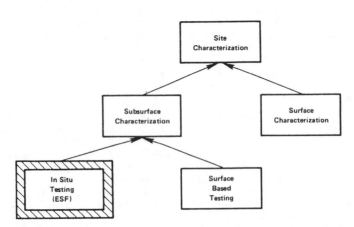

FIGURE 5. RELATIONSHIP OF IN SITU TESTING TO SITE CHARACTERIZATION

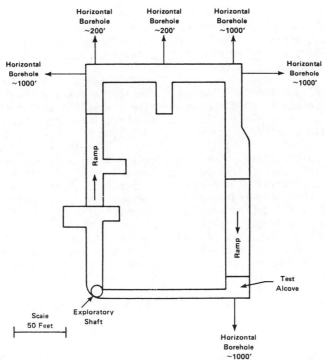

FIGURE 6. PLAN VIEW OF EXPLORATORY SHAFT FACILITY
(SITE SUITABILITY TESTING)

components
13. To evaluate in situ hydrological information
14. To develop data that can be used to quantify repository environment with respect to regulatory requirements such as working environment, air quality, dust, gases, brine pockets, temperature etc.

The data required to satisfy these requirements is to be produced in three inter-related test phases: Site Suitability, At-Depth Testing, and Test and Evaluation Facility.

The site suitability phase of testing will permit monitoring of the response of the rock to shaft drilling and excavation. The data gained from this phase will be used to help resolve technical issues and contribute to license applications, performance assessment, and repository design.

This phase of testing has been divided into the following four categories of in situ characterization: geological characterization, geoengineering characterization, hydrological characterization, and geochemical characterization. These tests will be conducted in a facility to be constructed from the exploratory shaft.

The purpose of the at-depth testing (ADT) is to further the process of repository site characterization and nominate the site for construction authorization as required by the NWPA. At-depth testing encompasses large scale tests in a time or space other than that included within the site suitability testing phase. The ADT tests are an expansion of site suitability testing and follow as soon as practical after the start of site suitability testing. For example, after tests have been

327

conducted in the boreholes which lie along the outer two drifts of the expanded facility, mining of the drifts will start. These tests are expected to run for three years or more.

The NWPA also authorizes the Department of Energy to construct a Test and Evaluation Facility (TEF) either colocated at a repository site or at an independent site. In the event that it is considered necessary to colocate a TEF, then it can be used for the characterization of the selected site in the following areas: geotechnical data for design verification; engineering data for site performance confirmation; and development and demonstration of technology for the operations of the repository. The decision to have a TEF is to be made at a later time and it will be based on the data needs.

The types of in situ tests to be performed and the data acquisition methodology is currently being developed. The test matrix to be developed will rely very heavily on the work performed so far and it will draw upon the experience gained at WIPP, Asse, Salt Vault, Avery, and other sites.

It is tentatively planned to perform tests to obtain in situ stresses using hydrofracturing, obtain closure or creep measurements using multipoint extensometers, and convergence recorders, obtain mining related stresses using borehole pressure cells and vibrating wire stress gauges, obtain thermal behavior of salt by placing heaters, obtain conductivity related information using thermocouples. Other major tests will include corejack tests similar to those at Avery, block heater tests, pillar stability tests. Methodology will be developed to obtain uncontaminated cores for salt mineralogy, geochemistry, water or brine chemistry, as well as mineralogical environment. The general geological environment will be determined by employing joint

and fracture surveys, mapping cores, performing geophysical logging, and by establishing instrumentation to obtain subsurface seismic environment.

In summary, in situ testing is considered a key activity in obtaining site specific data for the purposes of characterizing site/sites for the isolation of HLW. The data to be produced by in situ testing is expected to play a key role in validating numerical models for the purpose of predicting long-term behavior of the salt, provide information for the purposes of assessing the performance of the repository, provide data to be used for the purposes of developing repository layout and for the purposes of designing engineered barriers. Additionally, in situ testing will further advance the state instrumentation technology for material such as salt and provide a major data base addressing rock mechanics of salt.

References

[1] Committee on Waste Disposal, Division of Earth Sciences, Disposal of Radioactive Waste on Land, National Academy of Science - National Research Council, Publication Number 519, April 1957, p. 6.

[2] Bradshaw, R.L., Perona, J. J., and Blomeke, J.O., Demonstration Disposal of High Level Radioactive Solids in Lyons, Kansas, Salt Mines: Background and Preliminary Design of Experimental Aspects, ORNL-TM-734, January 10, 1964.

[3] St. John, C.M., Aggson, J.R., Hardy, M.P. and G. Hocking, Evaluation of Geotechnical Surveillance Techniques for Monitoring High-Level Waste Repository Performance, NUREG/CR-2547, U.S. Nuclear Regulatory Commission, Washington, D.C., 1982, pp. 1-20.

[4] Krause, W.B., _Avery Island Brine Migration Tests; Installation, Operation, Data Collection, and Analysis_, ONWI 190(4). Office of Nuclear Waste Isolation, Battelle Memorial Institute, Columbus, Ohio, December 1983.

[5] Matalucci, R.V., Christensen, C.L., Hunter, T.O.; Moleske, M.A., Munson, E.E., _Waste Isolation Pilot Plant (WIPP) Research and Development Program: In Situ Testing Plan, March 1982._ SAND81-2628 National Technical Information Service, Springfield, Virginia.

[6] Ubbes, W.F., _Conceptual Test Plan for Site Confirmation Testing at an Exploratory Shaft in Salt._ ONWI-493, September 1983. Office of Nuclear Waste Isolation, Battelle Memorial Institute, Columbus, Ohio 43201.

In situ tests and instrumentation in the WIPP facility*

L.D.TYLER
Experimental Programs Division, Sandia National Laboratories, Albuquerque, NM, USA

ABSTRACT

The WIPP (Waste Isolation Pilot Plant) is a DOE R&D Facility for the purpose of developing the technology needed for the safe disposal of the United States defense related radioactive waste. The in situ test program is defined for the thermal-structural interactions, plugging and sealing, and waste package interactions in a salt environment. An integrated series of large scale underground tests address the issues of both systems design and long term isolation performance of a repository.

INTRODUCTION

The Waste Isolation Pilot Plant (WIPP) has been authorized by the U. S. Congress in 1979 as a facility with a mission "... for the express purpose of providing a research and development facility to demonstrate the safe disposal of radioactive waste resulting from the defense activities and programs of the United States exempted from regulation by the Nuclear Regulatory Commission."[1] The WIPP is under construction in the salt beds of southeast New Mexico. The facility is to be used to permanently isolate transuranic (TRU) waste generated by the U. S. Defense program, and to provide an "underground laboratory" in which the concepts for safe disposal of defense high level waste (DHLW) and TRU waste in salt will be validated and demonstrated. The construction at WIPP began in July 1981, with two shafts and about two miles of underground drift excavated at the proposed facility depth of 655 m, Figure 1. The facility horizon is located in a very thick evaporite sequence of the Salado Formation, a small portion of which is shown by the stratigraphic section in Figure 2. Currently, the shaft excavation continues with the enlarging of two of the existing three shafts. The underground excavation is now focused on the development of the rooms for rock mechanics and simulated waste experiments. This paper will provide an overview of the in situ test activities of the R&D Program.

*This program supported by the U. S. Department of Energy under
  Contract #DE-AC04-76DP00789
**U.S. DOE facility

The WIPP R&D Program provides the technical basis for systems design
and safety and environmental assessments for the WIPP and future defense
waste repositories.[2] The initial phase of the technical development in
the WIPP Program involved laboratory testing and small scale field testing.
The results of these tests and the results of similar testing done over the
past two decades have provided the information to define the phenomena that
accompany placement of radioactive waste into salt. The WIPP R&D Program
is progressing into a phase in which in situ experiments on a realistic
scale and in an actual environment can be accomplished. This phase of in
situ testing will serve to validate codes, models and designs and to confirm, on a large scale, the results of laboratory experiments. The initial
in situ tests in WIPP will be conducted in a simulated fashion for both
non-heat and heat-producing waste. The effects of heat from defense high
level waste will be simulated using electrical heaters in actual emplacement canisters to reproduce the thermal effects anticipated. These simulated experiments will be followed in 1989 by a selected suite of experiments
using radioactive waste. The WIPP in situ simulation experiments are divided into three broad areas: thermal/structural interaction (TSI), plugging
and sealing, and waste package interaction experiments. These program
areas are described in the following paragraphs, and their locations in the
WIPP are shown in the underground layout, Figure 3. Approximately 4000
channels of data will be collected for the in situ simulated experiments
for a "testing" period of at least seven years.

Thermal/Structural Interaction TSI Experiments

An important issue in repository development concerns the underground
openings, and whether the structural response of the drifts, rooms, and
shafts can be predicted. Development of a repository in salt should not
only provide for stable rooms and entries during the operational period for
waste emplacement and possible retrieval, but also sufficient long term
rock-deformation to ultimately encapsulate and isolate the waste. Repository design is different from the typical mine design because of the emphasis on stability, not economic extraction, and because the heat generated
by the waste requires knowledge of room response not previously considered
in design. Given an underground heat loading, use of a numerical simulation technique is required to predict behavior of the repository rooms and
surrounding formation for several thousands of years.

The prediction of both short-term stability of the rooms and the
long-term encapsulation of waste requires reliance on adequate material
(constitutive) models, material parameters and numerical computer codes to
predict the effects of stress and heat on salt. The TSI in situ tests are
designed to provide the data base needed to validate and to improve the
predictive technique which can model the behavior of salt at ambient and
elevated temperatures for repository design and long term isolation
assessments.

The WIPP TSI In Situ Tests

WIPP program needs were defined after careful evaluation of past
laboratory and field testing in salt. The remaining issues for the

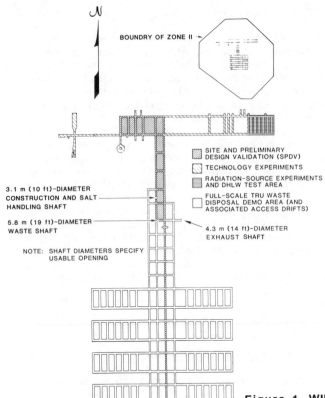

Figure 1. WIPP Underground Layout

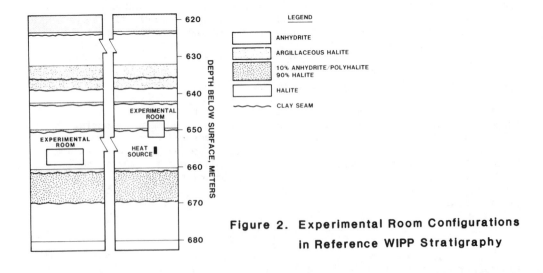

Figure 2. Experimental Room Configurations
in Reference WIPP Stratigraphy

further development of the predictive technique required an integrated series of large scale underground tests [3].

The TSI underground tests consist of the following six major individual tests:

1. 18-W/m^2 Mockup.
2. DHLW Overtest
3. Geomechanical Evaluation
4. Heated Pillar
5. In Situ Stress
6. Direct Shear of Clay Seams

The first four tests require excavated rooms and the in situ stress test requires some controlled excavation, Figure 3. The experiments have been laid out to ensure that the heaters in rooms A and B to the east of the facility and the experiment rooms G and H, to the west are in the same high-purity salt sequence. (Figure 2)

Two of the tests use large in situ rock masses with specified geometries to provide a check on the adequacy of the constitutive models developed from small laboratory tests. The geomechanics test (Room G) and heated pillar test (Room H) are these tests. The analysis of the more complex repository room configurations in two other tests use the proper constitutive models and the computer codes in calculating the room response. The 18-W/m^2 Mockup (Room A, A1, A2, and A3) and DHLW Overtest (Room B) experiments provide these tests. The last two are special tests to provide calculational inputs that can best be obtained or verified from in situ tests. The in situ direct shear of clay seams and in situ stress are such special tests.

The 18-W/m^2 Mockup (Room A) provides in situ results involving all the structural and thermal interactions expected in an actual repository. The configuration is shown in Figure 4. This test takes place in three rooms, each with uniform cross sections of 5.5m x 5.5m, separated by two 18 m thick pillars. The center room (A2) is the mockup of the reference repository configuration with simulated canisters. The thermal loading in the room is obtained from 28 0.47-kW simulated-waste canister-heaters 0.6m in diameter by 2.7m long, emplaced in holes 5.5 m deep in the floor of the room. Four 1.41 kW guard heaters are used at each end of the room to eliminate end effects. The outside guard rooms (A1 and A3) simulate thermal load through a simplified array on the room centerline. The thermal loading in these rooms is obtained from 1.41 kW heaters in hole 5.5 m deep in the floor spaced 3.4 m apart. This thermal load of 18-W/m^2 gives a maximum salt temperature of about 350 K (77 C) near the heaters at 3 years. The vertical closure of the A2 drift is predicted to be 0.6 m at 3 years.

Instruments and gages planned for the room consist of: (1) temporary and permanent manual vertical and horizontal closure stations, established concurrently with excavations; (2) permanent floor and ceiling displacement stations, composed of paired remote reading extensometers in opposed 15 m deep boreholes in floor and ceiling, (3) paired remote reading opposed horizontal extensometers in 15 m deep boreholes in the ribs; (4) permanent remote reading floor-ceiling stations; (5) remote reading stress gages, (6) inclinometers, and (7) thermocouples. The thermocouple arrays are to

333

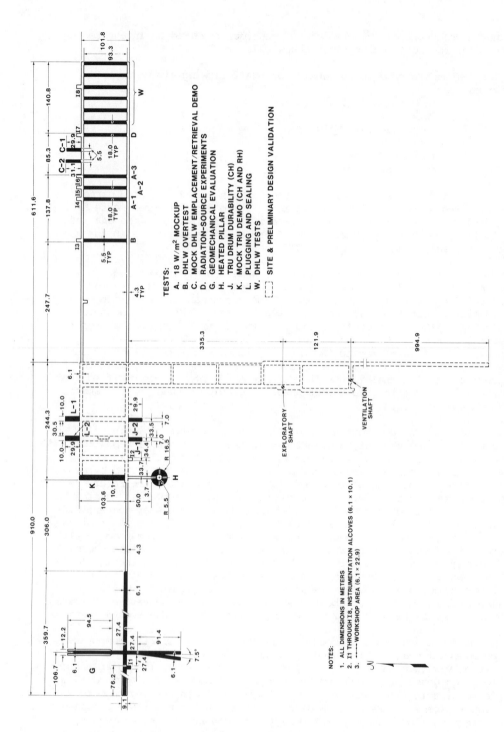

FIGURE 3. Layout of In Situ Tests for WIPP

TESTS:
A. 18 W/m² MOCKUP
B. DHLW OVERTEST
C. MOCK DHLW EMPLACEMENT/RETRIEVAL DEMO
D. RADIATION-SOURCE EXPERIMENTS
G. GEOMECHANICAL EVALUATION
H. HEATED PILLAR
J. TRU DRUM DURABILITY (CH)
K. MOCK TRU DEMO (CH AND RH)
L. PLUGGING AND SEALING
W. DHLW TESTS
   SITE & PRELIMINARY DESIGN VALIDATION

NOTES:
1. ALL DIMENSIONS IN METERS
2. I1 THROUGH I8, INSTRUMENTATION ALCOVES (6.1 × 10.1)
3. ——— WORKSHOP AREA (6.1 × 22.9)

determine the farfield temperatures around the room and the nearfield temperatures around several heaters. Also, the mechanical influence of a clay seam will be monitored by relative extensometer measurements of the rib-and-pillar surfaces at designated locations. Measurements will continue for at least 3 years.

The DHLW Overtest (Room B) will provide a data base for thermal/ structural calculations at a higher temperature where the deformations are accelerated to give, in a reasonable test time, the large strains and deformations that would occur in the reference test only after a long time period. In addition, in situ behavior of room backfill will be measured because the plans are to backfill the hot room after 3 years and then remove the backfill after 7 years of testing. The test configuration (Fig. 5) consists of a uniform cross section room with a heater emplacement array of 17 1.8 kW heaters spaced at intervals of 1.5 m center-to-center along the axis of the room. Two 4 kW guard heaters are placed at each end of the room to eliminate end effects. The instrumentation for the first 3 years of this test will be similar to that used in Room A. Additional special gages to measure room closure and backfill stress and compaction will be installed before the room is backfilled. Results from these additional gages and core sample tests will be obtained throughout the backfill compaction period. The predicted vertical closure of the overtest drift is 0.82 m at 3 years.

Two tests, the Heated Axisymmetric Pillar (Room H) and the Geomechanical Evaluation (Room G) are designed to be as representatives of two dimensional geometry as possible. These tests form the basis for verifying constitutive behavior. The most significant technical issue to be resolved by these verification tests is whether the behavior of a large volume of salt in a natural setting is predictable from models based on site specific laboratory data. Resolution of this issue is essential in validating numerical analyses for design and performance predictions for bedded salt repositories.

The Heated Axisymmetric Pillar experiment tests the heat dependance of the constitutive model. Because the geometry is axisymmetric, code analysis requires very little geometrical abstraction. The test configuration consists of a drift 3.7 m wide x 3 m high x 50 m long to provide access to an annular room 11 m wide x 3 m high surrounding the test pillar, which is 5.5 m in radius. The surface of the pillar is to be uniformly heated by a resistive blanket heater with an energy output of about 135 W/m^2.

Instruments will be installed to measure vertical and horizontal drift-closures, deformation of and stress in the salt pillar and around the drift up to 15.2 m into the salt. Thermocouples will be installed to measure temperature of the salt in the pillar and drift. Deformation and fracturing will be monitored using acoustic-emission triangulation. Measurements will continue for at least 3 years.

The Geomechanical Evaluation tests consist of three isolated, long drifts of various roof spans. These drifts are shown in Figure 3 at the west end of the R&D facility. Because of length of the drifts these drift configurations can be modeled accurately with two-dimensional codes. The changes in roof span in these drifts involve different volumes of material and, in a bedded sequence, different combinations of materials and clay

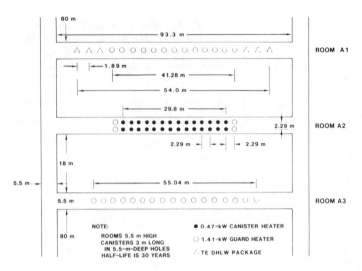

Figure 4. Plan View Schematic of the Heater Emplacements for the 18-W/m² RRC Mockup Test

seams. Portions of the test involve more complex geometries. The wedge pillar is designed to give information on the quasi-static yielding and fracture of a pillar. Also, the simple intersection formed by the cross-cut is to provide data for future 3D code calculation.

The development of the Geomechanical Evaluation test rooms is in four phases. In Phase 1, a main drift 360 m long and 3 m high by 6.1 m wide is driven from east to west in the Room G area. Phase 2 begins 1 year after the main drift is completed with the driving of a north south cross-rift 238.5 m long and 3 m height by 6 m wide. Also, in Phase 2, 61 m length of the west end of the main drift is widened to a 9.1 m roof span. Phase 3 will begin immediately after Phase 2 with the development of the wedge pillar that starts 27.4 m south of the intersection in the cross drift and is driven at a 7.5° angle to the cross drift. Phase 4 starts 1 year after completion of Phase 2 and involves widening a 91.4 m long section of the north end of the cross drift to give a 12.2 m roof span.

Instruments and gages planned for this test are grouped into six major stations; three stations are associated with the long 2D drifts, two stations with the 3D intersections and one station with the wedge pillar. Instruments and gages include individual anchor bolts (single point exten-someters) for manually measuring horizontal and vertical salt displacements around the drifts, coupled with multipoint extensometers for remote measure-ments of these displacements; temporary and permanent manual floor-to-ceiling closure stations; temporary and permanent manual rib-to-rib closure stations; remote-reading vertical and horizontal closure gages; stress gages and pressure cells. The prompt installation of temporary closure stations is critical in this test for early measurements.

The wedge pillar has inclinometer holes in addition to the typical instruments and gages listed above. This pillar will also be instrumented with geophones for detecting acoustic emissions during pillar fracture.

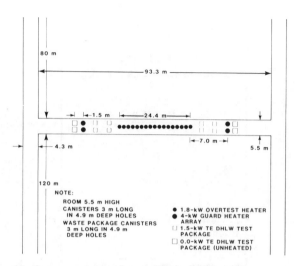

Figure 5. Plan View Schematic of Heater Emplacements for the Overtest

The In Situ Stress and Direct Shear of a Clay Seam are two tests that address very specific problems. In the first, the in situ stress is being defined using hydraulic fracturing methods. The in situ stress is an important boundary condition in any calculation. In the second, the shear strength of a large area of clay seam is to be determined.

The in situ stress measurement was in a long (182 m), 100 mm diameter borehole drilled along the Room G entry drift. Several hydrofrac tests were dye-marked along the length of the deep borehole. The fractures will be exposed and studied during drift excavation. Instruments include high-resolution caliper gages, a borehole viewer, an impression packer, a portable recorder and pressure cells.

The direct shear of a clay seam will be tested by cutting around a 1 x 1 m block in place in a wall or floor containing a representative clay seam. Flatjacks will be installed in slots cut around the block so that shear and normal stresses can be applied. Displacement of the seam will be measured as a function of applied stress.

WIPP Waste Package Performance Tests

Much of the perceived success of a nuclear-waste repository depends on the ability of the DHLW package to physically contain its enclosed waste radionuclides over hundreds, and perhaps, up to a thousand years. The waste package, the assembly of the waste form, and the engineered barrier materials, should also chemically retard or restrict the release rate of radionuclides for even longer, about 10,000 years, in the event the geologic barriers are breached.

An extensive body of data on the performance of HLW package material exists from laboratory experiments in overtest and the relevant environments of a salt repository [4]. The in situ testing of simulated DHLW and

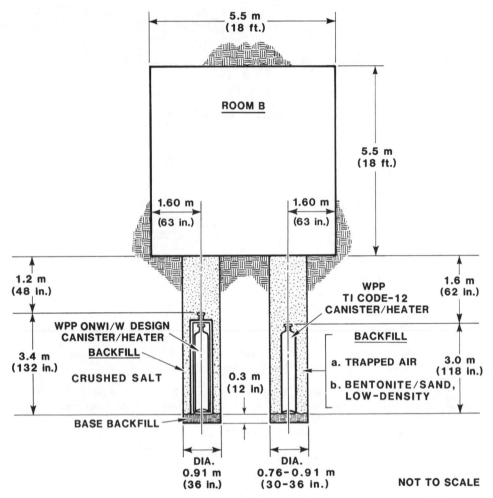

Figure 6. Cross-Section View of Simulated DHLW Emplacements in WIPP Room B

TRU packages in the WIPP are to be done under expected reference and overtest conditions. These tests are a valuable precursor to the actual radioactive waste test by extending the state-of-the-art technology beyond the laboratory. The tests are divided into two categories: Simulated DHLW Package Performance Tests and TRU Waste Package Performance Tests.

The Simulated DHLW Package Performance Tests are to be conducted at a reference repository condition in Room A (Fig. 4) and at an overtest condition in Room B (Fig. 5). Six electrically heated waste canisters will be emplaced in Room A1 to test canister and backfill materials under an expected repository environment. All canisters will be .6 m in diameter and 3 m long. Three canisters will be made of mild steel and three canisters will be made of mild steel with a 2.2 mm thick TiCode overpack. The canisters are placed outboard of the guard heaters in Room A1 and will

338

be operated at 0.47 KW power. Two types of backfill materials will be tested: crushed salt and a low density bentonite/sand backfill. One canister will have brine injected into the emplacement hole to test the canister material.

Twelve waste packages are emplaced in Room B to be tested at an accelerated test condition and a thermal load (1.5 KW/ canister), approximately three times greater than the reference condition. Some of the canisters will be tested with intentionally created defected or degraded canisters and some with artificially injected quantities of brine. Eight of the packages contain electrical heaters; the other four packages contain canisters filled with nonradioactive DHLW glass. The same backfill materials as tested in Room A1 also will be tested in the overtest environment. The canister material to be tested includes SS304L canisters with a mild steel overpack and TiCode-12 canisters. The instrumentation for the tests are thermocouples for the canister materials, backfills and nonradioactive glass and pressure gages for the backfill. The canisters will be tested for 3 to 7 years and will undergo extensive post test material testing.

The TRU Waste Package Performance tests will be conducted in the J rooms shown in Figure 3. The in situ testing of simulated/nonradioactive CH TRU waste drum and RH TRC waste canisters in the WIPP is necessary to evaluate and demonstrate the waste package physical integrity. Testing will be conducted under near reference conditions and severe overtest (accelerated aging) conditions of temperature and brine intrusion. Tailored backfill also will be tested to determine their effectiveness as engineered barriers. These tests allow the TRU Waste Package to be tested in an actual repository environment before radioactive waste is emplaced in the WIPP facility. The backfills being tested are granular bentonite/crushed salt (30 wt % / 70 wt %) and crushed salt . The predominant results of the tests will be in the area of corrosion resistance and backfill applicability for meeting the EPA standards.

WIPP Plugging and Sealing Program

The Plugging and Sealing Program provides for the development and demonstration of the technology required to establish plugs and seals in man-made penetrations (boreholes, shafts, drifts, storage rooms) to ensure long term isolation of nuclear waste. Basically, the program is designed to test and demonstrate the effectiveness of the various engineered barriers proposed for sealing vertical penetrations into the underground waste disposal facility and for isolating the underground rooms from each other [5]. The test program will quantify the permeability of the salt beds and other layers, especially where altered by the excavation, as a function of time as the walls of the room close upon the backfill. Microcracking due to excavation and stress relief will heal in a salt environment. The rate at which the healing occurs must be established. Both cement grouts and natural (salt) plugs will be tested for use at the WIPP. The Plugging and Sealing Program is currently in its initial development stages. The current field activities include: (1) in situ permeability tests conducted in horizontal boreholes drilled into pillars from 4 opposite faces of the same pillar on the same axis to determine the range of permeabilities that can be expected; (2) the plug test matrix for developing emplacement methods for candidate plug and backfill materials and for obtaining in situ cured samples for laboratory assessment of geochemical stability; and (3) demonstration and verification of capabilities to install, test and monitor in situ plugs in the AEC-7 drill hole.

SUMMARY

The WIPP R&D Program has entered into the in situ test phase. This currently addresses nonradioactive tests for Thermal Structural Interaction. Plugging and Sealing, and Waste Package Interactions. The TSI experiments and Simulated DHLW tests are currently being emplaced and will be operational by the end of this calendar year, 1984. The remaining experients are scheduled to be operational next year. The testing program is designed to provide the technology for developing the design and long term assessment of waste isolation for the WIPP and future defense waste repositories.

REFERENCES

1. PL9b - 164, December, 1979, U. S. Congressional Authorization B.11.

2. Matalucci, R. V., C. L. Christensen, T. O. Hunter, M. A. Molecke, and D. E. Munson, "Waste Isolation Pilot Plant (WIPP) Research and Development Programs: In Situ Testing Plan, March 1982," SAND81-2628, Sandia National Laboratories, Albuquerque, NM (1982)

3. Munson, D. E. and R. V. Matalucci, "Planning, Developing, and Fielding of Thermal/Structural Interactions In Situ Tests for the Waste Isolation Pilot Plant (WIPP)", Waste Management '84 Conference, Tucson, AZ, (March 1984).

4. Molecke, M. A., J. A. Ruppen and R. B. Diegle, "Materials For High Level Waste Canisters/Overpacks In Salt Formations", Nuclear Technology, Vol. 53, No. 3, (December, 1983).

5. Christensen, C. L., S. J. Lambert and C. W. Gulick, "Sealing Concepts for the Waste Isolation Pilot Plant (WIPP) Site", SAND81-2195, Sandia National Laboratories, Albuquerque, NM (1982).

Design and instrumentation of in situ brine migration tests at the Asse Salt Mine

T.ROTHFUCHS
GSF, Institut für Tieflagerung-WA, Schachtanlage Asse, Remlingen, FR Germany

ABSTRACT

Rock salt contains small amounts of different water forms which are released to High Level Waste (HLW) boreholes as a result of heat induced migration through the surrounding salt formation.

A special cold trap system was developed for the measurement of release rates and also the integral amount of released water. In order to evaluate these measuring data some additional instrumentation as for example seismoacustic monitoring, and gas pressure measurements were applied.

The layout of in situ experiments at the Asse Salt Mine, the technical and physical principles of the measuring methods and some typical results are discussed and presented.

1. INTRODUCTION

Rock salt formations are being considered as suitable sites for final nuclear waste disposal. Especially with regard to the disposal of heat producing high level radioactive waste (HLW) rock salt is favorable because of its high thermal conductivity and visco-plastic creep behavior under heat and stress. These material properties will avoid high waste temperatures and the development of fractures.

However, natural rock salt contains small amounts of different water forms and it has been observed in laboratory [1] and field tests [2,3] that this water migrates toward the heat source (or waste package) due to temperature and pressure gradients. The released water can accelerate corrosion of the waste package thus leaching radionuclides from the waste matrix.

The water may occure in the form of crystalline water of hydrated minerals like Polyhalite ($K_2MgCa_2(SO_4)_4 \cdot 2H_2O$) and Kieserite ($MgSO_4 \cdot H_2O$) as absorbed water on the grain boundaries or as saturated brine inclusions.

The average water content of Older Halite (Na2ß) at the Asse Salt Mine ranges from 0.04 to 0.25 weight % [1].

In order to estimate the arrival of this water at the waste package, field tests under representative in situ conditions are of high importance.

2.   BRINE MIGRATION

Two different mechanisms of brine migration must be taken into account
with regard to the measurement of heat induced water release.

Vapor Migration - This is motion of water vapor which is generated by evaporation water from brine in the salt porosity or from hydrated minerals.
This motion is assumed to be proportional to pressure gradients and to be
dependent on the permeability of the salt. Motion of the water can be described according to Darcy's law

$$V = - \frac{K}{\mu} \nabla P_V$$

where

V = velocity of the water vapor (cm/sec)
K = permeability of the salt (cm^2)
μ = viscosity of the water (centipoise)
P_V = water vapor pressure (atm)

Liquid Inclusion Migration - This is motion of small inclusions of saturated brine located within the salt crystals or on the grain boundaries.
This motion is known to be proportional to the local temperature gradient
and increases in velocity at higher temperatures. Jenks [4] published the
following equation for the velocity of liquid inclusions as a function of
temperature and temperature gradient

$$V = \nabla T_S \; 10^{(0.00656T - 0.6036)}$$

where

V = inclusion velocity (cm/yr)
∇T_S = temperature gradient in the salt (°C/cm)
T = temperature of the salt (°C)

3.   DESIGN OF EXPERIMENTS

The design of in situ simulation experiments should represent the
expected repository conditions as close as possible. With regard to the investigation of brine migration it is necessary to differentiate between
short-term and long-term effects.

It is possible to design the experiments to produce maximum possible
amounts of released water by enforcing both migration mechanism. These
experiments may be called "Accelerated Brine Migration Tests" (ABM). It is
also possible to design experiments to produce reduced amounts of water by
restraining one of the above mentioned mechanisms. These experiments may be
called "Restrained Brine Migration Tests " (RBM). Both types of experiments
have advantages and disadvantages.

ABM-Test - this type of experiment has the advantage to simulate the short
term phase immediately after waste emplacement into the borehole which is

342

of interest with regard to safty considerations during the operational phase of the repository. This experiment is designed to have

- high thermal gradient (representing a single borehole - enforcing liquid inclusion migration)

- low water vapor pressure in the borehole (representing an open borehole - enforcing vapor migration)

- reduced radial stress at the borehole wall (representing an annulus between waste canister and the borehole wall)

A disadvantage is, that it is impossible to differentiate between the migration mechanisms because of composed effects.

RBM-Test - this type of experiment has the advantage to simulate the long term phase after waste disposal which is of interest with regard to long term safety analysis. This experiment is designed to have

- reduced thermal gradients (representing overlapping temperature fields of several boreholes - restraining liquid inclusion migration)

- water vapor pressure built up in the borehole (representing a sealed borehole - restraining vapor migration)

- high radial stress at the borehole wall (representing a closed annulus between waste canister and borehole wall)

A disadvantage is, that only a very small amount of released water is produced, thus increasing the testing period and introducing measuring difficulties. By combining different design parameters of ABM and RBM-tests it is also possible to investigate the contribution or the importance of the different migration mechanisms.

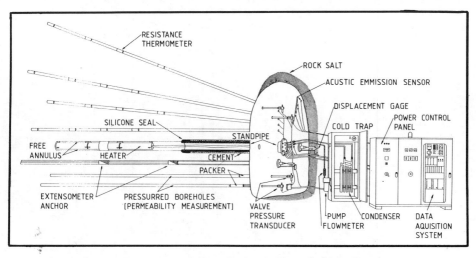

Figure 1: Arrangement of Temperature Test 5 (ABM Test)

Typical examples of the layout of both types of experiments are shown in figures 1 and 2.

Figure 1 shows the test set up of the so called Temperature Test 5 which was performed in an unlined horizontal borehole at the 775 m-level of the Asse Salt Mine. This heating test represents a single borehole ABM-test having a free initial annulus of 4 cm between the heater and the borehole wall. The water release to the borehole is measured by continous condensation of the water in a cold trap (see section 4.1) thereby keeping the water vapor partial pressure in the borehole at a low level.

Figure 2 shows the configuration of one of the four brine migration tests that are conducted on the 800 m-level at Asse Mine as a joint US/FRG effort [5] . This experimental set up represents a RBM-test. The overlapping temperature field of several boreholes is simulated by the use of eight "Guard Heaters" surrounding a central heater. Hereby, the temperature gradient at the central borehole wall is reduced as desired in a RBM-test. The annulus between the borehole wall and the heater is filled with a porous medium for keeping the annulus open for gas and water collection and for maintaining mechanical stress at the borehole wall. The central borehole is completely sealed and the released gases (e. g. water

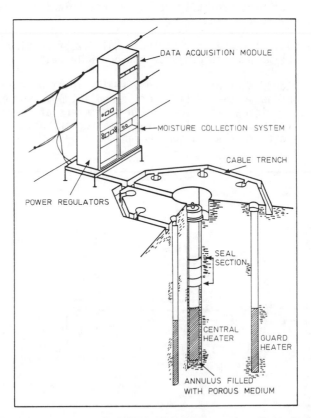

Figure 2: Arrangement of US/FRG Brine Migration Test (RBM Test)

344

vapor) are permitted to built up pressure. The water vapor concentration in the borehole atmosphere is determined by periodically performed gas analysis and a final cold trap measurement at conclusion of the test.

4. INSTRUMENTATION

The following section presents an overview of typical and most important instrumentation that has been used for brine migration investigations at the Asse Salt Mine. It also includes some information about the background of the measuring techniques.

4.1 Measurement of Water Release

For the combined measurement of heat induced water and gas release during in situ testing a special measuring system was to be developed that permitted the simultaneous determination of release rates and of the total amount of released components. In order to avoid measuring errors it is necessary to seal the borehole atmosphere from ambient air. Systems with open cycles were known from previous experiments [6, 7]. Here, the moisture contained in the borehole atmosphere was exhausted on a periodical basis to a dessicant canister where the water was absorbed into chemicals and measured by weighing. A disadvantage was, that other gases produced in the experiment were diluted and carried away.

A more advantageous system was developed for experiments at Asse. The system consists of a closed cycle connected to the borehole. The borehole atmosphere is used as a carrier gas which is continously pumped through the borehole, through the collection apparatus, and returned to the borehole. Part of the closed cycle is a cold trap where the moisture is condensed at reduced temperatures. Other gases existing in the borehole are not carried away and they can be determined by periodical gas analysis. An advantage of collecting the water coninously is, that a measurement of the rate of released water can be obtained as well as the total amount of released water.

Figure 3 shows the principle of a closed cycle apparatus with a cold trap which was used for Temperature Test 5 (see figure 1). In this case Helium was used as the carrier gas. The cold trap consists of a modified refrigerator in which a calibrated glas tube is placed as a water collector. The temperature inside the collector is digital selectable taking a temperature hysteresis of + 0.5° C into account.

The gas flow of approximately 2 l/min is initiated by a small membrane pump and is controlled by a flow meter. A calibration of the cold trap apparaturs resulted a measuring error of approximately + 6 % and a detection accuracy of 0.5 ml [8] .

4.2 Measurement of Salt Permeability

As mentioned in section 2 migration of water vapor may occur according to Darcy's law. In order to estimate migration rates it is necessary to know the permeability of rock salt under in situ conditions. Measurements of the permeability in the vicinity of a heated borehole were also performed during Temperature Test 5. Two parallel boreholes (P1, P4, Figure 1 and 3) at radial distances of 0.3 m and 0.8 m from the heater borehole wall were pressurized

with Neon up to 8 bars. The permeability was determined by use of the modified equation for radial Darcy flow of gases [9, 10].

$$K = \frac{\mu \dot{M} RT}{\pi h (P_B^2 - P_e^2)} \ln (r_e/r_B) \quad ; \quad \dot{M} = \frac{V}{RT} \frac{P_1 - P_2}{t_0 - t}$$

where

K = permeability (cm^2)
μ = viscosity of the gas (10^{-8} bar sec)
$\dot{M}$ = mass flow rate (Mol/sec)
R = Gas constant (bar cm^3 K^{-1} Mol^{-1})
T = absolute temperature (K)
h = length of borehole (cm)
P_B = average borehole pressure (bar)
P_e = pressure inside the pore space (bar)
r_B = radius of borehole (cm)
r_e = effective radius (cm)
V = volume of the borehole (cm^3)
P_1-P_2 = borehole pressure decrease (bar)
t_0-t = measuring duration (sec)

   Additional to these measurements gas samples were taken periodically
from the heater borehole to determine if Neon had passed through the salt.
The salt permeability was determined in this experiment to be smaller than
10^{-7} Darcy.

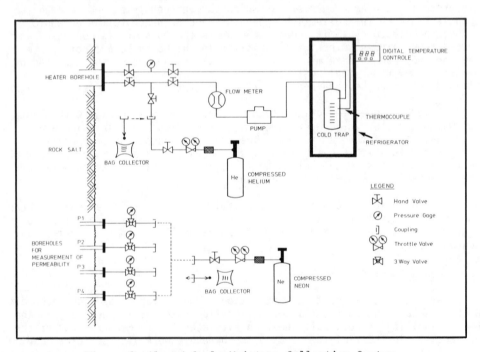

Figure 3: Closed Cycle Moisture Collection System
used during Temperature Test 5

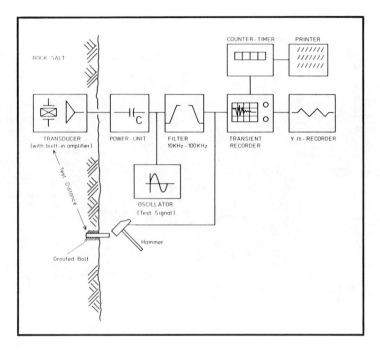

Figure 4: Block Diagram of Seismo Acustic Emission Measurements

4.3   Measurement of Seismo Acustic Emission (SAE)

The measurements of SAE rates and of seismic wave velocity are known to be suitable methods do estimate changes of rock mass structures [11]. With regard to the measurement of heat induced water migration through the salt formation it is of interest if changes of SAE rates can be used to explain increasing or decreasing water release rates or in other words increasing or decreasing permeability of the rock mass.

Figure 4 represents a block diagramm of the seismoacustic apparatus which was used during Temperature Test 5. This apparatus permitted counting of seismoacustic signals as well as the measurement of elastic wave velocities. A seismoacustic transducer is located in a borehole near the heated part of the rock mass. A variable filter is used to eliminate noise created by mining activities. Each signal is displayed on the screen of an osciloscope and the number of signals is summarized over 2 hour intervals. The number of signals of each interval is output on a printer.

For the measurement of seismic wave velocities a grouted bolt was used as a reference point for hammer beat measurements. The transient recorder was triggered by a 1.5 vdc signal which was generated by the hammer beat. The travel time of the induced elastic wave to the seismoacustic transducer was determined from a y/t-plot of the stored data including the trigger signal as well as the transducer signal.

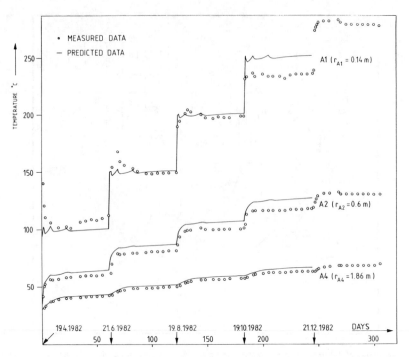

Figure 5: Temperature Increase During Temperature Test 5

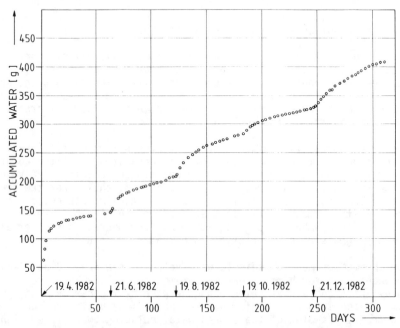

Figure 6: Accumulated Water - Temperature Test 5

5.   TYPICAL RESULTS AND CONCLUSIONS

Figures 5 through 8 show typical results obtained during Temperature Test 5. The borehole was heated stepwise to a final maximum temperature of 270° C. Figure 5 shows predicted and measured temperatures at the midplane of the heated section and Figrue 6 represents accumulated water versus time. Figure 7 shows the gas pressure (Neon) in borehole P1 which was pressurized over the complete heating period to signalize possible and significant changes of the permeability of the rock mass. Figure 8 finally presents the results of counting of seismoacustic signals.

It can be seen that the steep temperature and subsequent increase of stresses at the beginning of each heating step is accompanied by an increase of water release rates and the rate of seismoacustic signals. Since no gas pressure drop (except when gas sampling was performed) occured during the complete heating phase and since no Neon was detected in the heater hole it must be concluded that no significant changes of the permeability occured during the test. However, when the heaters were shut off at termination of test a sharp increase of SAE rates was observed and 12 hours later a sudden and complete pressure loss was observed in borehole P1.

From these measurements it can be concluded that a number of various investigations or measuring techniques are necessary to be performed to get an understanding of the mechanism of heat induced brine migration in salt. On the basis of in situ measurements, theoretical considerations and numerical

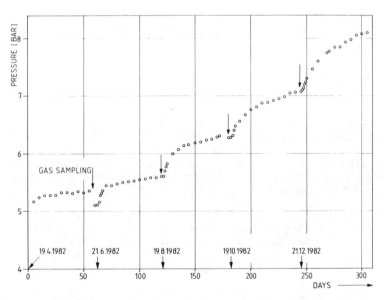

Figure 7: Gas Pressure Increase in Borehole P1 - Temperature Test 5

349

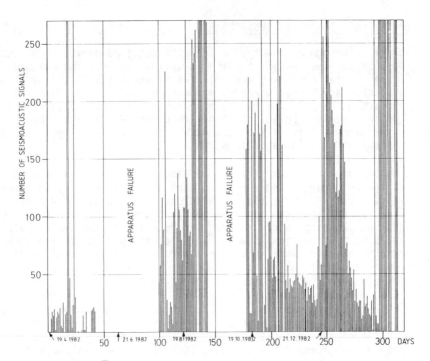

Figure 8: Number of Seismo Acustic Signals - Temperature Test 5

model development it will be possible to estimate the importance of brine migration and possible consequences.

6.   REFERENCES

[1]   Jockwer, N.:"Untersuchungen zu Art und Menge des im Steinsalz des Zechsteins enthaltenen Wassers sowie dessen Freisetzung und Migration im Temperaturfeld endgelagerter Abfälle, Gesellschaft für Strahlen- und Umweltforschung mbH München, Institut für Tieflagerung-WA, GSF-T 119, Juni 1981

[2]   Bradshaw, R. L. et al: " Project Salt Vault: A Demonstration of High Activity Solidified Wastes in Underground Salt Mines", Oak Ridge National Laboratory, ORNL-4555, April 1971

[3]   Rothfuchs, T. et al: "Simulationsversuch im Älteren Steinsalz Na2β im Salzbergwerk Asse - Temperaturversuchsfeld 4, Gesellschaft für Strahlen- und Umweltforschung mbH München, Institut für Tieflagerung, Abschlußbericht im Rahmen der Verträge 015-76-1 und 058-78-1 WAS D des Indirekten Aktionsprogramms der Europäischen Gemeinschaften

[4]   Jenks, G. H.: "Effects of Temperature, Temperature Gradients, Stress, and Irradiation on Migration of Brine Inclusions in a Salt Repository", Oak Ridge National Laboratory, ORNL-5526, 1979

[5]   ONWI-245, GSF-T 118: "Asse Salt Mine Brine Migration Test Program, Gesellschaft für Strahlen- und Umweltforschung mbH München and Office of Nuclear Waste Isolation, April 1981

[6]   RSI-0083, "Experimental Plan for In Situ Synthetic and Natural Brine Movement Studies at Avery Island", RE/SPEC, Inc., July 24, 1979

[7]   SAND 79-2226, "Salt Block II: Description and Results", June 1980

[8]   Rothfuchs, T. : Vorversuch zum Temperaturversuch 4 - Versuchsbericht, Gesellschaft für Strahlen- und Umweltforschung mbH München, Institut für Tieflagerung-WA, Interner Bericht, Februar 1979

[9]   Pirson, S. J.: Oil Reservoir Engineering, Second Edition McGraw-Hill Book Company, Inc., New York 1969

[10]   Katz, D. L. et al: "Handbook of Natural Gas Engineering, McGraw Hill Book Company, Inc., New York 1959

[11]   Blake, W. et al: "Microseismic Techniques For Monitoring the Behaviour of Rock Structures, U.S. Bureau of Mines, Bulletin 665, 1974

Basis and methods for in-situ measurements on liberation and generation of volatile components concerning the disposal of high-level radioactive waste in rock salt

N.JOCKWER
GSF, Institut für Tieflagerung, Braunschweig, FR Germany

ABSTRACT

Rock salt formations which are considered for disposal of radioactive wastes contain besides the main mineral halite the minor minerals polyhalite, anhydrite and kieserite as well as traces of clay, bitumen and carbonates. The water in the rock salt derives from the hydrated minerals, may be adsorbed at the crystal boundaries together with gases such as CO_2, H_2S and hydrocarbons, or form inclusions. By elevated temperature and radiation these volatile components will be liberated and further be generated.

Methods have been developed for the determination of the volatile components which are liberated already at natural mine temperatures, generated and liberated at elevated temperatures, and which are generated by γ-radiation (radiolysis).

Rock salt from the North German salt domes which is foreseen for the disposal of high level radioactive wastes consists, besides the main mineral halite, of the minor minerals polyhalite, anhydrite and kieserite in the range from 0,1 to 5 wt %. In some small layers the concentration of the minor minerals reaches up to 30 wt %. As a result of the hydrated minerals rock salt contains small amounts of water. 55 % of the investigated samples have a water content of less than 0,1 wt % and 75 % less than 0,2 wt %. The amount of water adsorbed on the crystal boundaries and in fluid inclusions is rather low as compared to the water derived from the hydrated minerals. Besides the minor minerals and water, rock salt contains traces of further solid components, such as clay, bitumen and carbonates as well as gases, such as CO_2, H_2S, CH_4 and further hydrocarbons. These gases will be liberated into the ventilation air or into the sealed disposal boreholes at natural mine temperatures already. Further heating and γ-radiation of the rock salt will lead to liberation and generation of further gas components, such as H_2; O_2; O_3; HCl; Cl_2; SO_2; H_2O as a result of thermal cracking and radiolysis. The amount of these components depends on the mineralogical composition and the stratigraphic layer of the rock salt.

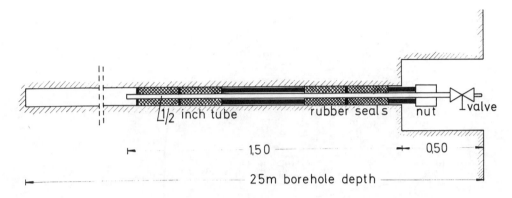

Figure 1: Principal drawing of a mechanical gastight packer
installed in a vertical borehole in Temperature
Test Field 6

The liberation and generation of the volatile components may
pressurize the sealed emplacement borehole, enforce the
corrosion of the containment and lead to the migration of
liberated radionuclides. Therefore, the gases in the surroundings of an emplacement borehole with high-level radioactive
wastes are important to the long-term safety concept. Laboratory investigation indicated that gas components which are already present in the rock salt, such as CO_2, H_2S and hydrocarbons will be liberated even below mine temperature. Above
mine temperature further components such as HCl will be generated by thermal cracking or chemical reaction. γ-radiation up
to 10^7 rads of salt samples with different mineralogical
composition indicated that components like H_2, O_2, Cl_2,
SO_4 and chloric hydrocarbons are generated by radiolysis.

In the Asse Salt Mine we are investigating the in-situ gas
generation and liberation at different conditions:

      a) at natural mine temperature
      b) at elevated temperature
      c) at elevated temperature with γ-radiation.

In order to determine these components liberated from the
rock salt, horizontal or vertical boreholes have been drilled
dry and sealed gastight with mechanical rubber packers. Figure 1
is a principal drawing of a packer installed in a vertical
borehole in Temperature Test Field 6. It consists of four
independent rubber seals which are spanned and pressed against
the borehole surface by a nut at the top. A tube of 1/2 inch
diameter goes down through the rubber seals into the borehole,
it is sealed at the top with a gas-valve from which the gas
samples are to be taken. To avoid corrosion and adsorption of

353

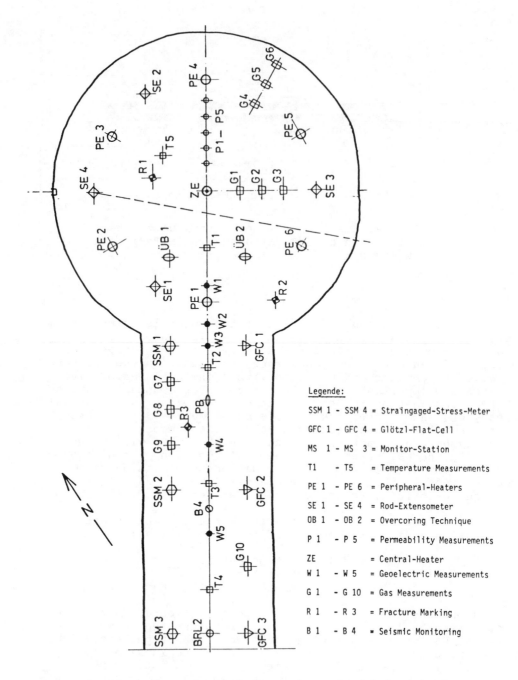

Legende:

SSM 1 - SSM 4 = Straingaged-Stress-Meter

GFC 1 - GFC 4 = Glötzl-Flat-Cell

MS 1 - MS 3 = Monitor-Station

T1   - T5   = Temperature Measurements

PE 1 - PE 6 = Peripheral-Heaters

SE 1 - SE 4 = Rod-Extensometer
OB 1 - OB 2 = Overcoring Technique

P 1 - P 5 = Permeability Measurements

ZE   = Central-Heater

W 1 - W 5 = Geoelectric Measurements

G 1 - G 10 = Gas Measurements

R 1 - R 3 = Fracture Marking

B 1 - B 4 = Seismic Monitoring

Figure 2: Plan view of Temperature Test Field 6 with the
borehole G1 - G10 for determination of gas
generation and liberation

the gas components the packer and the gas-valve consists of
sea-water resistent stainless steal. The packer and the valve
itself are gastight, but in the early beginning we had problems
to get the surface between the rubber seals and the wall of the
borehole gastight. We managed this problem by cleaning the
borehole surface from dust with a brush and by putting a thin
film of vacuum grease on the rubber seals.

After installation of the packers the boreholes have been
evacuated three times to a pressure less than 10^{-2} mbar and
flooded after each evacuation with very pure nitrogen. By eva-
cuating and flooding the borehole the starting point of the
borehole atmosphere was very well known, falsification by the
air was comparatively low and determination of gas components
with low concentration was much easier. To avoid migration of
gas components from the ventilation air of the mine into the
borehole, to indicate the tightness of the borehole and to take
gas samples very easily, the sealed borehole was pressurized to
1,5 bar.

In order to take the gas sample and to bring them into the
laboratory for analyzation, gas sampling bags and gas sampling
bulbs made of glass, teflon or coated steel have been used. At
present the gas determinations are carried out at different
laboratories. The disadvantage of this method is that the time
between taking the samples and analyzation last at least some
days. Adsorption or reaction of components might be possible,
which leads to falsification. In the future we intend installing a gas determination laboratory with a gas chromatograph
and mass spectrometer in the direct vicinity of the test fields
to minimize the effect of the sampling bags and the time between taking the samples and analyzation.

Figure 2 shows the plan view of Temperature Test Field 6 in
which in a central borehole (ZE) and 6 peripheral boreholes
(PE) 7 electrical heaters with a total heating capacity of
120 kW are installed. Besides the temperature induced stress,
strain, seismicity and generation of fracture, the generation
and liberation of gases up to 300° C will be investigated (for
details of the test field see contribution of W. Kessels). The
temperature test field is located in the Stassfurt halite which
contains thin layers of anhydrite. These layers are penetrated
by free saturated brine and gases like H_2S, CO_2 and hydrocarbons. All penetration components are liberated at natural
mine temperature already, but liberation is enforced by
elevated temperature. To determine liberation and generation of
gas components versus time and temperature, 10 boreholes with a
diameter of 56 mm and a depth of 25 m have been drilled in
different zones of this field. The boreholes G1; G2 and G3 are
within the heated area, the temperature of which will rise up
to 300° C, whereas with G4, G5 and G6 which are directly outside, the temperature will be between 200 and 250° C. G7; G8
and G9 are between 5 and 7 m and G10 about 13 m away from this
area. The temperatures will rise up to 100 or 50°C. All bore-

holes have been sealed 9 months prior to start of heating, therefore we know very well gas liberation and the saturation level of the different gas components at natural mine temperatures. Before the start of heating, boreholes G2, G5 and G8 will be rinsed and flooded again with very pure nitrogen. Later on this procedure will be repeated every month. With this method it might be possible to determine the total amount of the liberated and generated gas components. All other boreholes will only be disturbed by taking the gas sample. At the beginning of the heating gas samples will be taken twice a week from all 10 boreholes, later on it will be reduced.

Temperature Test Field 6 in the Asse salt mine is installed completly and heating will start in the near future and will last for about one year. During the whole time gas samples will be taken from all boreholes.

ACKNOWLEDGEMENTS

This investigation is being made within a research contact with the European Atomic Energy Community and Gesellschaft für Strahlen- und Umweltforschung mbH München.

REFERENCES:

Anthony, T.R.; Cline, H.E.: "Thermal migration of liquid droplets through solids", Journal of Applied Physics, 42, No. 9, 3380 - 3387 (1971).

Anthony, T.R.; Cline, H.E.: "The thermomigration of biphase vapor-liquid droplets in solids", Acta Metallurgica, 20, 247 - 255 (1972).

Aufricht, W.R.; Howard, K.C.: "Salt characteristics as they affect storage of hydrocarbons", Journal of Petroleum Technology, 733 - 739 (1961).

Borchert, H.: "Ozeane Salzlagerstätten", Bornträger Verlag, Berlin-Nikolassee (1959).

Braitsch, O.: "Entstehung und Stoffbestand der Salzlagerstätten", Springer Verlag Berlin, Göttingen, Heidelberg (1962).

Herrmann, A.G.: "Die Phasenanalyse als Schnellverfahren zur Analysierung von Salzgesteinen", Bergakademie, 201 - 207 (1956).

Herrmann, A.G.: "Praktikum der Gesteinsanalyse", Springer Verlag Berlin, Heidelberg, New York (1975).

Herrmann, A.G.: "Geowissenschaftliche Probleme bei der Endlagerung radioaktiver Substanzen in Salzdiapiren Norddeutsch-
lands", Geol. Rundschau 68, 1076 - 1106 (1979).

Herrmann, A.G.: "Geochemische Prozesse in marinen Salzablage-
rungen: Bedeutung und Konsequenzen für die Endlagerung radioaktiver Substanzen in Salzdiapiren", Z. dt. geol. Ges., 131,
433 - 459 (1980 a).

Herrmann, A.G.: "Geochemische und mineralogische Grundlagen für
die Endlagerung radioaktiver Substanzen in Salzdiapiren Nord-
deutschlands", Fortschr. Miner. 58, 2, 169 - 211 (1980 b).

Herrmann, A.G.: "Grundkenntnisse über die Entstehung von
marinen Salzlagerstätten", Der Aufschluß, 32, 45 - 72
(1981).

Hofrichter, E.: "Zur Frage der Porosität und Permeabilität von
Salzgesteinen", Erdöl-Erdgas-Zeitschrift, 92, 3, 77 - 80
(1976).

Jockwer, N.: "Laboratory investigation on the water content
within the rock salt and its behavior in a temperature field of
disposed high level waste", Scientific basis for nuclear waste
management, Boston (1980).

Jockwer, N.: "Untersuchungen zu Art und Menge des im Steinsalz
des Zechsteins enthaltenen Wassers sowie dessen Freisetzung und
Migration im Temperaturfeld endgelagerter radioaktiver Abfälle", Dissertation, Technische Universität Clausthal
(1981).

Jockwer, N.: "Transport phenomena of water and gas components
within rock salt in the temperature field of disposed high
level waste", Workshop on near-field phenomena in geologic
repositories for radioactive waste, Seattle (1981).

Jockwer, N.: "Gas production and liberation from rock salt
samples and potential consequences on the disposal of high
level radioactive waste in salt domes", Scientific basis for
nuclear waste management, Berlin (1982).

Jockwer, N.: "Laboratory investigations on radiolysis effects
on rock salt with regard to the disposal of high level radioactive wastes", Scientific basis for nuclear waste management,
Boston (1983).

Kessels, W.: "Thermomecanical calculations and constructional
criteria for an in situ test on the cracking of salt induced by
high thermal power", CEC/NEA Workshop on disign and instrumentation of in situ experiments, Brussels (1984).

Roedder, E.; Belkin, H.E.: "Fluid inclusions in salt from the Rayburn and Vacherie domes", Louisiana, US Geological Survey Open File Report, 79 - 1675 (1979 a).

Roedder, E.; Belkin, H.E.: "Application of studies of fluid inclusions in permian Salado salt, New Mexico, to problems of siting the waste isolation pilot plant", Scientific basis for nuclear waste management, $\underline{1}$, 313 - 321 (1979 b).

Tollert, H.: "Beiträge zur Porosität im Steinsalz", Kali und Steinsalz, $\underline{4}$, 2, 55 - 60 (1964).

Seismic transmission measurements in salt

B.HENTE & G.GOMMLICH
GSF, Institut für Tieflagerung, Braunschweig, FR Germany

ABSTRACT

Seismic transmission measurements offer a nondestructive possibility to investigate inaccessible areas of rock in situ. Therefore even qualitative or relative results, e.g. changes with time, are valuable. Rock parameters which affect seismic velocities (compressional or shear waves) are porosity, density of microcracks, pressure and the type of rock. The instrumentation which has been used for such measurements in the Asse mine is described as well as some newly developed components to be used in future experiments.

RESUME

Des mesures de transmission sismique offrent une possibilité non-destructive d'investiguer des parties de roches in situ inaccessibles. Ainsi des résultats même qualitatifs ou relatifs, comme changements avec le temps, sont valables. Les paramètres des roches qui influencent les vitesses sismiques (ondes de compression et de cisaillement) sont la porosité, la densité de microfissures, la pression et bien entendu le type de la roche en question. Les instruments qui ont été employés pour de telles mesures dans la Mine de l'Asse sont décrits ainsi que des éléments récemment développés qui seront utilisés dans des essais futurs.

1. INTRODUCTION

The propagation velocities of seismic waves (compressional and shear mode) in rock depend - in addition to the rock type - on several parameters, among them porosity, density of microcracks and pressure.

Although the possibilities for the interpretation of velocity measurements are limited - see for instance [1] and [2] for theoretical and laboratory based accounts, a series of in-situ-measurements has been performed in the Asse to obtain an idea of the order of magnitude of possible effects. Since seismic transmission measurements would be a non-destructive tool to investigate inaccessible areas of rock, even qualitative results are valuable.

In situ determination of seismic wave attenuation would be another challenging task in this context, but experimental difficulties as the reproducible coupling of transducers in situ still require some pre-investigations.

2. MEASURING CONFIGURATIONS

Many configurations are imaginable. The experimentally easiest one is transmission from wall to wall in a pillar. Other possibilities are wall- or floor-borehole and borehole-borehole. If at least two receivers are lined up with the source, the propagation time between them can be determined absolutely independently of a source trigger pulse. Moreover, changes in the shape of the wavelet can be detected (attenuation measurements).

If the area under investigation contains regions of considerably lower velocities as compared to the surroundings (extreme degree of fissuring, poorly condensed backfill), the interpretation of the measurements becomes difficult because indirect seismic rays may arrive earlier than the direct ones. In such cases the area must be covered by a dense fan of rays (several source and/or receiver locations). Preferably the seismic signal should be recorded simultaneously by a number of receivers.

3. SEISMIC SOURCES

The easiest way of excitation is a hammer blow on a metal anvil which is in good contact to the rock. The electric contact between hammer and anvil can be used to trigger the signal recording (time reference). At older walls or floors the coupling of the signal to the rock may be very poor by this method, because wall-parallel cracks may have formed some 10 cm behind the wall (or floor). In such cases we use drill holes of at least 50 cm depth with a metal rod in it which replaces the anvil. The hammer mostly used has a weight of 0.5 kg.

P-wave spectra due to this source have their maximum amplitudes between 1 and 10 kHz. Signal-stacking (see section 5) of five hammer-blows is sufficient to penetrate about 80 m of rock salt. In the beginning we used a 16 kg hammer, which had the disadvantage of low signal frequencies and of destroying the rock beneath the anvil.

The availability of a borehole source has been a strong requirement. Electric sparker sources, which could be purchased from the shelf, require a brine-filled borehole. Since this could influence the dynamical properties of the surrounding salt and would also render the handling in horizontal boreholes more difficult, we do not use them. In addition, from seismic investigations of another company in the Asse using sparker sources, we could observe that the excitation of shear waves was insufficient.

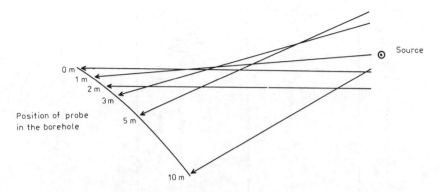

Fig. 1: Apparent ray directions as determined from observations in the borehole

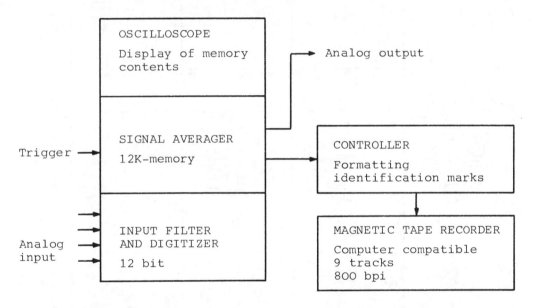

Fig. 2: Block diagram of data acquisition and averaging system

Consequently, we have constructed our own borehole source using pneumatic components. Pulses in two different directions can be generated. The diameter of the probe is 116 mm. Unfortunately, we could test it in the laboratory only so far, because the data acquisition and averaging unit (see section 5) had to be sent back to the manufacturer for repairs.

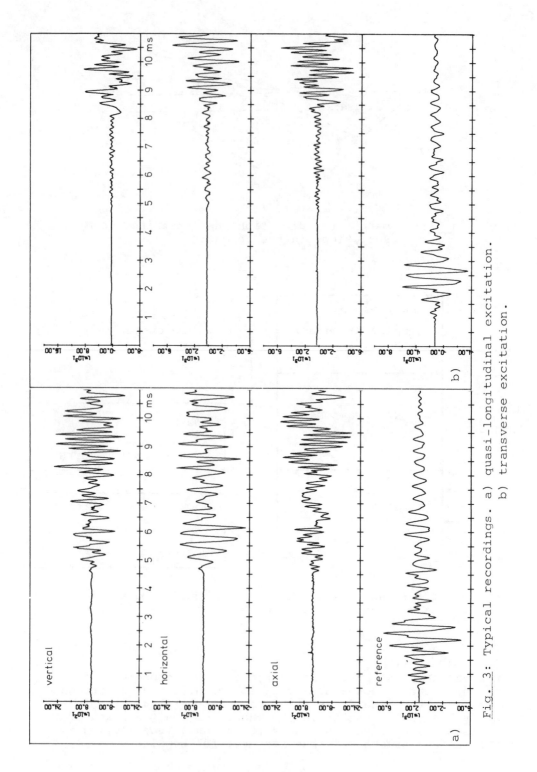

Fig. 3: Typical recordings. a) quasi-longitudinal excitation.
b) transverse excitation.

362

4. TRANSDUCERS AND COUPLING

We use piezoelectric (quartz) accelerometers with built-in amplifiers. Their sensitivity is 1000 mV/g (g = gravitational acceleration), their frequency response is flat $\pm$ 1 dB from 1 to 3000 Hz and $\pm$ 3 dB up to 5 kHz. The cylindrical transducers have a diameter of 19 mm and a height of 36 mm.

For measurements at the wall or floor three of these components are mounted perpendicular to each other on a metal cube, which in turn is fixed to an anchor cemented into the rock. Presumably due to interference from the anchor this method of coupling does not reproduce the acceleration components as they must be expected from the measuring configuration, i.e., the determination of wave directions would lead to incorrect results.

For measurements in boreholes a probe has been developed in which three accelerometer-components are mounted. The probe body consists of refined steel, has a length of 28 cm and fits to borehole diameters from 66 to 80 mm. It can be positioned in the borehole by special rods and is fixed to the borehole walls by two compressed air driven pistons.

Unlike the anchor-mounting this type of coupling provided satisfactory directional measurements as demonstrated in Figure 1.

In assessing these results the many possible sources of mistakes must be taken into account: calibration of the transducers, transverse sensitivity, cross-talk, dynamics of the probe body and the coupling to the rock.

5. DATA ACQUISITION AND AVERAGING

The storage and the in situ part of the data processing are shown by a block diagram in Figure 2. Sampling rate and input filter can be selected in accordance with the measuring needs. The trigger, based on the first electrical connection of hammer and anvil, determines the zero-time for the signal propagation. Its 90 % rise time is about 20 ns.

Four channels of data are stored in the memory providing 1024 sample points for each channel. Memory contents can be displayed in situ. In the stacking mode the growing of the signal out of the noise can be viewed immediately after each repeat of the hammer blow. When the displayed signals seem acceptable they can be recorded on digital tape for further evaluation.

As laboratory components have been used for this system, special care had to be taken to protect them from the hostile environment in the mine, e.g. dust and heat. Nevertheless, after about two years, the A/D-converter of the averager became susceptible to heat, i.e., always after half an hour of operation digitization errors occurred.

6. EXAMPLES AND SOME RESULTS

Figure 3 shows an example for a transmission measurement wall-borehole. In Figure 3a the excitation was close to the longitudinal direction (rod/anvil pointing towards receiver); in Figure 3b transverse. Shear wave arrivals can be determined unambiguously only in the latter case.

For propagation paths in mainly virgin rock salt compressional wave velocities of 4.5 to 4.6 km/s have been obtained, as was expected. Paths closer to the walls led to values of about 4.45 km/s, and close to heaters 4.13 km/s have been measured.

REFERENCES

[1] Walsh, J.B.: "The Effect of Cracks on Poisson's Ratio", J. Geophys. Res., 70 (20), 5249 - 5257 (1965).

[2] Wagner, F. C. and Engler, R.: "Zum unterschiedlichen mechanischen Verhalten von Spalten und Klüften in Gesteinsproben", Gerlands Beitr. Geophys., 92, 486 - 492 (1983).

Microseismic monitoring in salt

D.FLACH & B.HENTE
GSF, Institut für Tieflagerung, Braunschweig, FR Germany

ABSTRACT

Sudden stress releases in rock salt, which may be due to high
mechanical or thermal load, can be detected and located by a
passive seismic array. The requirements to be met by such a
system are explained, and some experiences with a system, which
is in operation in the Asse since 1980 covering the whole mine,
are described. A second array of a smaller scale is being
installed about a new thermal test field. An outline of its
design is given.

RESUME

Des détentes de contrainte subites dans le sel gemme qui peuvent
être causées par une haute charge mécanique on thermique peuvent
être détectées et localisées par une disposition passive de
mesure sismiques. Les exigences aux quelles un tel système doit
répondre sont expliquées. Quelques expériences faites avec un
système qui fonctionne dans la Mine de l`Asse depuis 1980 et qui
la recouvre entièrement, sont décrites. Une seconde disposition
sismique à une échelle plus petite est un train d`être installèe
au-dessus d`un nouveau champ d`essai thermique. Un apercu de sa
conception est donné.

INTRODUCTION

Openings in the rock or local heating lead to stress accummu-
lations in the rock, which in turn may cause sudden stress releases. If these events, which indicate the forming of new fractures or movement along existing fractures, are small, they are
called microseismic activity. Large events with visible damage
would be called rockbursts.

By seismic methods such events can be detected, counted, located, and its energy release and dimensions can be estimated.
Seismic monitoring is used in certain particularly endangered
production mines as an early rockburst warning system for many
years, see ref. [1] for a review. In the following a microseismic monitoring system, which is operated in the Asse mine
since 1980, will be described. The results, which have been obtained in the Asse so far, are reported in [2].

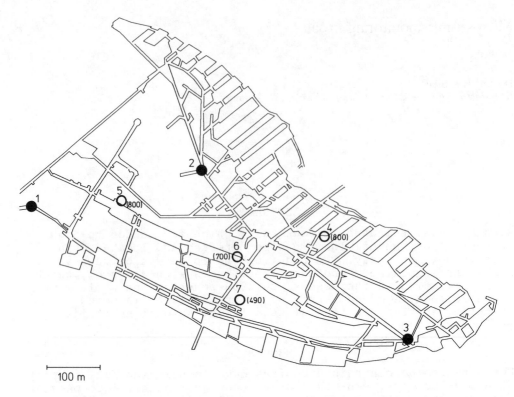

<u>Fig. 1</u>: Distribution of geophones in the Asse. Shown is
the 750-m-level, where geophones 1, 2 and 3 are
installed. The levels of the others are indicated
in brackets.

REQUIREMENTS

Source location is possible, if there are at least 4 transducers
(geophones or accelerometers). They must not be arranged in a
co-planar configuration and their coordinates must be known. In
selecting the transducer locations the local noise level should
be kept as low as possible.

The automatic event recording must start early enough for
unambiguous determination of the first arrivals in the
recordings.

INSTRUMENTATION

Originally there have been vertical geophones at 7 locations
in the Asse-mine. Recently, these have been complemented by
horizontal components at 4 of these locations. The distribution
of the geophones in the mine is represented in Fig. 1.
Three-dimensional transducers are installed at points 1, 2, 3,
and 7.

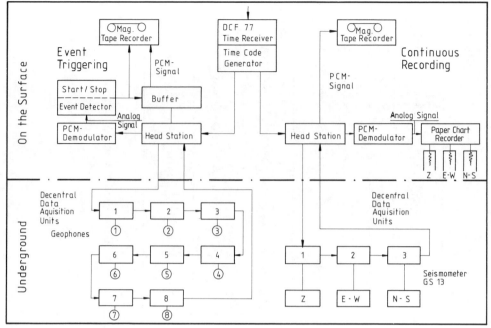

Fig. 2: Block diagram of the monitoring system

The geophones are cemented into horizontal boreholes of 1 to 2 m depth. They have a resonant frequency of 4.5 Hz, linear characteristics up to 300 Hz and a damping rate of 56 % of the critical value.

The data acquisition system uses a special PCM (Pulse Code Modulation) -technique, which requires a ring of cable connecting all stations and the head station at the surface rather than cable links from each transducer to the surface. See Fig. 2 for a block diagram. The signals are digitized at a sampling rate of 1482 Hz and with a resolution of 12 bit. The data are transmitted to the surface station continously. If a certain signal level is exceeded at two channels within a given short time window, a magnetic tape recorder (also PCM) is started. The data are delayed by a storage by 3 s before they are recorded. Thereby signal arrivals and eventual forerunners do not get lost. A time code, which is derived from a receiver of DCF 77 (German time code transmitting station), is added to the recorded signals.

When the horizontal components were added, another, independent system of the same kind was installed. Therefore the original acquisition system, which was designed for 8 channels, could remain in operation. For exact synchronisation of the horizontal and vertical recordings a redundant horizontal transducer was included in the vertical system.

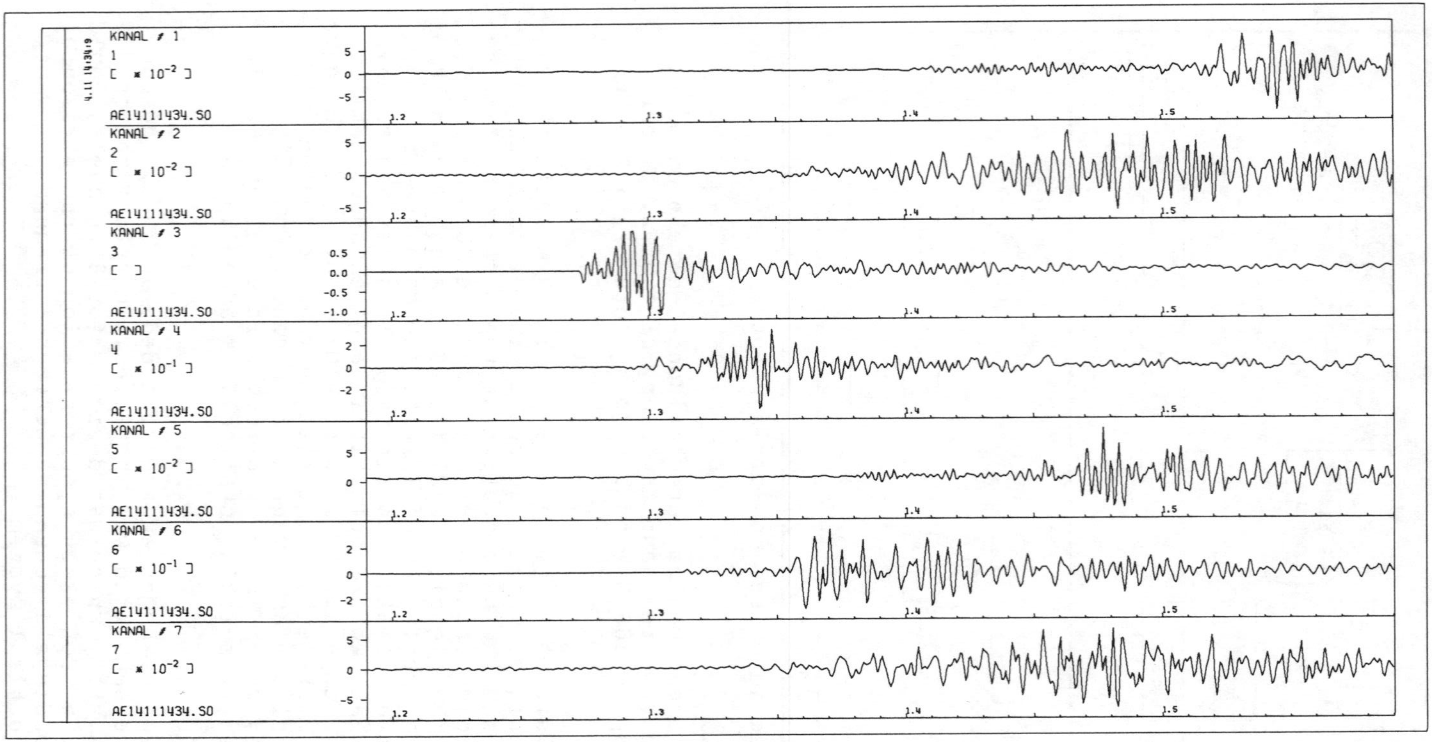

Fig. 3: Registration of a microseismic event. Times are given in s.

Another, similar monitoring system is installed about a new heater experiment. This array is of smaller dimensions (local monitoring), and it is designed for signal frequencies up to 10 kHz. Because of this high data rate an underground transient storage is used which stores the signals of a microseismic event in real time and transmits them to the surface at a slower rate.

The PCM-tapes can be read by a computer. Then plots can be obtained at arbitrary scales and the data are prepared for further signal analysis (filtering, correlation).

OBSERVATIONS

A typical microseismic event as observed by the 7 vertical geophones is shown in Figure 3. The channel numbers here correspond to the station numbers in Figure 1. The event was located about 100 m west of point 3 in Figure 1.

Spectra of spontaneous events observed by the large array (whole mine monitoring) decrease clearly below the upper frequency limit of the monitoring system (300 Hz). The local array covering a broader bandwidth has not been in operation until the time of writing, but higher frequencies are expected here due to the smaller distances.

The first arrival times are not always as well defined as in the example given in Figure 3. Gradually increasing amplitudes or ambient noise in many cases would not support automatic signal detection and processing. Because of the low rate of microseismic events in the Asse, one can do without fully automated monitoring however.

The error of source location is typically 20 m for artificial events (blasts). For locations not well within the array the error may be greater.

REFERENCES

[1] Cook, N.G.W.: "Seismicity Associated with Mining",
    Eng. Geology, 10 (2 - 4), 99 - 122 (1976).

[2] Hente, B.: "Microseismic Monitoring of a Salt Mine",
    Proc. Field Measurements in Geomechanics,
    pp. 1371 - 1379, Zurich

Thermomechanical calculations and constructional criteria for an in-situ-test on the cracking of salt induced by high thermal power

W.KESSELS
GSF, Institut für Tieflagerung, Braunschweig, FR Germany

ABSTRACT

In the Asse salt mine an in-situ-test for the heating of salt with a total heating capacity of 120 kW will be carried out over a period of approximately one year. The chief aim of this test is the investigation of possible cracking due to the thermal load on the salt. Furthermore, questions on the gas liberation, brine migration, existing compression gradients and permeability changes of the salt are being considered. Following a short presentation of the test set up, the results of thermomechanical calculations and estimations are introduced. The results of homogeneous rocks show that macroscopic cracks in unworked rocks are not to be expected during the heating phase.

According to the prototype model cracking may occur after several months in case of sudden shutdown of the heaters. Here the calculations in the centre of the heating area show a rather sharply defined region with cracking. In the case of a continuous shutdown of the heaters over a longer period of time fracturing of the salt rock will not occur.

1. INTRODUCTION

The experiment which I want to present is now being carried out in the Asse salt mine. The thermal power of this test was switched on in April of this year. The reason for performing this test is shown in Figure 1. Here a heater was implemented in the rock salt for a brine migration test [1]. Thermally induced stresses and the low viscosity of the heated salt here crushed the heater. In more detail you can see this in Figure 2. The investigations which are presented in the following are made to verify thermomechanical numerical programmes [2]; [3]. The correctness of such programmes is very important for a check on the thermal stresses caused by a realistic high-level waste (HLW) disposal in a large area of a salt mine. For the thermal visco-elastic loading of the salt we cannot find parameters which en-

able us to extrapolate the investigations of a small heated area
to a larger one in the sense of a Fourier or Rayleigh number.
The first number can be applied only in the case of thermoelastic materials [4].

2. THE TEST SITE AND ITS EQUIPMENT

In Figure 3 you can see a horizontal cross section with several boreholes for different investigations. There are stress
measurements with Sandia stress meters, with Glötzl cells, and
with flat jacks [5] which were developed at the Ift and described in my previous presentation. Complementary to the stress measurements we carry out stress and displacement measurements by
marked extensometer. On the casing of the central borehole we
measure the pressure with AWID flat jacks. Electrical conductivity measurements are carried out in four boreholes. Here we can
measure changes in the resistivity if the salt cracks or if the
brine migrates through the pressure or thermal gradient. A review on the electrical conductivity measured in situ in the Asse
salt mine is given by Kessels et al. [6]. The migration of the
brine and gas is the other topic of the investigations, as you
can see on the marked boreholes for gas analysis and permeability measurements (Fig. 3). Experiences of this are given by
Jockwer [7]. Last not least we carry out seismic measurements in
the active and passive mode. We want to measure time and location of the fractures and to determine the forms of the wave-
lets. This is of importance to identify the origin of mechanisms
[8]. In-situ informations can also be obtained on the change of
material parameters, especially in the heated range.

The vertical cross section for the extensometer measurements
is shown in Figure 4. The heaters are installed at 15 to 25 m
depth. The extensometer, which was installed with an angle of
45° with respect to layering, is imposed to measure a possible
displacement along a layer.

3. ANALYTICAL ESTIMATE OF THE INFLUENCE BY THE
VARIATIONS OF THE DIFFERENT MATERIAL PARAMETERS

When planning an experiment it is important to estimate the
dependence of the investigations considered on the various material parameters. Because we have a dependence on various material parameters we cannot check this up by means of a time and
money consuming programme.

It is, of course, clear that an analytical check can only
show the dependence qualitatively. Here we have taken a point
source with periodical variation. The solutions for the temperature and for the radial component of the tensor are given in
equation 1. (For elastics the solution is given by Nowacki [9].

371

$$T_1 = \text{Re}(T) = \frac{Q}{4\pi\kappa R} \cos\left(\omega t - R\sqrt{\frac{\omega}{2\kappa}}\right) \exp\left(-R\sqrt{\frac{\omega}{2\kappa}}\right) \ ,$$

$$\sigma_{RR} = \frac{Qm_o\mu_o}{\pi R^3\omega}\left\{\exp(-R\sqrt{\gamma/2})\ [(1+R\sqrt{\gamma/2})\sin\ (\omega t - R\sqrt{\gamma/2})\right.$$

$$\left. +R\sqrt{\gamma/2}\cos\ (\omega t - R\sqrt{\gamma/2})]-\sin\ \omega t\right\}$$

R = distance from the point source

ω = angular frequency

t = time

κ = thermal diffusivity

$m_o\mu$ = $\dfrac{1+\nu}{1-\nu}\ \alpha\mu_o$

μ_o = sliding modulus

ν = Poisson ratio

Q = power amplitude

α = thermal coefficient of expansion

γ = ω/κ

Here we had to take the complex moduli according to the principal correspondence of elastic and viscoelastic equations [10]. The measured quantity is then the real component of equation 1. The complex moduli are taken according to the Maxwell model.

$$\text{Re}(\sigma_{RR}) \sim \text{Re}(m\mu) = m_o\mu_o\ \frac{\omega^2}{\omega^2\ \beta^2}$$

$$\beta^2 = \frac{3K_o\ (\eta/\mu_o)}{4\mu_o + 3K_o}$$

η = viscosity

K_o = modulus of compression

The incorrect assumption in this model is the linearity between stress and the time derivation of the strain with a temperature independent viscosity.

In the analytical calculations we have taken the following material parameters as references.

$$E = 25 \ (Gp_a) \ ; \ \gamma = 0,25 \ ; \ Q = 120 \ kW \ ; \ \alpha = 0,45 \cdot 10^{-4} \ (K^{-1})$$

$$\kappa = 2,6 \cdot 10^{-6} (m^2/sec) \ ; \ R = 10(m) \ ; \ \mu = 10^{17} \ (\frac{N \cdot sec}{m^2})$$

$$\omega = \frac{2 \ \pi}{2 \ (a)} \ ;$$

The most important variations of the material parameters are to be found for the viscosity. A realistic stress-strain relation is given by Albrecht et al. [11] and shown in equation 4.

$$\dot{\varepsilon} = 0,419 \cdot e^{-\frac{7047 \ K}{T}} \cdot \sigma^{5.0} \frac{bar^{-5}}{d}$$

A better but more complicated dependency for secondary creep is treated by Langer [12] or Hunsche [13].
Figure 5 shows the viscosity as a function of stresses and temperature. It is calculated in accordance to equation 4.
For better understanding the relaxation times are also given.
Here we can see that the characteristic relaxation times in the heated range as in case of the high-level-waste disposal (500 K temperature and 100 bar deviatoric stresses) are in the order of a few hours.

In the colder areas around the HLW-disposal field we have relaxation times in the order of some years. For our experiment this means that we can expect only small shearing stresses in the heated area. Consequently, the heated area is similar to a fluid-filled sphere under high pressure.

In spite of this high non-linearity dependence the described analytical solution for a periodically time-dependent heat source can qualitatively describe the behaviour on the heated salt.

In Figure 6 the time- and distance dependency of the temperature is shown. The negative temperature is caused by the negative heat production in the centre at periodical variation. We can see that at 10 m the temperature is in a realistic order of magnitude and amounts to about 500 bar above the undisturbed level.

Figs. 7 and 8 show the distance and the time-dependence of the radial and tangential components of the stresses. Here, too, the stresses are in an order of magnitude of 500 bar and are therefore realistic. Nonrealistic is, of course, the tensile stress of the salt. Gessler [14] has found tensile stresses less than 25 bar.

We can see that in the cool phase tensor stresses occur in the radial direction and compressive stresses in the tangential direction. In the heating phase it is reverted. Next we look at Figure 9 which shows the displacement in an expected magnitude of some centimeters. What an extensometer measurement looks like in practice can be seen in Figure 10. This extensometer measurement was carried out by Rothfuchs [15] in the Asse salt mine. Because the extensometers are installed from the gallery we have not only the displacement caused by the heater in an infinite salt, but there we also measure the creep into the gallery. We can see this by the extensometer measurements made during the first 150 days.

The normal convergence of the gallery is also shown here. The extensometer measurements are very important for the numerical verification of this experiment. This is the best mechanical measurement which we can perform at present.

Now we estimate the stress amplitude in the dependency of the viscosity and of the elasticity modulus (Fig. 11). We can see that there is only a small range of viscosities where we have a viscoelastic behaviour with the given parameters. For viscosities of less than 10^{16} Nsec/m^2 we have only insignificant stresses. The conclusion is that here we have heated fluid and from 10^{18} Nsec/m^2 we have heated an elastic body, but this is only correct for the given parameters.

At different distances from the heated point sources we have different ranges of viscous behaviour as shown in Figure 12. In Figure 13 we see that for great time periods the stresses decrease in spite of a high energy input into the salt. That means that the slope of the power input is important for the stresses occurring in a high-level-waste disposal.
A slow filling of the high-level-waste disposal field is therefore best for avoiding high stresses.

4. NUMERICAL THERMOMECHANICAL CALCULATIONS

After this linear viscosity calculation we treat a numerical nonlinear calculation from Albers and Schlich [16] performed for the experiment presented here. In Figure 14 we can see the time-dependent temperature for three points shown in the model at the left of the illustration. The cylinder is taken to have constant heat production. The total power is 120 kW. A detailed calculation shows that we cannot heat with a power of 120 kW for a whole year because in this case the heater will reach too high a temperature. For this reason we have limited the surface temperature of the heater to 400 °C. This leads to a decrease of power to about 80 kW after one year.

The model for the mechanical calculation is shown in Figure 15. We have used a non-hydrostatic boundary condition in the calculation. This is also measured near the experimental site by Feddersen [17], but here the horizontal stress field component was lower than the vertical component, as shown in Figure 16.

In Figures 17 and 18 we can see that the maximum stresses in the centre are reached after a month. The stresses in the centre are then relatively stable with a slow decrease. A comparison of the two stress components shows that there are no differences after one year. This means that in the area of high temperature the relaxation time is so short that we can regard the salt as being a fluid. If we switch off the heater after one year we get a sharply bounded area with tensile stresses. This is very important for the seismic monitoring of the boundary of the area where fractures occur.

5. DISPLACEMENTS ON JOINTS

Last we will briefly treat the displacements on joints in the salt. These are fractures which we expect to locate with extensometer measurements and with a seismic array. We can explain such displacements as shown in Figure 19.

The heater in the origin of the marked coordinate system produces a stress field with a compressive stress in e_r-direction and a tensile stress in e_ϕ-direction. So we have, together with the superimposed geological stress field, a tangential and a normal stress on the joint. If the quotient of both is greater than the coefficient of friction we have a displacement.

$$\tau = \frac{3}{4} \, \sigma_{rr} \, \sin 2 \, \Psi \qquad \sigma = P + \frac{1}{4} \, \sigma_{rr} \, (1 + 3 \, \sin 2 \, \Psi)$$

$$\mu = \frac{\frac{3}{4} \, \sin 2 \, \Psi}{-\frac{P}{\sigma_{rr}} + \frac{1}{4} \, (1 + 3 \, \cos 2 \, \Psi)} \qquad \sigma_{rr} = - \frac{a}{r^3}$$

Figure 20 represents the quotient of tangential and normal stresses along two joints for a heater test with 60 kW and for a surrounding hydrostatic stress of 170 bar.
Here we can see that at a perpendicular layer distance of 10 m from the heated center, a joint will be displaced if the coefficient of friction is less than 0.3.

We think that this experiment in this sense is good for verification of the validity of the numerical programmes in their theoretical basis and in their numerical accuracy.

REFERENCES

[1]   Rothfuchs, T.: "Thermophysical In-Situ-Investigations on HLW-Disposal in Rock Salt", GSF/IfT Annual Report, Commission of European Communities, 1983.

[2]   Wallner, M..; Wulf, A.: "Thermomechanical calculations
         concerning the design of a radioactive waste repository in
         rock salt", ISRM Symposium, Aachen. Publ. A.A. Balkema,
         Rotterdam, for Deutsche Gesellschaft f. Erd- und Grundbau,
         Essen, 1982.

[3]   Albers, G.; Ehlert, C.: "Zur Entwicklung von Rechenmo-
         dellen zur Abschätzung der thermomechanischen Auswirkungen
         eines Endlagers für radioaktive Abfälle im Salzgebirge",
         ISRM Symposium, Aachen. Publ. A.A. Balkema, Rotterdam, for
         Deutsche Gesellschaft f. Erd- und Grundbau, Essen, 1982.

[4]   Kopietz, J.; Eisenburger, D.; Koss, G.: "Geophysikalische
         Untersuchungen an Salzformationen im Hinblick auf die End-
         lagerung hochradioaktiver Abfälle in Geologischen Körpern
         (Geothermik)", Abschlußbericht zum Forschungsvorhaben
         SR 173, BGR, Hannover, 1983.

[5]   Kessels, W.; Flentge, I.; Kolditz, H.; Muth, M.: "AWID,
         ein absolut messendes Druckkissen nach dem Kompensations-
         verfahren im Druck", Report GSF/IfT, T-174, 1983.

[6]   Kessels, W.; Flentge, I.; Kolditz, H.: "D.c. Geolelectric
         Soundings for Determining Water Content in the Salt Mine
         Asse, (FRG)", Report No. T 166. To appear in: Geophys.
         Prospecting, 1983.

[7]   Jockwer, N.: "Untersuchungen zur Art und Menge des im
         Steinsalz enthaltenen Wassers sowie dessen Freisetzung und
         Migration im Temperaturfeld endgelagerter radioaktiver Ab-
         fälle", GSF/IfT-Bericht T 119, 1981.

[8]   Hente, N.; Gommlich, G.; Flach, D.: "Mikroseismische Über-
         wachung von Bergwerken zur Tieflagerung radioaktiver Ab-
         fälle", Glückauf-Forschungsheft 2, 1982.

[9]   Nowacki, W.: "Thermoelasticity", International Series of
         Monographs on Aeronautics and Astronautics, I: Solid and
         Structural Mechanics, 1962.

[10]   Christensen, R. M.: "Theory of Viscoelasticity", Academic
         Press, 1971.

[11]   Albrecht, H.; Meister D.; Wallner, M.: "Bestimmung geo-
         technischer Kannwerte von Salzgesteinen", 3. Nat. Tagung
         über Felsmechanik, Aachen, 1978.

[12]   Langer, M.: "Felsmechanische Probleme bei der Errichtung
         von Speicherkavernen", ISRM Symposium, Aachen, Publ. A.A.
         Balkema, Rotterdam, for Deutsche Gesellschaft f. Erdu.
         Grundbau, Essen, 1982.

[13]   Hunsche, U.: "Ergebnisse der Untersuchungen zum Festig-
        keits- und Fließverhalten von Steinsalz", 5. Zwischen-
        bericht zum Forschungsvorhaben SR 138, Bundesanstalt f.
        Geowissenschaften und Rohstoffe, Hannover, 29 S., 1980,

[14]   Gessler, K.: "Vergleich der einaxialen Zugfestigkeit mit
        der Dreipunkt-Biegezugfestigkeit und unterschiedlichen
        Spaltzugfestigkeiten", Kali u. Steinsalz, 8, Nr. 12, 1983.

[15]   Rothfuchs, T.; Schwarzianeck, P.; Feddersen, H.: "Simula-
        tionsversuch im Älteren Steinsalz Na2β im Salzbergwerk
        Asse", GSF/IfT Abschlußbericht, 1983.

[16]   Albers, G.; Schlich, M.: "Thermomechanische Belastungen
        eines Salzstocks im Umfeld von Einlagerungszonen bzw.
        Versuchsanordnungen", 1. Halbjahresbericht GSF/IfT zum
        EG-Vertrag, 1983.

[17]   Feddersen, H.-K.; Frohn, C.: "Absolutspannungsmessungen im
        nichtaufgeheizten Gebirge", 1. Halbjahresbericht GSF/IfT
        zum EG-Vertrag, 1983.

Figure 1: Survey of a crushed heater in the heated salt

Figure 2: Detailed picture of a heater crushed by the heated salt

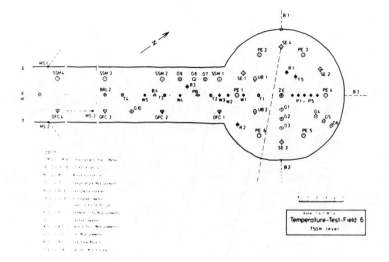

Figure 3: Horizontal cross-section of the test site TV6 in the Asse salt mine

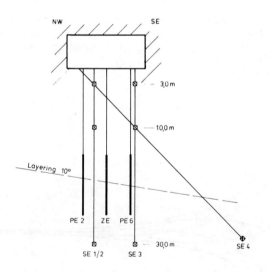

Figure 4: Vertical cross-section of the test site TV6 in the Asse salt mine with extensometer boreholes

379

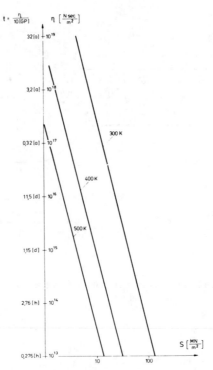

Figure 5:   The viscosity of salt for different deviatoric
stresses and temperatures. It is calculated by a
linearization of the nonlinear creep law of Hunsche
(1981)

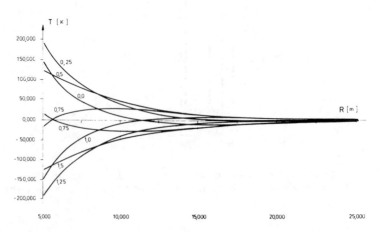

Figure 6:   The time (in years) and distance dependency of the
temperature for the given parameter

380

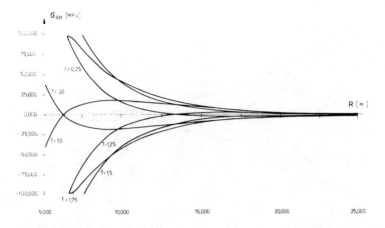

Figure 7: The time (in years) and distance dependency of the radial stresses for the given parameter

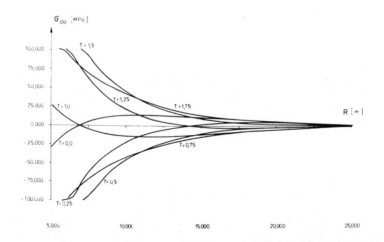

Figure 8: The time (in years) and distance dependency of the tangential stresses for the given parameter

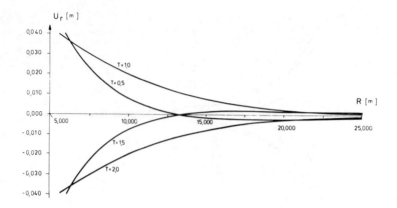

Figure 9:   The time (in years) and distance dependency of the displacement for the given parameters

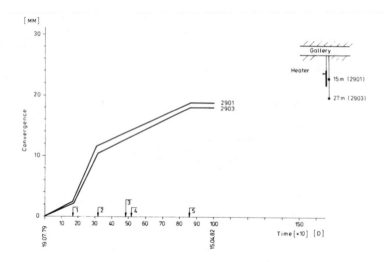

Figure 10:   Extensometer measurement during a heater test in the
Asse salt mine carried out by Rothfuchs et al. (1983)
1. Heater switched on with 100 % power
2. Heat load amounting to 60 % of the starting power
3. Heat load amounting to 57 % of the starting power
4. Heat load amounting to 44 % of the starting power
5. Heater switched off

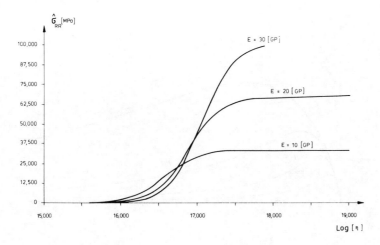

Figure 11: The amplitude of the radial stress component in its dependency from the viscosity and the elastic modulus

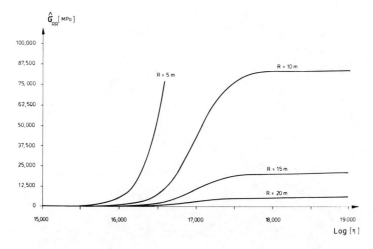

Figure 12: The amplitude of the radial stress component in its dependency from the viscosity and distances

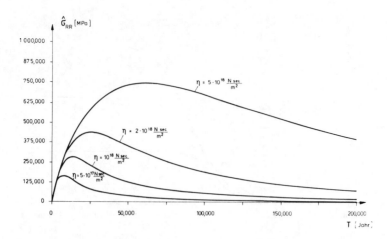

Figure 13: The amplitude of the radial stress component in its dependency on the periodic length and the viscosity

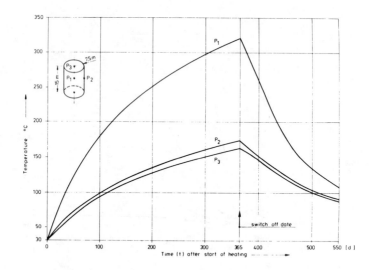

Figure 14: Temperature versus time at homogeneous heating with 120 kW in the cylinder shown at marked points

384

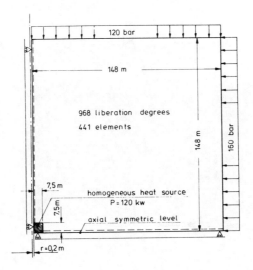

Figure 15: Model for the thermomechanical calculations

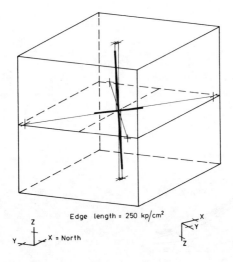

Figure 16: Results of a triaxial measurement in the Asse salt
mine (carried out by Firma Interfels)

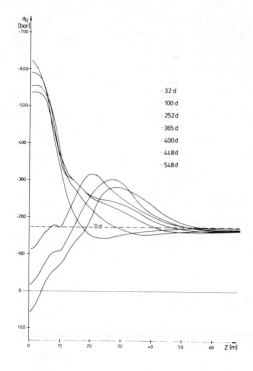

Figure 17:
The distance dependency
of the radial stress com-
ponents at different times

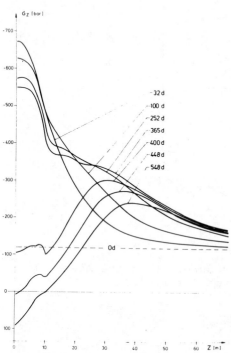

Figure 18:
The distance dependency
of the tangential stress
components at different
times

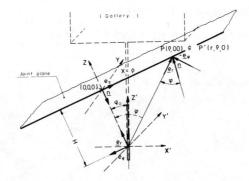

Figure 19: Model for displacements along joints under a spherically heated range

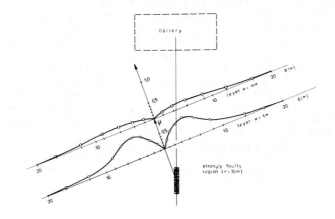

Figure 20: The quotient of tangential and normal stress along two joints for a heater test with 60 kW and for a hydrostatic stress of 170 bar

Testing of an absolute measuring flat jack according to the compensation method (AWID)

W.KESSELS

GSF, Institut für Tieflagerung, Braunschweig, FR Germany

ABSTRACT

A measurement principle for a flat jack is being introduced by means of which stress measurements may be carried out without requiring any material paramaters of the flat jack for the measurements. No calibration measurements are needed for this absolute measuring flat jack. A temperature dependence does not exist. A leap in the electrical resistance of the flat jack, which is brought about by two metal sheets separating when being pumped up with hydraulic oil at reaching of external pressure, is evaluated. Besides theoretical considerations on the mode of operation of the flat jack, laboratory tests were performed in an autoclave as well as in a tube filled with salt grit under a uniaxial press. The error which arose during measurements in the oil pressure autoclave was due to the magnitude of the reading accuracy of the measuring devices, i.e., it was smaller than 0.3 % at 150 bars. This is to be considered as the error which may be assigned to the flat jack.
Finally we treat an arrangement of six flat jacks for determining stress tensors in the heated area of the Temperature Test - Field (TV 6) in the Asse salt mine.

1. INTRODUCTION

For the disposal of high-level-waste (HLW) in geological formations knowledge of the thermomechanical stresses is of particular importance. This is related to the whole area of the disposal field and also to the pressure on the casing of the HLW.

A measurement of these stresses is difficult because the temperature field and the stress field vary with time [1]. So we get time dependent internal stresses near the equipment installed for making measurements in a borehole. These stresses are caused by the different thermomechanical behaviour of the equipment and of the geological formation. For this reason we have developed a flat jack for direct and temperature independent stress measurements at the Institut für Tieflagerung in Braunschweig. This is called AWID (Absolut-Widerstandssprung-Druckmeßkissen) [2]. The principal function of the flat jack is to determine the point of separation of two thin metal sheets by a step function. Natau determined this function from a strain measurement [3]. Here we take an electrical resisivity for the signal function. The advantage of this is insensitivity as to mechanical, chemical, or physical load [4].

The flat jack was tested by laboratory experiments. Now it is installed in the heated area of the Temperature Test Field 6 (TVF6), which is described in this section. Here we want to measure the pressure on the casing of the central borehole by using three flat jacks. At some distance from this we measure the stress tensor with six flat jacks. For this we have carried out some statistical and numerical calculations to find the best arrangement. A similar treatment for overcoring measurements is to be found in Gray et al [5].

2. PRINCIPLES

In Figure 1 the construction principle of an AWID flat jack is shown. The metal sheets A/B are soldered at the edges so that they are leak-proof for oil. Two of the four contacts are for constant current input and two for potential measurements. If the pressure between A and B reaches the loading on the outside of the flat jack we get a leap in the potential measurement. The pattern of the current lines before and after the leap is shown in Figure 1; it explains the resistivity change.

Figure 2 shows a first AWID measurement under hydraulic oil pressure. Here we have used two metal sheets with different thicknesses (2.4 mm and 0.2 mm). The stiffness of this was so high that we could not get a sharp leap. But we get the same resistivity when the hydraulic oil pressure on the outside is reached. The flat jack is deformed from measurement to measurement. We can see this by comparison of measurement I and II, which were made at the same pressure. II was carried out later.

In Figure 3 the complete measuring technique is shown. A motor drives the oil pump with constant speed. The pressure of the pump is recorded on the X-coordinate of the recorder and the resistivity on the Y-coordinate.

3. LABORATORY INVESTIGATIONS

Laboratory measurements are carried out in the high pressure autoclave and under a press in salt grit. The principle of autoclave measurement is shown in Figure 4. In Figure 5 is shown a measurement made according to the latest state of engineering. Here we can see a sharp leap when the hydraulic pressure at the outside is reached. The error of the AWID measurements has the same level as we find with the usual pressure measurements. The pump is driven by a motor as is shown in Figure 3. Important for the accuracy is the velocity of pumping. Too high velocities give too high pressures at the point of the resistivity leap. This is caused by the hydraulic flow resistance between the two metal sheets. Measurements which investigate this are shown in

Figure 6. If the velocity of the motor is too high the leap occurs at a too high pressure. In Figure 7 we can see the linear dependency of the velocity of revolution on the difference between measured and true pressure. The accuracy of the shown experiments in the hydraulic autoclave is the accuracy of the AWID flat jack.

Other measurements were carried out under a press which is shown schematically in Figure 8. Here salt grit was filled into a greased tube. The salt grit was pressed by the two stamps. The function of the flat jack was tested up to 300 °C. Here no deterioration in accuracy was found. Further moistening of the salt with brine did not diminish accuracy.

In the measurements in salt grit we cannot find as sharp a leap as in the measurements in the high pressure autoclave. In Figure 9 such a measurement is shown. Here we also get a high accuracy. At a pressure of 350 bars, however, we have a leakage. This is compensated by a higher pumping velocity. The first measurements were carried out with a hand pump. This resulted in more erroneous results, as can be seen in Figure 10. These results, however, have errors of less than 5 % and can be considered as good. In Figure 10 we can see that there was high friction between salt and tube. In Figure 10, in a second cycle, we measured more than 100 bars with the flat jack, using a stamp pressure of 50 bars.

4. INSTALLATIONS IN THE MEASUREMENT FIELD

When determining the stress tensor particular problems occur through use of the flat jack in a small field area. Three of the most vital problems are to be characterized as follows:

1. The stress field which is disturbed by the installation of the flat jack must reassume the unperturbed state of stress after surface contact of the rock with the measuring device through relaxation processes.

2. Time-dependent strongly variable stress and temperature fields can build up time-dependent variable internal stresses around the flat jacks.

3. The arrangement of the flat jacks must enable determination of the whole stress tensor with a high degree of accuracy.

Points 1 and 2 are not of such great importance for the intended measurements in Temperature Test 6, as the determination of the stress tensor is carried out in the area with a maximum temperature. Here the relaxation rate of the salt is so short that a very fast force contact has to be expected.

These short relaxation rates favour the decrease of the perturbation stresses which are to be found in the immediate vicinity of the flat jack.

390

The thin, altogether only 0.4 mm thick flat jacks prevent the buildup of large internal stresses due to their only slight stiffness.

Point 3 is to be considered in more detail in the following.

The aim here was to find an optimum arrangement for six flat jacks in order to determine the stress tensor. A statistical evaluation with a number of more than six flat jacks is not to be carried out in this case in order to keep the measuring efforts at a reasonable level.

In order to carry out an optimum calculation with variations of all free parameters it is found that the 18 free parameters to be included in the optimization (three components of the six normal unit vectors of the flat jacks) would call for an unwarranted calculational effort.

Seen from a more general point of view an optimum arrangement of the flat jacks is to be found under the aspect that the stress tensor to be determined is completely unknown. It may be assumed that the flat jack determines the normal component of the stress vector and thus a directional quantity.

This spatial orientation of the flat jack permits the conclusion that a spatially oriented information can be determined by the flat jack. As it was assumed, however, that the stress tensor to be determined is unknown, it appears sensible to consider the area in its entire information content as isotropic.

Therefore, the optimum arrangement for the six flat jacks is the one in which the same solid angle is assigned to each normal vector of a flat jack. Thus a regular polyhedron arrangement is automatically obtained in the case of the six flat jacks of a dodecahedron, as is illustrated in Figure 12.

For an additional revision of this flat jack arrangement, set up according to symmetry considerations, the "evaluations programme" was developed which permits a tensor determination according to equation 1:

$$P_i = (\hat{\sigma} \, \underline{n}_i \,) \cdot \underline{n}_i$$

P_i = pressure at the flat jack

$\hat{\sigma}$ = tensor of stress $\qquad\qquad$ (1)

$\underline{n}_i$ = normal unit vector of the flat jack

This programme was then enlarged into a statistics programme in order to determine the influence of fluctuations of various factors.

An error was introduced by means of a Gaussian distribution with a random number generator into the stress values near the

391

flat jacks, calculated for a given stress tensor. These errors
were introduced independently for each of the six flat jacks.
Figure 13 shows a distribution of errors as well as the fluctuations of the determined stresses of the principal axes. The distribution is ascertained from 100 evaluations. In order to simulate an unknown position of the stress tensor it was rotated by
the angle beta (Figure 12) in the range of 0° up to 90°.

During this rotation the position of the X-axis in direction
of the smallest main axial stresses was maintained. This calculation was then plotted as a function of various angles alpha
(see Fig. 12). Figure 14 illustrates standard deviations determined above the angle alpha which are to be considered as errors. It is shown that the error of the main normal stresses determined exhibits a distinct minimum in Z- and Y-direction at
30 °. This emphasizes the correctness of the previous symmetry
considerations. For sigma$_X$ there was a minimum at 15°. The arrangement of the flat jacks was turned only around the X-axis.
The X-axis of the flat jack arrangement was fixed in the direction of the smallest main normal stress.

Of particular interest is that the error of alpha = 20° changes only slightly up to alpha = 40°, i.e., the arrangement of
the flat jacks supplies relatively stable results as compared to
small deviations from the nominal range of alpha = 30°.

Figure 15 shows the mean values of the main normal stresses
plotted in 100 calculations. A deviation of the mean values from
the reference stresses characterizes a non-linear error dependency. The errors (statistical deviations) propagate within a
range of 15 - 60°, mainly in a linear direction.

REFERENCES

[1]   Wallner, M.; Wulf, A.: "Thermomechanical calculations con-
      cerning the design of a radioactive waste repository in
      rock salt", ISRM Symposium, Aachen. Publ. A.A. Balkema,
      Rotterdam, for Deutsche Ges. f. Erd- und Grundbau, Essen
      (1982).

[2]   Kessels, W.; Flentge, I.; Kolditz, H.: "Patentanmeldung
      über eine Vorrichtung zur Bestimmung eines Spannungszu-
      standes", amtliches Aktenzeichen P 3403521.4 (1984).

[3]   Natau, O.: "Ein praktisch verformungsfrei messendes Element
      zur Bestimmung von Spannungen im viskoelastisch-plastischen
      Gebirge", Sitzungsberichte des 2. Kongresses der Inter-
      nationalen Gesellschaft für Felsmechanik, Belgrad, 1970,
      Bd. 4 (1970).

[4] Kessels, W.; Flentge, I.; Kolditz, H.; Muth, M.: "AWID, ein
     absolut messendes Druckkissen nach dem Kompensationsver-
     fahren im Druck", Report GSF/IfT, T-174 (1983).

[5] W.M. Gray, N.A. Toews: "Analysis of Accuracy in the Deter-
     mination of the Ground-Stress Tensor by Means of Borehole
     Devices," Status of Practical Rock Mechanics, Proceedings
     of the Ninth Symposium on Rock Mechanics held at the Colo-
     rado School of Mines, Golden, Colo., April 17 - 19, 1967.

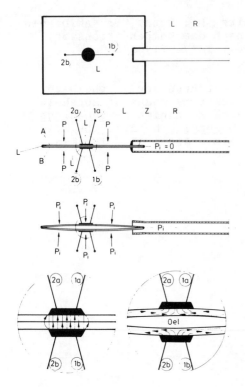

Fig. 1 [QL]
The AWID Flat jack during the beginning and after the completion of the measuring process.

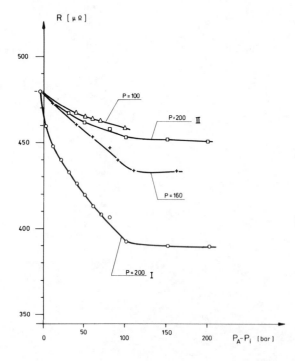

Fig. 2 [QL]
Measuring pressure with an AWID flat jack comprising one thick and one thin metal sheet, for different pressures. Measurement II was carried out after measurement I.

P_A: pressure in autoclave

P_i: pressure in flat jack

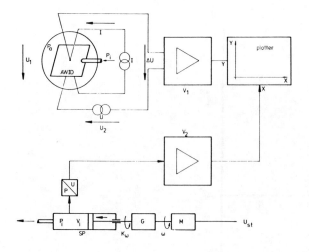

P_a - external pressure

P_i - internal pressure on the AWID flat jack

$V_{1,2}$ -   DC - amplifier

G - gearing

M - DC - motor

SP - high-pressure screw press

Fig. 3   Schematic representation of arrangement for measuring with the AWID flat jack.

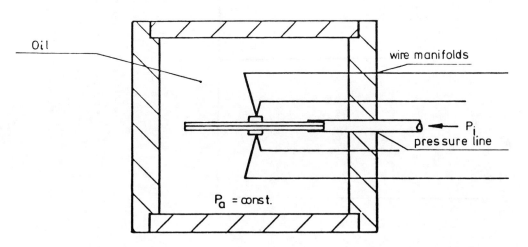

Fig. 4   Schematic representation of arrangement for testing the AWID flat jack in an oil-pressurized autoclave.

U 2a, 2b

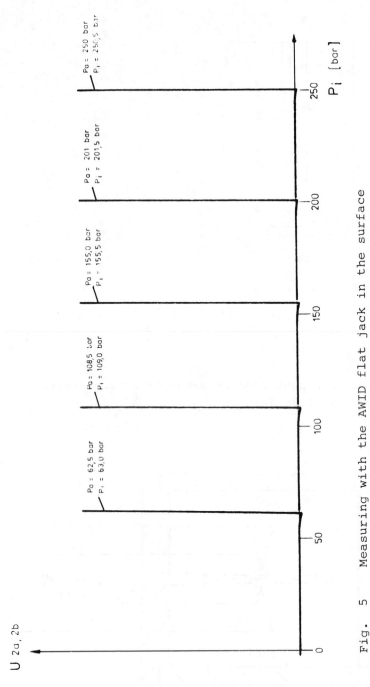

Pₐ = 62,5 bar
Pᵢ = 63,0 bar

Pₐ = 108,5 bar
Pᵢ = 109,0 bar

Pₐ = 155,0 bar
Pᵢ = 155,5 bar

Pₐ = 201 bar
Pᵢ = 201,5 bar

Pₐ = 250 bar
Pᵢ = 250,5 bar

0   50   100   150   200   250 P_i [bar]

Fig. 5   Measuring with the AWID flat jack in the surface
pressure apparatus of the IfT, with low pump speed
and with suppressed zero point at the beginning of
the measurement.

P_A: pressure adjusted in the autoclave

P_i: pressure determined with the flat jack

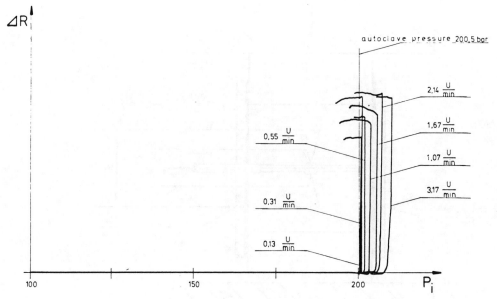

Fig. 6 Measurement with AWID flat jack in oil-pressurized autoclave, with constant pressure inside the auto-clave, for different pumping speeds, as characterized by the rotation speeds of the pump motor.

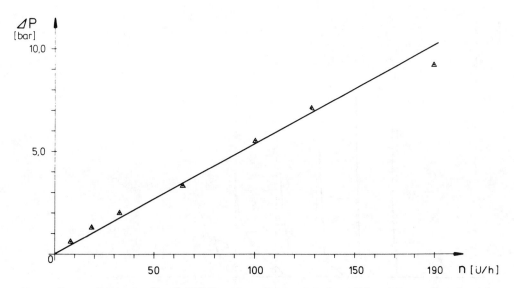

Fig. 7 Error P of AWID measurements in oil-pressurized auto-clave at 100 bars, as a function of the rotation speed of the pump motor.

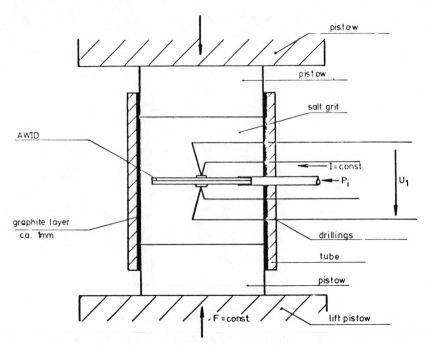

Fig. 8 Schematic representation of arrangement for testing the AWID in salt grain in a uniaxial press.

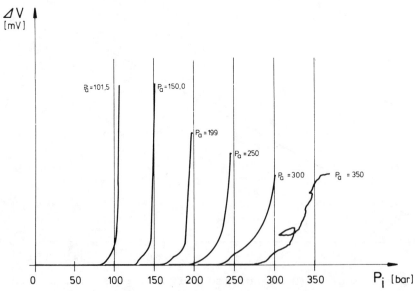

Fig. 9 Measuring with the AWID in salt grain, using a motor-driven pump.

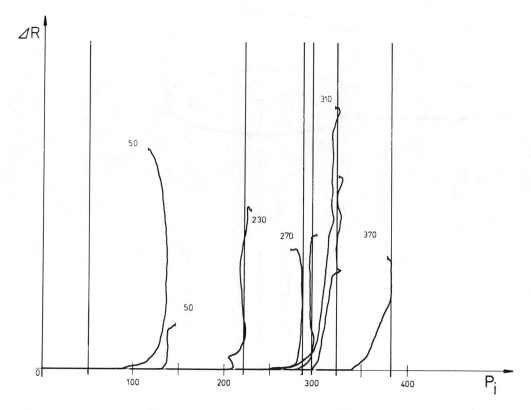

Fig. 10   Testing the AWID in salt grain in the uniaxial press
of the IfT. The measurement was carried out with a
hand pump. The especially high pressures of more than
100 bars for an adjusted external pressure of 50 bars
resulted from the fact that for this measurement the
measuring cycle was run through the second time,
and the salt grain was locked up in the tube.

399

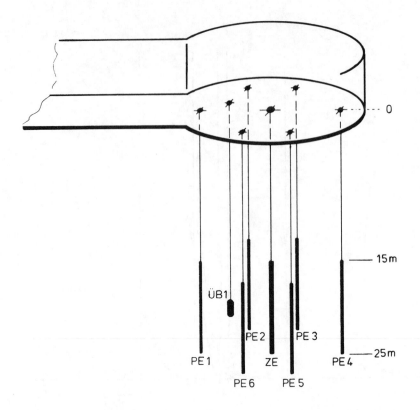

Fig. 11   Measuring site for determining the stress tensor
with the AWID flat jack (ÜB 1) in Temperature Testing
Area 6 in the Salt Mine Asse.

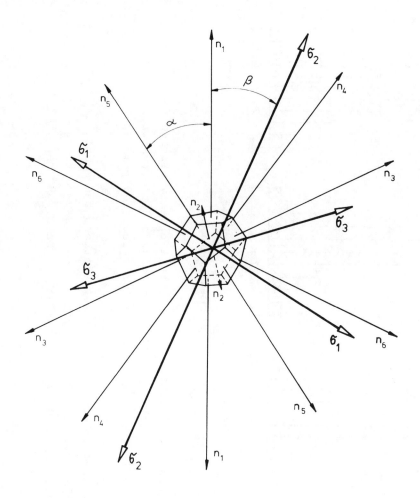

Fig. 12   Optimal arrangement of 6 flat jacks for determining
          the 6 components of the stress tensor. The surfaces
          of the dodecahedron characterize the ranges of
          influence of the flat jacks. Their unit normal vectors
          are designated with n_i. The principal axes of an
          assumed stress tensor are designated with sigma.

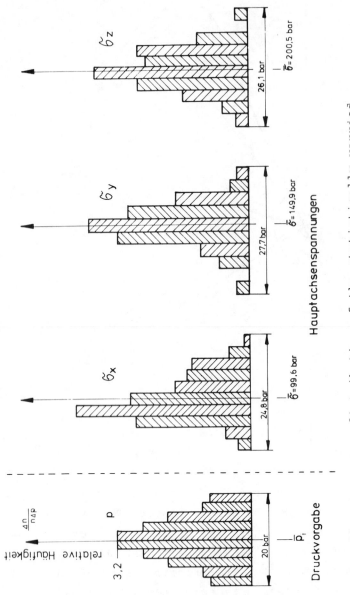

Fig. 13 Frequency distribution of the statistically varried adjusted pressures at the flat jacks and the resulting statistical distributions of the calculated principal stresses, with the flat jacks arranged in a dodecahedron array. In all calculations the magnitudes of the system of the principal axes (assumed to be without errors) were fixed (at 100 bars, 150 bars, 200 bars).

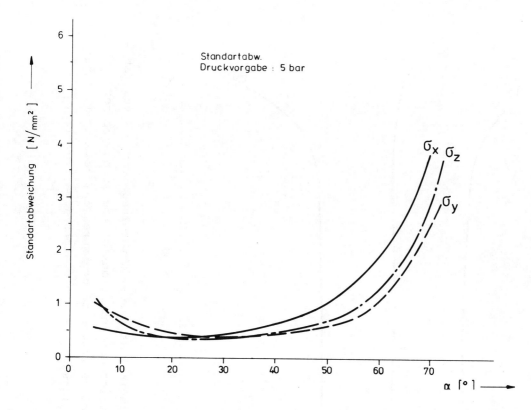

Fig. 14   Error (standard deviation) for a determination of
the stress tensor as a function of the angle alpha
(see Fig. 12), for an error in pressure measurement
with a standard deviation of 5 bars.

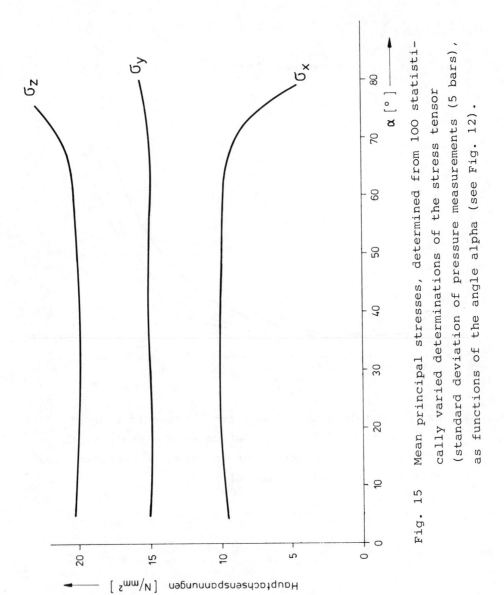

Fig. 15   Mean principal stresses, determined from 100 statistically varied determinations of the stress tensor
(standard deviation of pressure measurements (5 bars),
as functions of the angle alpha (see Fig. 12).

In-situ heating experiments for the production and analysis of critical stresses

J.KOPIETZ & D.MEISTER
Bundesanstalt für Geowissenschaften und Rohstoffe (BGR), Hannover, FR Germany

ABSTRACT

To study the conditions under which thermally induced fracturing in salt
formations may occur and how it can be reliably demonstrated, a large-scale
in-situ heating experiment is presently being set up by the BGR in the Asse
salt mine. Various geophysical and geomechanical methods of field measurements will be used to obtain the experimental data base for a numerical
analysis of critical stresses. Finite element calculations are used to
precalculate the temperatures, strains, and stresses, to establish design
criteria for this experiment, and for comparison with the measurements.
These investigations will be also used for improving and verifying of
numerical models used in the thermomechanical calculations.

1. INTRODUCTION

Discussions of possible failure scenarios for a flooded mine gave
rise to a series of thermomechanical in-situ experiments started several
years ago by the BGR in the Asse salt mine to study the problem of whether
the heat released from high-level radioactive waste might cause thermally
induced fracturing yielding pathways through which water or brine could in-
trude into the repository and come in contact with the waste canisters [1].

In the course of these heating experiments, a distinct horizontal
fracture was produced when the heater was turned off after a 50-day-heating
period at a heating level of about 60 % of the initial one [2]. This fracturing resulted from an extremely high rate of cooling of the heated rock
salt mass due to the abrupt end of heat production, which would not occur in
a real repository. However, experiments producing failure of the rock could
be an important aid for the validation of safety reserves with regard to the
limit of thermomechanical loading to be established for repository design.

Hence, a large-scale in-situ heating experiment of this kind is
currently being set up by the BGR in the Asse salt mine, to obtain a comprehensive experimental data base for an analysis of the geophysical and geomechanical conditions governing the development of critical stress.

Finite element calculations of the temperatures, displacements, and
stresses to be expected in the course of this experiment are used in establishing design criteria for this experiment, especially the heating pattern
to be applied. The interpretation of these investigations will be done

mainly by comparing the results obtained from the model calculations with those of the measurements, and this comparison in turn will be used to improve and verify numerical models on which the thermomechanical calculations are based.

2.   TEST FIELD

A single heater, as in the earlier test, is being used in the thermo-mechanical in-situ heating experiment presently being set up by the BGR in the Asse salt mine. For this experiment, a drift has been excavated at the 800-m level of the Asse mine (Stassfurt rock salt) about 20 m long, 6.6 m wide, and 5.1 m high. As shown in Fig. 1, the borehole configuration comprises about 20 observation boreholes around a central one containing the electrical heater. The boreholes are 30 to 35 m deep.

3.   HEATER DESIGN

The design of the electrical heater is based on commercial mineral-insulated heating cables sheathed with Inconel and with a conductor of nickel chrome. This cable, 6 mm in diameter, is mounted in loops on a cylindrical surface 100 mm in diameter and 20 m long within the steel casing of the borehole. The hollow space between the cables will be filled with glass beads to avoid convection of air in the borehole; in this way the thermal properties of vitrified high-level waste are roughly approximated. The heater is dimensioned for a safe maximum power output of 50 kW which corresponds to 2.5 kW per meter.

4.   MEASUREMENT PROGRAM

To obtain as comprehensive a data base as possible for a quantitative analysis of critical stress, the rock-mechanical and geophysical measurement program of this experiment has been considerably extended in comparison to those of the earlier, more qualitative experiments. The methods to be used were tested for the most part during a preliminary experiment. In accord-ance with the objectives of this large-scale experiment, the main emphasis has been put on rock-mechanical measurements of high accuracy.

4.1   Temperature Measurements
To monitor the temperature field in the course of the experiment and to obtain the necessary correction values for the instruments, temperature measurements will be performed in the boreholes and on the surface of the test field at about 60 points. Platinum resistance sensors (Pt 100) will be used to measure temperature (4-wire resistance measurements). This will be done to achieve a high reliability of the temperature data, which are an im-portant basis for the evaluation and interpretation of all the other data. The protecive mantle for the sensors consists of Inconel metal and that for the leads of Teflon. By calibration at the site, an accuracy better than 0.1 °C is obtained. The temperature in the heater assembly is measured with thermocouples (NiCu-Ni) which are mineral-insulated and sheathed with Inconel.

4.2   Heater Power
The thyristor-controlled input power of the electrical heater is measured with a commercial 3-phase power transducer circuit. In addition,

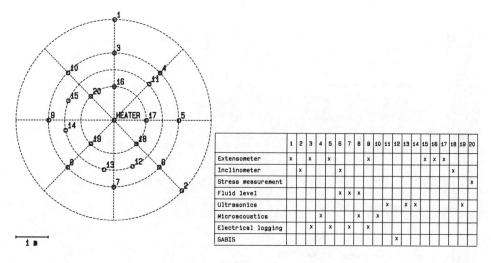

Fig. 1 Borehole arrangement in heating experiment TMV IV

the incremental input power (kWh) is measured with an eddy current meter in connection with a digital counter.

4.3   Measurement of Fluid Levels

The occurrence of thermally induced fracturing will be monitored by measuring the levels of oil in a number of the observation boreholes. Commercial piezoresistive pressure transducers will be used for this purpose, as in earlier experiments.

4.4   Displacement measurements

The deformation of the rock will be determined using convergence, extensometer and inclination measurements, as well as mine surveying. The convergence of the gallery is being measured between points in the roof and the floor, as well as between the two walls in several profiles. The measuring is done by hand using a measuring tape of invar steel. The accuracy is about ± 0.1 mm. The extensometer measurements record movements in the interior of the rock, especially in the thermally affected floor, but also in the roof and walls. Seven multiple-position rod extensometers (BGR system) were installed. The displacement of the extensometer rods with respect to a reference point at the extensometer head will be automatically measured by an inductive displacement transducer. In addition, a new single-position extensometer using glass fiber rods as shown in Fig. 2 has been installed for the purpose of instrument testing. The horizontal components of displacements will be measured in three boreholes with inclinometer using the principle of a pendulum. For the mine surveying, all of the extensometer heads will be used as reference points connected with a reference point outside the test area; the measurements will be carried out using a hydrostatic tube balance with inductive displacement transducers.

4.5   Stress Measurements

To analyse the initial state of stress in the unheated rock mass of the test field, short-term stress measurements have been made using the

407

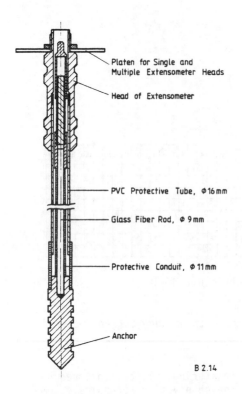

Platen for Single and
Multiple Extensometer Heads

Head of Extensometer

PVC Protective Tube, Φ16mm

Glass Fiber Rod, Φ9mm

Protective Conduit, Φ11mm

Anchor

B 2.14

Fig. 2
Single-position borehole
extensometer (Glötzl)

overcoring method with inductive displacement transducers (BGR system) and
dilatometer measurements [3]. The stresses induced by heating and subsequently cooling of the rock mass in the course of the heating experiment
will be determined by long-term measurements with hydraulic pressure cells
(type Glötzl). Figure 3 illustrates the methods schematically.

4.6   Ultrasonic Measurements and Acoustic Emission
In three boreholes of the test area, hole-to-hole ultrasonic measurements will be made to detect the formation of fissures and changes in
the acoustic parameters of the rock salt. The arrangement of the boreholes
has been selected so that measurement can be made in the immediate vicinity
of the heat source. The measuring system was developed by the BGR. The
precision of the traveltime measurements is about ± 5.0 µs. The acoustic
release of energy during the heating of the rock and the subsequent cooling
will be recorded by a 2-channel monitoring system, also developed by the
BGR. The event rate, the amplitude frequency, the duration of the events,
and the chronological sequence of events can be measured using this equipment, the schematic diagram of which is shown in Figure 4.

4.7   Geoelectric Borehole Logging
Geoelectrical borehole logging methods suitable for monitoring the
formation of fractures in salt rock are being developed by the BGR for use
in this heating experiment. A DC 2-electrode method has been developed
based on the measurement of the current fed at constant voltage into the
rock by a movable borehole probe. Figure 5 illustrates these measurements

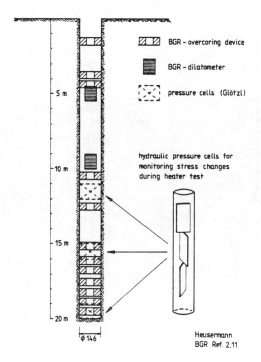

BGR – overcoring device

BGR – dilatometer

pressure cells (Glötzl)

hydraulic pressure cells for
monitoring stress changes
during heater test

Heusermann
BGR Ref. 2.11

Fig. 3
Stress-measurement devices

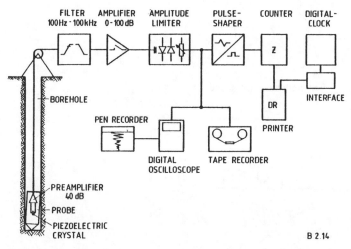

FILTER
100Hz - 100kHz

AMPLIFIER
0 - 100 dB

AMPLITUDE
LIMITER

PULSE-
SHAPER

COUNTER

DIGITAL-
CLOCK

Z

INTERFACE

BOREHOLE

PEN RECORDER

DR

PRINTER

DIGITAL
OSCILLOSCOPE

TAPE RECORDER

PREAMPLIFIER
40 dB

PROBE

PIEZOELECTRIC
CRYSTAL

B 2.14

Fig. 4 Block diagram for acoustic emission measurement

with the electrical borehole logs of a profile across the test field. These
logs clearly demonstrate several inhomogeneities encountered in the geological structure, probably associated with the presence of small varying amounts
of brine. AC methods currently being tested use high-frequency resonance
circuits, the frequency of which is altered by changes in the inductance or
capacitance of the circuit caused by fractures in the borehole.

409

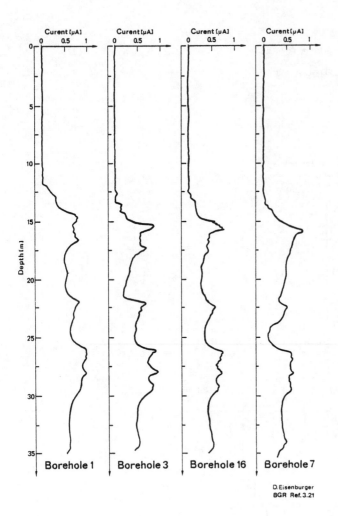

Fig. 5
Example of electrical
borehole logs

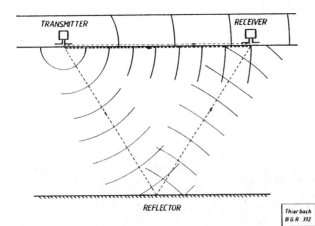

Fig. 6
HF measurement
method

4.8   Acoustic Borehole Scanning

An acoustic borehole scanning method developed by the Westfälische Berggewerkschaftskasse (WBK), which is an advanced version of the borehole probe known under the name of "Borehole Televiewer", is being applied in this experiment for the first time in salt formations. With this method -- called SABIS (Scanning Acoustic Borehole Image System) -- the traveltime and attenuation of the amplitude of a rotating ultrasonic beam are measured. By 3-D plots of the evaluation of traveltimes it is possible to portray the structure of the borehole wall and also to demonstrate borehole convergence.

4.9   HF Measurements

High-frequency electromagnetic radar measurements are being used to obtain additional information on the geological structure of the test site. This method, developed by the BGR, has been used for about 10 years to explore the interior of salt deposits [4]. Discontinuities in the salt, including pockets of brine can be located by pulsed electromagnetic radiation and by measurement of the traveltime and direction of the reflected pulses. The principle of the method of measurement is shown in Fig. 6.

4.10   Permeability Measurements

Permeability measurements will be carried out, as in earlier tests, with a procedure developed by the BGR based on the slug-test method. In a fluid between two packers sealing a borehole section, the pressure is suddenly increased. After turning off the pressure line, the decrease in pressure between the packers is recorded. The permeability is determined from this record by the use of model curves.

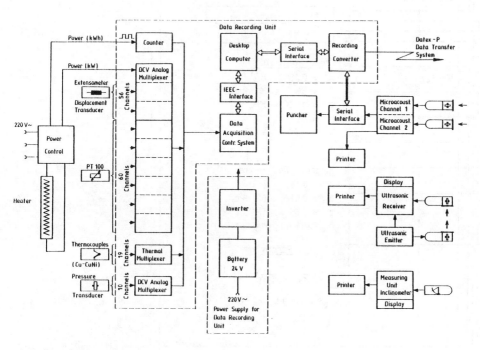

Fig. 7 Block diagram of the data aquisition

4.11 Data Aquisition

A commercial data aquisition system comprising a desktop computer with a tape cartridge drive (HP 85) and a control unit, including a scanner and a digital voltmeter (HP 3497A), is used to record geophysical and rock-mechanical data from about 150 monitoring points installed in the test field. The data are stored directly by the computer on magnetic tape and then transmitted at specified intervals by telephone line (Datex-P) to the central computer system at the BGR in Hannover. A block diagram of the data aquisition system is shown in Fig. 7. Ultrasonic and electrical borehole logging as well as the inclinometer measurements are recorded independently of this system.

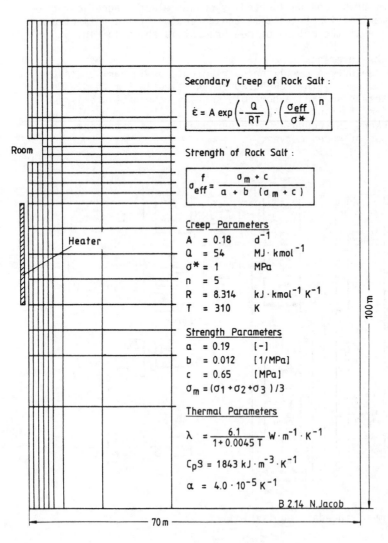

Room

Heater

Secondary Creep of Rock Salt :

$$\dot{\varepsilon} = A \exp\left(-\frac{Q}{RT}\right) \cdot \left(\frac{\sigma_{eff}}{\sigma*}\right)^{n}$$

Strength of Rock Salt :

$$\sigma_{eff}^{f} = \frac{\sigma_{m} + c}{a + b\ (\sigma_{m} + c)}$$

Creep Parameters

$A\ \ = 0.18$ $\ d^{-1}$
$Q\ \ = 54$ $\ MJ \cdot kmol^{-1}$
$\sigma* = 1$ $\ MPa$
$n\ \ = 5$
$R\ \ = 8.314$ $\ kJ \cdot kmol^{-1} K^{-1}$
$T\ \ = 310$ $\ K$

Strength Parameters

$a\ \ = 0.19$ $\ [-]$
$b\ \ = 0.012$ $\ [1/MPa]$
$c\ \ = 0.65$ $\ [MPa]$
$\sigma_{m} = (\sigma_{1} + \sigma_{2} + \sigma_{3}\ )/3$

Thermal Parameters

$$\lambda\ = \frac{6.1}{1 + 0.0045\,T}\ W \cdot m^{-1} \cdot K^{-1}$$

$$C_{p}3 = 1843\ kJ \cdot m^{-3} \cdot K^{-1}$$

$$\alpha\ = 4.0 \cdot 10^{-5}\ K^{-1}$$

B 2.14 N.Jacob

100 m

70 m

Fig. 8
FEM mesh and
input paramter

412

5. THERMOMECHANICAL MODEL CALCULATIONS

The non-steady-state temperature, displacement, and stress occurring during the large-scale heating experiment are being calculated using the finite element computer programs ADINAT and ADINA [5].

5.1 Thermomechanical Modelling

A rock section 700 m high and 70 m wide, including the drift and heater used in the large-scale experiment, was modeled for the FEM calculations. The FEM mesh shown in Fig. 8 consists of 661 nodes from 200 isoparametric elements of 8 nodes each. The regular design of the mesh distinctly reduces the effects of the geometry of the mesh on the stress and deformation. The calculations are carried out for the case of rotation symmetry. At the sides of the model only vertical displacements are possible and only horizontal displacements are possible at the bottom. A geostatic load of 20 MPa is applied at the top of the model. As initial condition, a hydrostatic state of stress is assumed for the rock by using a Poisson's ratio of 0.49. A Young's modulus of 25 000 MPa is used. The natural rock temperature is set at 37 °C. The heat input from the heater is distributed over 5 nodes (s. Fig. 8). Air ventilation in the drift is roughly modeled by keeping the surface temperature at a constant temperature of 32 °C. The table in Fig. 8 shows the constitutive laws and the values used in the calculations for the mechanical and thermal properties.

5.2 Initial Results

These initial calculations are for a power input of 900 W/m (totaling 18 kW) and a 110-day heating period preceded by a twenty-day period before excavation of the drift and a 20-day period following excavation, totaling

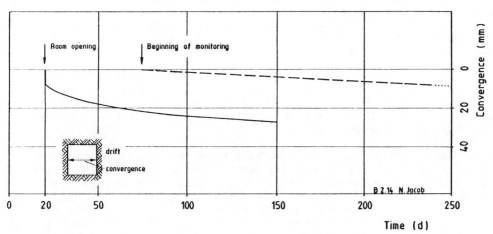

Convergence of drift
— — measured
——— computed

Fig. 9 Comparison of computed and measured drift convergence

413

150 days. A distinct deformation of the rock up to a distance of 15 m from the heater is clearly shown during the 110-day heating period.

The calculated convergence of the drift without the influence of heat is shown in Fig. 9 in comparison with the measured values, as an example for the cavity deformation. Convergences significantly different from this curve do not result from the calculations that include the influence of heat. An elastic convergence of about 8 mm can be seen in the figure immediately after excavation. After 130 days, the horizontal convergence amounts to about 27 mm; the rate of convergence is already constant at 0.0483 mm/d. Measurements begun 30 days after excavation show convergence that is a linear function of time. There is good agreement between calculated and measured rates of convergence.

Figure 10 shows an example of the development of effective stresses and the local safety factors defined by the following equation [6]:

$$\eta = \frac{\sigma_{eff}^{f} \text{ (according to material strength)}}{\sigma_{eff} \text{ (according to FE computations)}}$$

The definition of σ_{eff}^{f} is given in Fig. 8. σ_{eff} is given by the following equation:

$$\sigma_{eff} = [\ 1/6(\sigma_1 - \sigma_3)^2 + (\sigma_1 - \sigma_2)^2 + (\sigma_2 - \sigma_3)^2]^{1/2}$$

The figure illustrates effective stresses and local safety factors at six selected nodes. In general, the curves demonstrate a sizable decline in the bearing capacity of the rock in the initial phase of the heating period. But the safety factor remains greater than 1 with an increasing trend during the remaining heating period.

Due to the relatively wide scatter in the material strength coefficients obtained from laboratory testing of rock samples, it is being considered whether a method for calculating the liklihood of failure [7] should be used in addition to the above illustrated local safety factors for the analysis of critical stresses.

These initial results of the simulation of the heating phase reveal that no critical state of stress will occur at the points or in the sections investigated. The occurrence of near-field critical stresses cannot be excluded completely, but the calculations have not yet been sufficiently evaluated for the area immediately around the heater. The production of thermally induced critical stresses is expected to occur (according to the results of the earlier experiments) mainly in the cooling phase after the heater has been turned off. But no precalculation results for the cooling phase are as yet available.

After concluding the current calculations, additional runs with increased levels of the input power of the heater are planned.

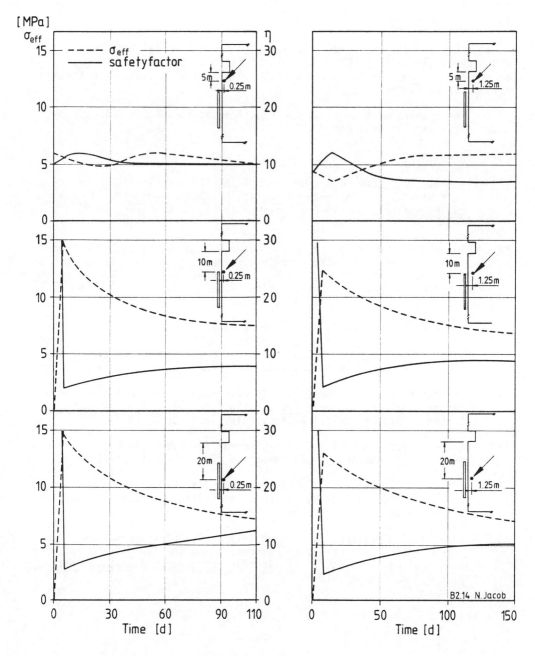

Fig. 10 Effective stresses and local safety factors as a function of time after beginning of heating

415

6.   ACKNOWLEDGEMENTS

This investigations are commissioned by the Federal Ministry of Research and Technology (BMFT). The test field in the Asse salt mine has been made available by the Gesellschaft für Strahlen- und Umweltforschung m.b.H. (GSF), München.

The authors wish to thank their colleagues for helpful discussions and figures used in this paper.

7.   REFERENCES

[1]   Kopietz, J., Jung, R.: "Geothermal In-situ Experiments in the Asse Salt Mine", Proc. Seminar on In-Situ Heating Experiments in Geological Formations, pp. 45-59, Ludvika/Stripa, OECD, Paris, 1978.

[2]   Eisenburger, D., Kopietz, J., Liedtke, L., Meister, D.: "Instrumentation and FE Calculations of In-Situ Heating Experiments Related to the Permanent Disposal of High-Level Wastes", Proc. Intern. Symposium Field Measurements in Geomechanics, ETH Zürich, Sept. 1983, pp. 1309-1319, A.A. Balkema, Rotterdam, 1984.

[3]   Heusermann, St., Pahl, A.: "Stress Measurements in Underground Openings by the Overcoring Method and by the Flatjack Method with Compensation", Proc. Intern. Symposium Field Measurements in Geomechanics, ETH Zürich, Sept. 1983, pp. 1033-1045, A.A. Balkema, Rotterdam, 1984.

[4]   Thierbach, R., Mayrhofer, H.: "Elektromagnetische Reflexionsmessungen in Salzlagerstätten", Proc. Fifth Internat. Symposium on Salt, Northern Ohio Geological Society, Cleveland, Ohio, 1978.

[5]   Bathe, K.J.: "A Finite Element Program for Automatic Dynamic Incremental Nonlinear Analysis", Cambridge, Massachusetts, 1978.

[6]   Liedtke, L., Kopietz, J.: "Thermomechanical Calculations Related to Thermally Induced Rock Loosening in an Underground Cavity", Computers & Structures, 17, pp. 891-902, 1983.

[7]   Liedtke, L., Meister, D.: "Stability Analysis of Underground Structures in Rock Salt Utilizing Laboratory and In-Situ Testing and Numerical Calculations", Solution Mining Research Institute Meeting, Manchester, England, 1982.

416

In-situ creep experiments under controlled stress in rock salt pillars
Design, instrumentation and evaluation

U.E.HUNSCHE & I.PLISCHKE
Bundesanstalt für Geowissenschaften und Rohstoffe, Hannover, FR Germany

ABSTRACT

The Federal Institute for Geosciences and Natural Resources of the Federal Republic of Germany plans to carry out a second in-situ creep experiment using a rock salt pillar containing a number of clay layers. The large- and small-scale deformations will be measured by several methods. The comparison of results with laboratory experiments and model calculations will increase the ability to extrapolate small-scale experiments to larger scales. A further objective is to develop new and improved techniques for in-situ experiments. The start of the experiment is scheduled for August 1984 in the Asse salt mine.

Within the scope of research work for the final deposition of radioactive waste in a rock salt formation, the mechanical behaviour of rock salt has been intensively investigated during the last several years. The laboratory experiments which have been carried out in the Federal Institute for Geosciences and Natural Resources (BGR) of the Federal Republic of Germany (FRG), deal mainly with the following fields:

- Creep behaviour of rock salt under different stress and temperature conditions and also the influence of origin. The following average creep law for steady-state creep has been developed on the basis of theoretical investigations and numerous creep experiments (1) (2) (3) :

$$\dot{\mathcal{E}}_s = A \cdot \exp\,(-Q/RT)\;\cdot\left(\frac{\sigma}{\sigma^*}\right)^{n}$$

A = 0.18 d^{-1} : structure factor

Q = 12.9 kcal $\cdot$ mol^{-1} : activation energy

R = 1.986$\cdot10^{-3}$ kcal$\cdot K^{-1}\cdot mol^{-1}$: universal gas constant

T   : temperature in K

σ : compressive stress

$\sigma^* = 1$ MPa   : normalizing stress

n = 5   : stress exponent

417

- Failure behaviour of rock salt under different stress and temperature conditions. The experimental results have been condensed in several formulas which are still in discussion.
- A new aspect being investigated is the influence of discontinuities (clay or anhydrite layers) on the mechanical behaviour of rock salt.

The practical knowledge gained and the deduced mechanical formulas have been used for numerous analyses and model computations for underground structures. Nevertheless, there still remain the problem of reliable extrapolation to large scales and long times.

This paper deals with the control and validation of the transfer of some laboratory experiments on the salt mechanics of large volumes under in-situ conditions specifically with the planning of an in-situ deformation experiment using a specially designed salt pillar. It has the following objectives:

- measurement of three-dimensional creep of a large volume of rock salt;
- investigation of the influence of clay layers on the deformation of the salt pillar;
- observation of small-scale creep and fracture behaviour in the vicinity of discontinuities;
- comparison of in-situ results with results from the laboratory;
- comparison of in-situ results with results from finite-element model calculations;
- development of new and improved techniques for in-situ experiments.

From these objectives it can be seen that this will be an experiment with aspects of both basic and applied research. The results of the experiment will have an influence on the design, monitoring and the safety assessment of underground openings in rock salt. The planning of the in-situ experiments and the accompanying laboratory experiments with these objectives was started in the second half of 1983.

One of the biggest problems associated with the in-situ experiment is to measure and to control the stress within the examined volume of rock salt. The reason is that the measurement of stress in rock salt is very problematic. Therefore, the only reliable possibility is to cut a pillar and measure and control the stress in a slit in the pillar using a large steel flatjack. The geometry of the pillar and the stress-producing system are shown in figure 1. The planned pillar will be 350 cm high and 150 cm on a side. It is connected at the roof and floor with the surrounding rock. The stress must be controlled by a simple system which is able to work very reliably under mine conditions for months and years.

It is an ideal situation that the BGR has already been successfully carrying out a large in-situ creep experiment for 2.5 years using a large rock salt pillar. This experiment shows many

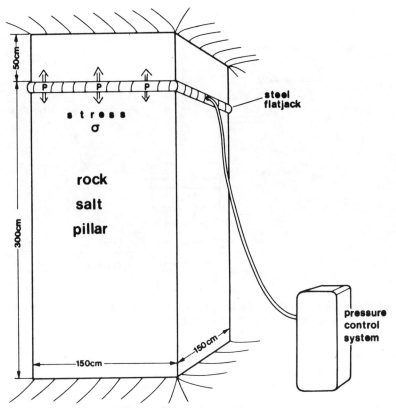

Fig. 1: Geometry of the rock salt pillar and the stress-producing
system of the planned experiment.

of the design and instrumentation features which are needed for
the new experiment. The main aim of this experiment is to compare
the creep behaviour of a rock salt pillar with the above-mentioned
creep law derived from small samples in the laboratory.

The experiment is being carried out in the Asse salt mine. For the
investigation of creep behaviour, stresses of 9 and 10 MPa were
used. The main result is that there is good agreement of the
steady-state creep behaviour of small laboratory samples with
that of the in-situ salt pillar. A complete description and evaluation have been published in (4).

Because of the importance for the new experiment, some more information will be given of the old one. Figure 2 shows a schematic of the simple hydropneumatic pressure unit. It has worked
very reliably for 2.5 years at pressures of up to 110 bars.
The precision of the stress control system was $\pm$ 3 bar ($\approx$3 %).
This is a good value, but because of the great stress exponent of
n = 5, a higher accuracy is desirable. Therefore, a new simple

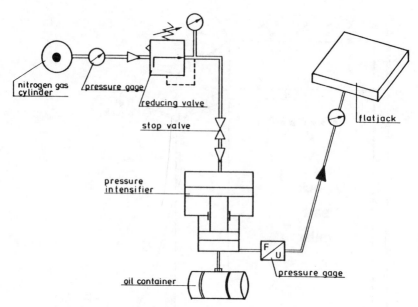

Fig. 2: Schematic of the hydropneumatic pressure unit for the first in-situ creep experiment.

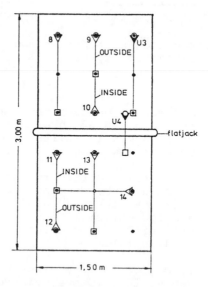

Fig. 3: Instrumentation for the deformation measurements in the first in-situ creep experiment. Shown is side 4 of the pillar.
8, 9, 10, ...   : LVDTs
U3 & U4   : dial gauges

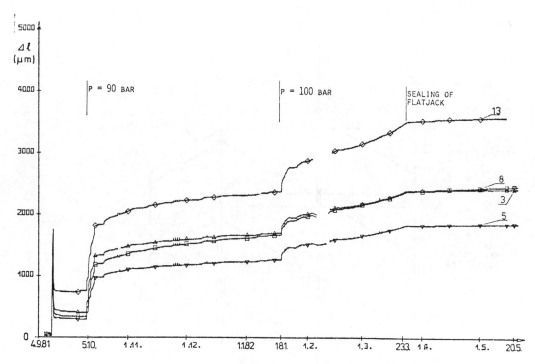

Fig. 4: Deformation-time curves of four vertical measuring intervals of the first in-situ creep experiment.
Measuring interval 1 m.

system has been developed in the meantime which regulates the
pressure within ± 0.1 bar (0.1 % at 100 bar).

Figure 3 shows the instrumentation for the deformation measurements. A total of 26 LVDTs and 6 dial gauges for vertical and
horizontal measurements were installed on the pillar. The
deformation-time curves of four vertical measuring intervals of
the first 8 months are shown in figure 4 as examples. The experiment also included stress measurements, multi-extensometer measurements (9 m deep) in the roof and floor and analogous creep experiments in the laboratory. In addition, two coherent optical
measuring methods (holography and the speckle technique) were tested for the first time below ground on the test pillar for measuring the deformation of an entire area on a no-contact basis.
Measurements and results of these two methods are published in
(5).

Coming back to the new experiment, it can be stated that it is to
some extent a refined repetition of the first one. The great
difference will be that it will also include the influence of
discontinuities.

421

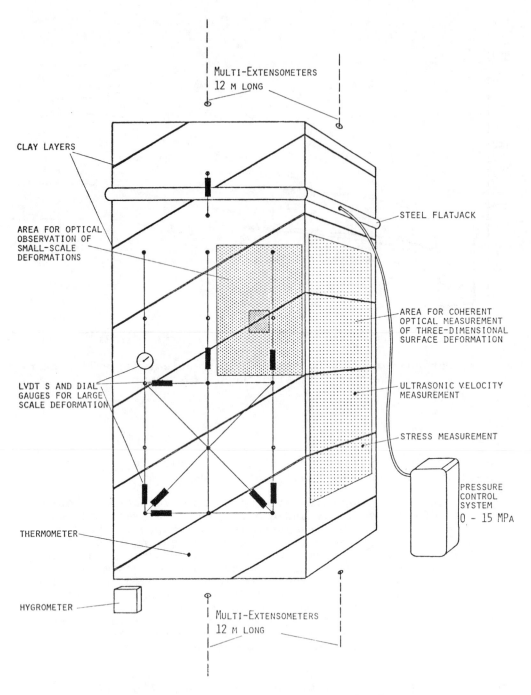

Fig. 5: Experimental features and instrumentation of the planned in-situ creep experiment.

The aims of the new in-situ test and the experience from the
first one led to the design shown in figure 5. The pillar will be
cut in the Asse salt mine at a depth of 750 m in a section where
the rock salt contains numerous clay layers (Na 2 TL). The experiment will consist of the following features:

- continuous measurement of large-scale deformation in vertical,
  horizontal, and diagonal directions by 30 LVDTs and 6 dial
  gauges;
- measurement of the large- and small-scale three-dimensional
  deformation during several time intervals by coherent optical
  methods;
- observation of small inhomogeneous deformation by special
  methods (e.g. grids, "brittle" paint);
- measurement of stress;
- measurement of ultrasonic velocities;
- measurement of temperature;
- measurement of humidity;
- measurement by eight multi-extensometers, 12 m long in roof and
  floor;
- system for pressure and stress control and regulation;
- analogous laboratory tests on samples from the neighbourhood of
  the pillar.

The data will be evaluated by analysis of the different large-
and small-scale deformation measurements and their comparison
with laboratory and modelling results.

It is rather certain that this large-scale deformation experiment
will increase our ability to extrapolate our laboratory results
to larger scales and will also help in the development of new and
improved techniques for in-situ experiments.

The cutting of the pillar in Asse salt mine is scheduled for June
1984 and the start of experiment for August. The test will run
for at least one year.

The experiment is sponsored by the BMFT under project no. KWA
5202 6.

References

(1) Wallner, M., Caninenberg, C., and Gonther, H.:
    Ermittlung zeit- und temperaturabhängiger mechanischer Kenn-
    werte von Salzgesteinen, Proc. 4. Int. Congr. Rock Mech.,
    Montreux 1979, 1, 313 - 318 (1979).

(2) Albrecht, H., and Hunsche, U.: Gebirgsmechanische Aspekte bei
    der Endlagerung radioaktiver Abfälle in Salzdiapiren unter
    besonderer Berücksichtigung des Fließverhaltens von Stein-
    salz, Fortschr. Miner. 58 (2), 212 - 247 (1980).

(3) Hunsche, U.: Results and interpretation of creep experiments
    on rock salt, Proc. First Conf. on the Mechanical Behavior of
    Salt, Nov. 1981, The Penn. State Univ., University Park (1982).

(4) Hunsche, U., Plischke, I., Nipp, H.-K., and Albrecht, H.: An in-situ creep experiment using a large rock-salt pillar, Proc. 6. Int. Symp. on Salt, Toronto 1983, in press.

(5) Beek, M.-A., Garbe, B., Geier, W., Hunsche, U. and Plischke, I.: Erfassung von Deformationen an einem untertägigen Salzpfeiler mit Hilfe kohärentoptischer Meßverfahren, in: Optoelektronik in der Technik: 6. Int. Kongr. - Laser 83, Springer Verl., Berlin, 170 - 181 (1984).

In-situ experiments on the time dependent thermo-mechanical behaviour of rock salt

L.H.VONS & J.PRIJ
Netherlands Energy Research Foundation (ECN), Petten

ABSTRACT

In a 300 m deep dry-drilled borehole in the ASSE-II mine, ECN has performed several experiments on the determination of the time dependent thermo-mechanical behaviour of rock salt. In this paper the following experiments will be described:

(1) Transient isothermal borehole convergence (performed at a total depth of 1050 m from December 1980 - April 1982).

(2) Measurement of the compression of the salt on a heater (at a depth of 1000 meters from June 1982 - August 1982).

(3) Borehole convergence due to a transient temperature field (at several depths from April 1983 - January 1984).

Special attention will be given to the measuring technique and data collection system. A discussion will be given of the applicability of results obtained.

1. INTRODUCTION

Since 1979 ECN is involved in in-situ experiments in the 300 m deep dry-drilled borehole in the ASSE-II mine in the FRG. The field of interest is the isothermal and thermo-mechanical behaviour of rock salt in a salt dome.
The results of these tests are serving as material input data for analytical simulation models of a repository in a salt dome which, in their turn, give basic information for the safety analsyis of the total concept. In this paper first the isothermal measurement techniques will be discussed and in the second part the design and measurement techniques of the heated probes which are used in the borehole is described.

2. THE ISOTHERMAL CONVERGENCE MEASUREMENTS

These measurements have been reported in [1]; therefore, only a brief description shall be given.
As soon as possible after the drilling of the hole the convergence measurement at the deepest point of this borehole was started. This measurement was continuously carried out up to the end of 1981 only with some breaks for diameter measurements at other levels and hole-wall inspection.
The choice of the type of measurement equipment was mainly determined by the short timespan available for the design, fabrication and purchase of the equipment parts. A video system was chosen. A TV-camera looks at six dials, a com-

FIG.1. CONVERGENCE MEASURING PART.

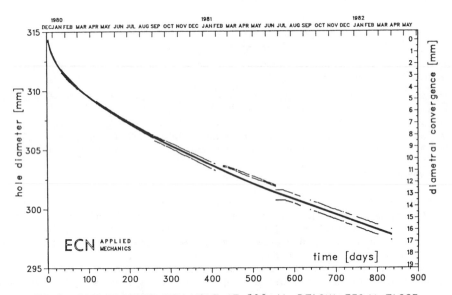

FIG. 2. CONVERGENCE ECN HOLE AT 292 M. BELOW 750 M. FLOOR

pass, two temperature devices and two watches (fig. 1). In this way three diameters and their directions are measured. The time and date can be read from the digital displays of the watches and the bore-wall temperature and the camera temperature are digitally displayed. This entire picture is sent to a monitor and a video recorder for displaying and storing of the measured data.
The system worked automatically by means of a time clock. The accuracy of these measurements is high (± 0.01 mm), especially the relative accuracy, which is of interest to analyse creep effects. Fig. 2 gives the results.
The diameter convergence measurements as a function of the depth in the hole were more complex. Because the creep rate had to be determined from the (small)

426

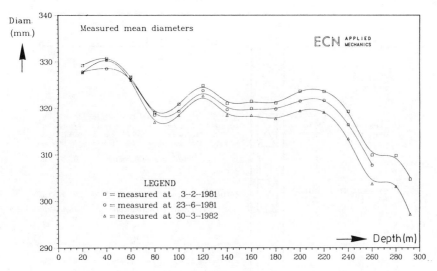

FIG. 3. BORE HOLE DIAMETER AS FUNCTION OF DEPTH.

difference in diameters measured at different times. Fig. 3 gives the results. The results in the upper part of the hole are disturbed due to some brine migration effects and due to an interaction between the stresses along the borehole and the stresses around the cavity.

3. HEATED EXPERIMENTS

3.1 Introduction

Two types of heated experiments are executed in the 300 m hole, namely: 2 pressure tests (HPP-I and HPP-II) and 5 free convergence tests (HFCP), fig. 4. The general aim of these tests is to verify computer program models based on the finite element method. From the onset it was decided that these tests should run automatically as was the case in the isothermal experiment. However, this type of test showed to be much more sensitive for mechanical and, especially, electrical faults. But after a certain "learning period" the experiments were run as planned. In fact only the first pressure test and the third free convergence test were partly a success.

The general lay-out of the equipment at the 750 m level is given in fig. 5. Inside the cabin the computer and the disc-unit are placed in a cupboard which is cooled by a refrigerator. The computer controls the heat input for the probe, alarms in case of an upset condition, presents the measuring data on monitor, prints and stores the information on a disc-unit. The experiment is watched at ECN-Petten by means of a telephone connection.

In case of a power failure of the electrical system the computer alarms the duty officer of the ASSE-mine, who will inform the ECN-personnel in Petten. If a power failure occurs, the power for the heaters in the probe is automatically reduced to zero. When the power failure is lifted, the heat input is automatically slowly brought back to the original level. These events (time, duration of failure) are also registrated.

From a data-transmission point of view it is obvious that an analog system can not be applied here. In the Cooled Instrumentation Box (CIB), therefore, all

427

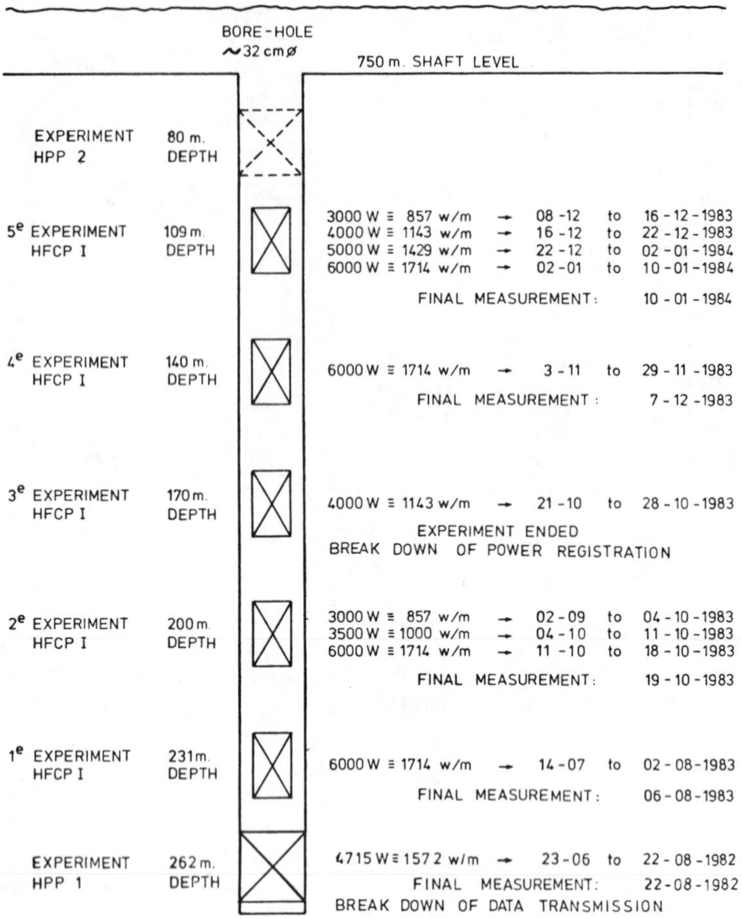

BORE-HOLE
~32 cm⌀

750 m. SHAFT LEVEL

EXPERIMENT HPP 2	80 m. DEPTH	
5ᵉ EXPERIMENT HFCP I	109 m. DEPTH	3000 W ≡ 857 w/m → 08 -12 to 16 - 12 - 1983 4000 W ≡ 1143 w/m → 16 -12 to 22 - 12 - 1983 5000 W ≡ 1429 w/m → 22 -12 to 02 - 01 - 1984 6000 W ≡ 1714 w/m → 02 -01 to 10 - 01 - 1984 FINAL MEASUREMENT: 10 - 01 - 1984
4ᵉ EXPERIMENT HFCP I	140 m. DEPTH	6000 W ≡ 1714 w/m → 3 - 11 to 29 - 11 - 1983 FINAL MEASUREMENT: 7 - 12 - 1983
3ᵉ EXPERIMENT HFCP I	170 m. DEPTH	4000 W ≡ 1143 w/m → 21 -10 to 28 - 10 - 1983 EXPERIMENT ENDED BREAK DOWN OF POWER REGISTRATION
2ᵉ EXPERIMENT HFCP I	200 m. DEPTH	3000 W ≡ 857 w/m → 02 - 09 to 04 - 10 - 1983 3500 W ≡ 1000 w/m → 04 -10 to 11 - 10 - 1983 6000 W ≡ 1714 w/m → 11 -10 to 18 - 10 - 1983 FINAL MEASUREMENT: 19 - 10 - 1983
1ᵉ EXPERIMENT HFCP I	231 m. DEPTH	6000 W ≡ 1714 w/m → 14 - 07 to 02 - 08 - 1983 FINAL MEASUREMENT: 06 - 08 - 1983
EXPERIMENT HPP 1	262 m. DEPTH	4715 W ≡ 1572 w/m → 23 - 06 to 22 - 08 - 1982 FINAL MEASUREMENT: 22 - 08 - 1982 BREAK DOWN OF DATA TRANSMISSION

FIG. 4 EXPERIMENTS IN ASSE II WITH HFCP 1 AND HPP.

the signals are digitalized. The CIB contains the following main electrical systems:

- power supply for the strain gauges and transducers;
- amplifier for the strain gauge and transducer signals;
- scanners;
- analog digital transformers.

To improve the reliability of the above mentioned electrical system, the equipment is split up in four independent signal groups and corresponding cables. The induced heating by the electrical equipment, mainly generated in the power packs, resulted in a max. temperature of about 50°C during the test (ambient temperature ~ 42°C). In order to improve the thermal conductivity in the gap between CIB and salt wall, spring clips of phosphor bronze are mounted at the outside of the CIB.

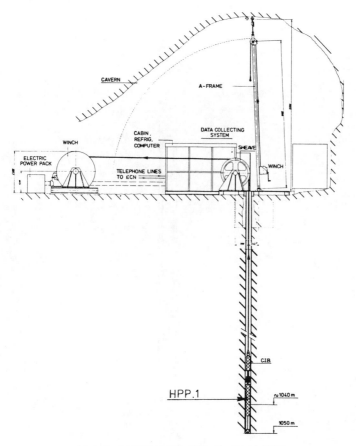

FIG.5 CAVERN AT 750 M. GALLERY

3.2   The construction of the HPP-I and HPP-II

The actual measuring system is based on the knowledge that pressure
loadings on a cylinder can be derived if sufficient radial displacements
are known. In fig. 6 the general construction and the lay-out of the measuring system of the HPP is given. From this figure it can be seen that the
HPP consists of a straight tube which is able to withstand the pressure
build-up of the salt formation due to the induced heat flow. The tube is
divided in 5 parts, the three parts in the middle are heated, the outer
parts are thermally insulated from the heated area. The application of
these outer parts was necessary in order to avoid the deformation of the
salt over the end of the heated area, which would lead to a lower value of
the maximum pressure and should have led to difficulties in the interpretation.
The division of the heated area in three parts is only done to minimize difficulties in the assembly phase.
The heating system consists of a thermo-coax wire which is placed in a groove
(double pitch), machined in the outer surface of the cylindrical wall. To guard
the wire against mechanical damage and chemical attack (brine), the wire is
covered with a lead string which is forced into the groove. At both ends of each

429

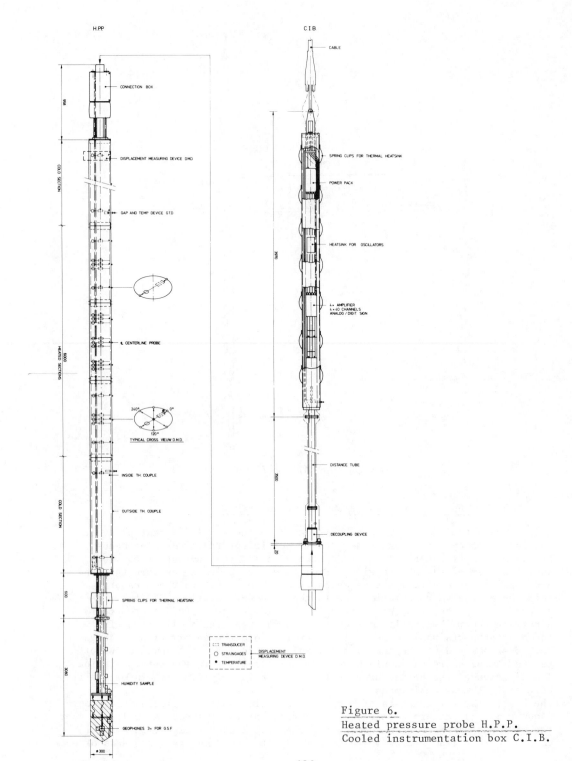

H.P.P.

- CONNECTION BOX
- DISPLACEMENT MEASURING DEVICE DMD
- GAP AND TEMP DEVICE GTD
- ℄ CENTERLINE PROBE

240°   0°

120°

TYPICAL CROSS VIEW D.M.D.

- INSIDE TH COUPLE
- OUTSIDE TH COUPLE
- SPRING CLIPS FOR THERMAL HEATSINK
- HUMIDITY SAMPLE
- GEOPHONES 3× FOR G.S.F

COLD SECTION

HEATED SECTION

COLD SECTION

⌀ 300

C.I.B.

- CABLE
- SPRING CLIPS FOR THERMAL HEATSINK
- POWER PACK
- HEATSINK FOR OSCILLATORS
- 4× AMPLIFIER
  4×40 CHANNELS
  ANALOG / DIGIT. SIGN
- DISTANCE TUBE
- DECOUPLING DEVICE

DISPLACEMENT
MEASURING DEVICE D.M.D.

- ⊏∷⊐ TRANSDUCER
- ◯ STRAINGAGES
- ● TEMPERATURE

Figure 6.
Heated pressure probe H.P.P.
Cooled instrumentation box C.I.B.

430

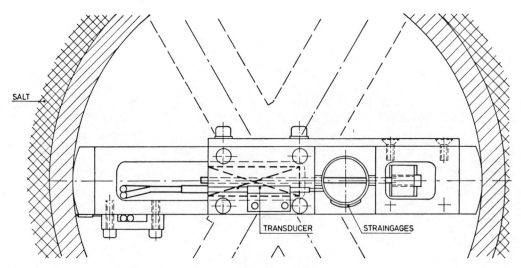

FIG. 7. DISPLACEMENT MEASURING DEVICE

heater part the wire is welded over a distance of 8 cm in the groove, this is done to avoid leakage of the 0-ring seal during the pressure test.

In order to obtain a high level of reliability of the pressure measurements at the elevated temperature of 200°C two different systems, strain gauges and transducers, are applied to measure the displacement Δd of the cylinder of the probe. They are mounted in the Displacement Measuring Device (DMD), fig. 7. The strain gauges are glued on the outside of the cylindrical ring which is compressed during the test due to the inwards directed displacement of the cylinder of the HPP. A full bridge strain gauge arrangement is applied to avoid temperature compensation problems and to obtain a relatively high input signal for the data transmission system.

The displacement transducer is of the induction type, the application of some ferritic material is then essential.

The temperatures are measured with thermocouples of the chromel-alumel type. The thermocouples are located:

- the center part of each of the 23 DMD's is equipped with one thermocouple for temperature control of the strain gauges and displacement transducer;

- 13 internal thermocouples are placed between the wall and a number of DMD's;

- 12 external couples are placed at the outside of the probe in such a way that direct contact with the salt surface is obtained. These couples are situated as close as possible to the corresponding internal couples in order to get an indication when contact between the probe surface and the salt formation has been established.

The DMD's were calibrated with a prescribed displacement system at temperature levels of ~20, ~150 and ~200°C. The strain gauges and the transducers showed minor temperature dependency in the applied temperature range. During the pressure test on the body of the HPP by means of hydraulic-jacket several DMD's showed hysteresis effects. It was shown, however, that these effects were reproducable. After transporting the whole assembly to the ASSE-mine and a short testing period, the first pressure test was started up with a heating power of

431

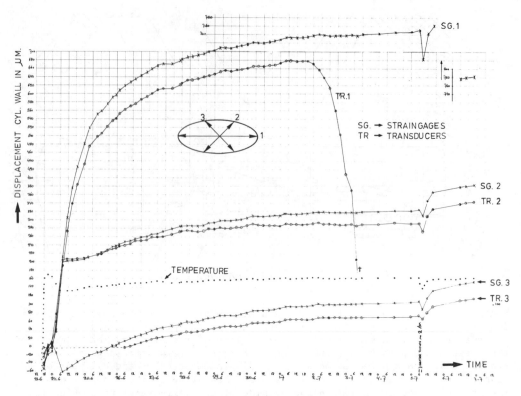

FIG. 8 PART OF MEASUREMENTS OF HPP.1 TEST.

~4715 W on the probe (Ø = 296 mm) and a radial gap of some millimetres between probe and the salt-wall.

In fig. 8 a part of the data of the first 12 days is given for the mid section of the HPP. During the first phase 23-6 → 25-6 all the strain gauges and transducers are showing an unclear picture. This phase is terminated after the probe has settled itself in the hole; the temperature of the HPP has reached a local maximum and the actual pressure build-up is starting. During the second phase 25-6 → 5-7 the deformation of the cylinder wall and the temperature are rising. In the measuring data small oscillations are visible, due to small disturbances of the heat input. The latter was caused by the large oscillations of the voltage of the power supply of the mine; later on stabilizers were installed to improve the situation. At the end of this phase a power break down of 2 hours occurred. After this period, due to the shrinking of the probe and the ongoing creep of the salt, the "thermal gap" is closed and by starting up again this "gap" will act as an extra deformation and causes a step in the pressure build-up. This effect was noticed by the four power interruptions during the test period of about 2 months. In the last phase, not depicted here, the deformations had almost a constant value with a tendency to decrease, which is in agreement with analytical simulation models. It has been shown that the

432

FIG.9. HFCP WITH MEASURING SYSTEM TIED UP JUST BEFORE
LOWERING THE PROBE.

pressure can be calculated as the mean value of the sum of the 3 deformations
of one level multiplied by a constant. This procedure leads to a maximum pressure of 350 bar. As can be observed in fig. 8 the strain gauges gave higher
values than the transducers but the difference remained almost constant in the
last phase of the test. The reason is probably a strong temperature dependency
of the transducer-circuit in the CIB for this was recently observed during the
calibration of the HPP-II at ECN. For the transducers an O-drift was found as
a function of the temperature of the CIB, therefore non-active transducers are
mounted now in the CIB. Due to the unexpected large a-symmetric displacements
the measurements of the single DMD's are only indicative. Therefore only levels
with 3 DMD's are applied in the planned HPP-II test. Finite element analyses
have shown that a max. pressure of about 400 bar can be expected. The difference with the measured one can be explained by the above mentioned differences
and model assumptions with respect to boundary conditions.

3.3 The construction of the HFCP

The HFCP consists of a straight tube with a diameter of about 18 cm in
order to create a sufficiently large gap between probe and wall surface

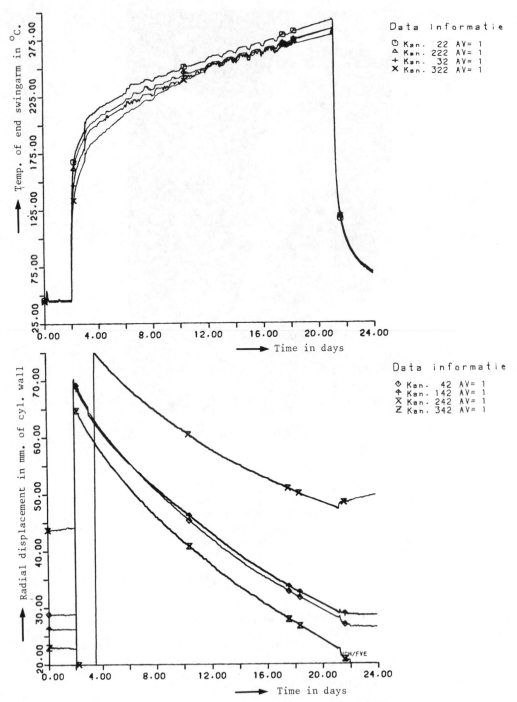

FIG. 10. PART OF MEASURED VALUES OF HFCP - TEST,
MID SECTION, DEPTH 231 M., AMBIENT TEMP. ~ 41 °C

($\Delta R \sim 50$ mm) to study the convergence of the heated salt, fig. 9.
The heating system is essentially of the same type as used in the HPP's. The
heated length is also 3 m, with unheated sections of 1.5 m on both sides. To
obtain a defined radial heat input into the salt anti-convection shields are
mounted between the sections of the HFCP. The displacement measuring system
consists out of a swing arm connected by a transmission rod system to a transducer. In this probe a single measuring system is applied because in case of a
disturbance the test is easily terminated and hoisted for reparation purposes,
this in contrast with the HPP tests. A number of 30 transducers are mounted on
the probe, divided in four groups as done in the HPP tests. Thermocouples were
used to measure the temperature of the body of the HFCP ($\sim 360^{\circ}$C max.) and the
area in the vicinity of the salt wall. Some results of the first test at the
231 m level (fig. 4) with a heat input relevant for a repository is given in
Fig. 10. The lower part of the figure gives some displacement values as a
function of time, the upper part the corresponding temperatures of the end of
the swing arms resting against the salt. The maximum salt temperature itself
at the end of the test is about 190°C. The displacement swing arm is for transport reasons in the hole tied up with a nylon wire against the HFCP-body. On
the desired depth the wire was blown by means of the heating of the probe.
The first 2 days were used to test the electronics without heating the HFCP.
Next the heating power was switched on and the process of convergence started.
After 20 days the heating power was shut off in order to avoid sticking of the
probe in the hole. It is of interest to observe that the process of convergence
stops immediately after power shut-off. The reason that "Kan 242" is out of
line with the other signals lies in the fact that shortly before starting up
"chart 3" had to be replaced without the practicle possibility of calibrating
that part of the system. After terminating the test, the HFCP was hoisted up in
steps in order to calibrate the signals of the charts against each other.
Next the probe was hoisted up to another level in the hole for the following
test.
After the end of the test series the displacements were measured again with another probe in order to check the last convergence results of the HFCP. Taking
into account that in the last applied measuring method, only the mean value of
the diameter was measured and the obvious questionable values of chart 3, the
results showed a satisfactory overall agreement, see fig. 11.
In the analytical work performed uptill now it has been shown that the global
deformation behaviour of the salt could be used to derive the temperature dependency of the creep law. Moreover the analyses confirm the measured phenomenon that after a power shut-off the convergence returns to the "cold" isothermal level. The analyses also give some indications that cracks could occur
after shut-off.
The complete evaluation of the measurements of the heated tests and the allied
analytical simulation work will be accomplished at the end of this year.

4.   CONCLUDING REMARKS

     The following concluding remarks are made here, firstly about the in-
strumentation for in-situ experiments and secondly concerning the applicability
of the results obtained.

   4.1   In-situ instrumentation

     From the isothermal experiments it can be observed that for the static
measurements the applied technique worked excellent. However, for future experiments it is thought to use an additional electrical measurement system in

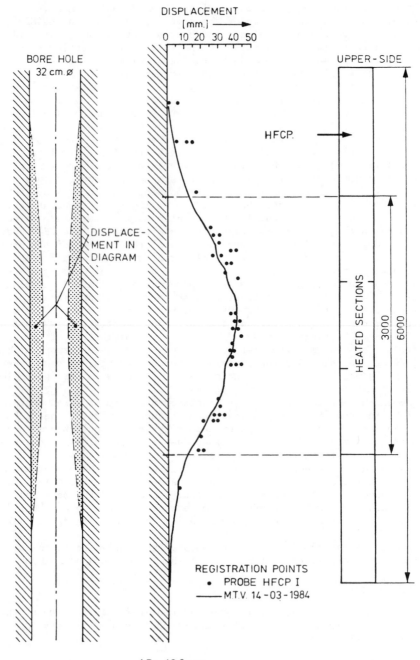

FIG. 11. COMPARISON BETWEEN HFCP-VALUES VERSUS THE
MEASURING TELEVISION VEHICLE MEASUREMENTS.

436

combination with the applied technique in order to make the data processing easier. In-situ calibration of such an electrical system is advisable. The applied system for convergence measurements along the length of the hole has to be improved. The system used in the 300 m hole equipped with one measuring clock rotating over the inner surface of the wall proved to be a tedious affair and a too small basis to check creep effects etc. To improve the situation the system should be extended with several measuring clocks and should be started directly after the drilling of the hole.

Pressure measurements at elevated temperatures have been proven to be a complicated affair. The applied method gives satisfactory results, however, a desired improvement should be to calibrate the system during the in-situ test. More than three measuring points at one level are advisable in order to improve the accuracy of the final result. The best method in this respect is to measure all displacements of the inner surface of the cylinder by means of a movable "DMD"-system. However, this is only applicable for holes with a larger diameter than the hole under consideration because in that case the displacement is larger. A pressure method based on bags, filled with liquid, and being compressed by rock salt is very useful in isothermal situations. However, the application in heated experiments is not very attractive because thermal expansion effects give rise to control and interpretation problems. The experience with the probe for the heated free convergence did hardly give way to any essential improvement. From all experiments executed uptill now it appears, however, that an in-situ calibration possibility is essential for a flexible and accurate measuring system.

4.2 Applicability of the results

The experiments performed uptill now could be used to derive the constitutive properties of rock salt. The isothermal convergence measurements at 314 K were used [2] to derive a secondary isothermal creep law based on the equivalents according to Von Mises.

$$\dot{\varepsilon}^{cr}_{eq} = 8.8 * 10^{-11} \sigma^{5.5}_{eq} \; [day^{-1}]$$

The convergence of the heated salt is used to derive the temperature dependency of this creep law resulting in:

$$\dot{\varepsilon}^{cr}_{eq} = 8.8 * 10^{-11} \sigma^{5.5}_{eq} \; (exp \; (-\frac{e}{T} + \frac{e}{314}))$$

The constant e was found to fall into the range of 7500 – 8250 which is equivalent to activation energies of 62 – 69 kJ/mol. These constitutive properties are in good agreement with results of laboratory experiments. Another application of the result is the direct measurement of some parameters which are relevant for the design of a repository. The maximum compression of the salt on a waste container can be taken from the HPP results without transformation.

A further important application of the results is the use in the validation of computer codes. All types of experiments are analysed in detail with the finite element program GOLIA [3] and uptill now the deviations between analyses and measurements are within the accuracy of the measurements.

437

REFERENCES

[1]   Doeven, I. et al: "Convergence measurements in the dry-drilled 300 m
        borehole in the ASSE-II salt mine",(EUR 8670), ISSN: 0379-4229,
        CODE: EARRF 5(2) 121-324 (1983).

[2]   Prij, J., Mengelers, J.H.J.: "On the derivation of a creep law from iso-
        thermal borehole convergence", ECN-89. (1981).

[3]   Prij, J. et al: "Finite element program for the analysis of the
        transient thermo-mechanical behaviour of a salt dome", Contract with
        CEC, ECN 83-19 (1983).

Acoustic crosshole measurements of cataclastic thermomechanical behaviour of rocksalt

J.P.A.ROEST & J.GRAMBERG
Laboratory for Rock Mechanics, Delft University of Technology, Netherlands

ABSTRACT

A plan will be shown of the measuring of the cataclastic effects in rock salt, surrounding a heated part of a deep borehole.
The test will be carried out in free convergence at a depth of 800 m. A picture of the instrumentation will be shown. Model tests are carried out, in order to investigate the cataclastic thermomechanical behaviour of rocksalt around a borehole during a transient period of heating and subsequent cooling. At a scale of 1 : $4\frac{1}{2}$ acoustic crosshole measurements are carried out under conditions of heating and compression comparable to the proposed deep borehole test.
The result will be shown and the relationship between cataclastic effects and acoustic velocity differences will be discussed.
The work is carried out with support of the Commission of the European Communities.

1. INTRODUCTION

For the reposition of H.L.W. a method has been planned by the Netherlands Energy Research Foundation.
This method consists among others of a system of very deep vertical bore holes in rock salt, drilled in the dry way. For reposition the H.L.W. will be put into canisters. These canisters will be placed into the bore holes; the H.L.W. will produce a certain amount of heat.

A test bore hole has been drilled in the Asse II rock salt mine in Germany. In this about 300 m deep test bore hole the heating by the H.L.W. is simulated by an electric heating and measuring device with a heated section of 3000 mm length. During such a heating experiment the structural changes in the environmental rock salt will be surveyed by the Delft Laboratory of Rock Mechanics. To this end a method has been developed, making use of acoustic cross hole measurements, see fig. 1. The differences in the measured acoustic velocities will form an indication for the loosening of the structure (decompaction) or tightening (compaction). An important point will be the detection of eventually occurring open cracks and crevices during the heating process.

The acoustic method, to be applied has already been tested with good results in a gallery wall of the Asse II mine, at a depth of 850 m.

A next stage of investigation was the simulation of the acoustic effects of local heating of a bore hole. This has been carried out in the laboratory to a scale of 1 : $4\frac{1}{2}$. The results will be discussed in this paper.

439

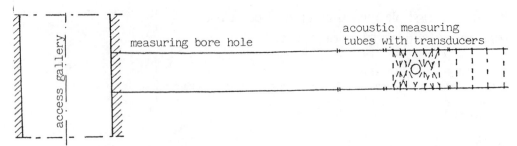

Fig. 1. Measuring cataclastic effects in environmental rocksalt of a heated vertical test bore hole by crosshole measurements.

2. ACOUSTIC CROSS HOLE MEASUREMENTS WITH HEATING ELEMENT TO A SCALE OF
1 : $4\frac{1}{2}$. FREE CONVERGENCE

2.1. Description of the test 'RSV'

In the laboratory a number of tests were carried out with rock salt cubes or blocks, provided with a central bore hole of 66 mm diameter, simulating a repository deep bore hole. A heating element, consisting of an electric heating spiral, was installed within the bore hole, free from the wall. The block, consisting of two halves, was placed in the 3 x 3,5 MN triaxial compressive machine, see fig. 2. The relevant pressure was applied, see fig. 3, traject A. The pressure was kept constant during the major part of the test. (trajects B,C,D,E). After about half an hour (traject B) the heating was switched on. After a period of upheating (traject C) the temperature was kept constant during about six hours. Then the heating was decreased gradually. As a result the temperature of the wall of the bore hole decreased gradually as well (traject E). Then the pressure was released (traject F). The whole test lasted $12\frac{1}{2}$ hours.

During the test the temperature was measured in four locations in the block. The course of the bore hole wall temperature as well as the course of the temperature of the block surface is shown in fig. 3-a.

In each steel plate of the compressive machine five transducers had been installed, see fig. 2 and fig. 3-e. During the test a number of acoustic cross hole measurements have been carried out, using a transient recorder for storage of the seismograms. The results of the measurements 3-1, 3-2, 3-3,3-4 and 3-5 are given in fig. 3-c.

It must be observed here, that the emphasis of this kind of measurements has to be put rather on the differences between the velocity data than on the velocities themselves, in their relation to the unknown absolute values.

2.2. Results of the test 'RSV'

Traject A. During the loading up to 27.7 MPa the acoustic velocities increase very slightly, indicating some compaction (fig. 3-c). Some convergence is measured (fig. 3-d).

Traject B. During one half hour of constant triaxial pressure, effects of the same kind are observed.

Traject C. During the period of upheating the velocity values decrease relatively much, about 2% ;
the 3-3 line more than the other lines (fig. 3-c). This is indicativefor decompaction, probably the result of micro cataclastic-plastic flow towards the

Fig. 2. One of the halves of the rock salt cube(with 1 cm grid)
with bore hole and heating element in the 3 x 3,5 MN
compressive machine ; the right side plate shows the
acoustic measuring points provided with contact mass.

cavity. The flow near by the cavity (line 3-3) is more intensive than at
greater distance (line 3-5). At the same time the convergence increases sharply, about $2\frac{1}{2}$ mm or about 4% of the bore hole diameter, see fig. 3-d.

Traject D. The temperature of the bore hole wall is kept constant at
254°C. The temperature of the block surface increases slowly up to 58°C and
then remains constant.

The acoustic velocities decrease only slightly, indicating an overall
slight increasing decompaction, i.e. loosening of the structure. The decompaction of the near by region, indicated by the line 3-3, is still more than
the decompaction of the more distant regions, e.g. lines 3-5 and 3-1. This is
most understandable, because the material flow to the bore hole is most intensive from the near by region.

The convergence grows more gradually about another 4%, fig. 3-d.

Traject E. The heat supply is diminished gradually and finally the e-
lectric heater is switched off, while the pressure is maintained at 27.7 MPa.
Within $1\frac{1}{2}$ hour the bore hole wall temperature falls down to about 53°, fig. 3-a.

The acoustic velocities show a remarkable effect. The influence of the
near by region on line 3-3 shows a very sharp fall, fig. 3-c. This indicates
probably the forming of radial tensile cracks at the bore hole wall. The more

441

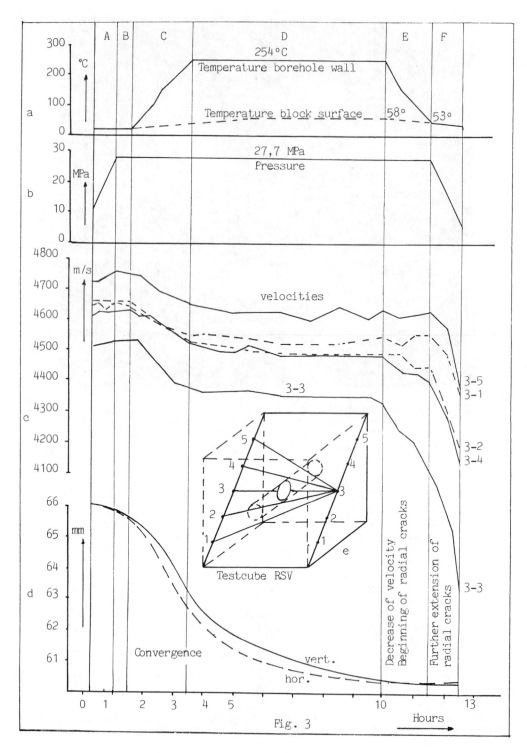

Fig. 3

442

distant regions show only a moderate fall, indicating an overall decompaction (loosening) of the material by shrinkage.

The convergence almost does not change. This indicates, that the shrinkage as a result of temperature fall has to be digested completely by the environmental rock salt.

Traject F. In this traject the pressure is lowered to 5.5 MPa, whereas the temperature remains almost constant. At the moment of lowering of the pressure the acoustic velocities affected by the more distant regions (lines 3-4, 3-2 and 3-1, 3-5) show a dramatic fall, whereas the sharp fall of line 3-3 is continued.

This illustrates that the lowering of the loading pressure increases the decompaction, fracturing included. Obviously the intensity of these phenomena increase from near by the bore hole towards the more distant regions. As is shown in the figures 4 and 5 radial tensile cracks have opened. They have been made visible by staining. Concentric cracks may occur as well, but not so evidently.

By this stage of the test it is shown clearly that the developed method of acoustic cross hole measurements will be appropriate to indicate delicate structural changes as well as the occurence of open cracks and crevices.

It may be emphasized here, that the above explained phenomenon of cracking by sudden temperature fall is not relevant to the H.L W. reposition problem. For, such a fast temperature fall will never occur under reposition circumstances, nor a sudden stress release.

Under these circumstances the long term effect will work : the property of plasticity of rock salt will prevent the forming of tensile stress and therefore of tensile cracks.

3.   THE EFFECT OF A SIMULATED CANISTER AT LABORATORY CONDITIONS.
      LIMITED CONVERGENCE.

This test has been carried out twice. The first test was interrupted as a result of a mechanical defect. The results, however, were indicative for the expected effect. The second test, 'RSX', showed the special effect in a better way.

3.1. Description of test 'RSX'

A block rock salt of the same size, 300 mm x 300 mm x 300 mm, provided with a bore hole, diameter 66 mm has been used. A steel cylinder simulated the canister. The convergence would be limited by the rigid cylinder. ⌀ 61,5 mm

Just as has been done in the RSV test, first the pressure was applied, this time up to 33 MPa, and then the temperature was raised gradually to 190°C at the bore hole wall. It was tried to keep this temperature constant during $7\frac{1}{2}$ hours, but it rose a small amount up to 210°C.

After $7\frac{1}{2}$ hours the temperature was lowered to 110°C. Then the test took unexpectedly an abrupt end.

3.2. Results of Test 'RSX'

The trajects A' and B' in figure 6 are quite similar to the trajects A and B in fig. 3, however, the convergence is more detailed, see fig. 6-d. The Traject C'is similar as well, showing considerable decompaction of the environmental rock salt.

Traject D' : At the beginning of this traject the converging rock salt reaches the rigid steel cylinder. At first it will touch the cylinder at some points only, indicated by point P on line 3-3, fig. 6-c. From now the convergence is hindered, but it will continue until the annular space is filled up completely, see fig. 7.

Beginning in point P the originally decreasing velocity in the near by region is increasing again, see line 3-3, fig. 6-c.

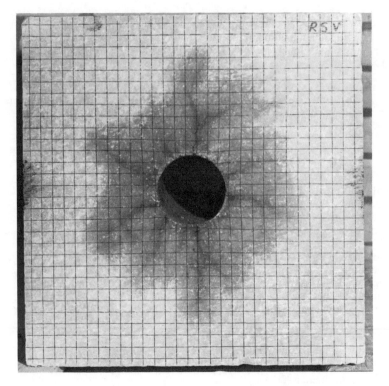

Fig. 4. Radial cracks and zones of intensive decompaction shown up by
staining; they are the result of a combination of relatively
fast temperature fall and the lowering of triaxial compressive stress.

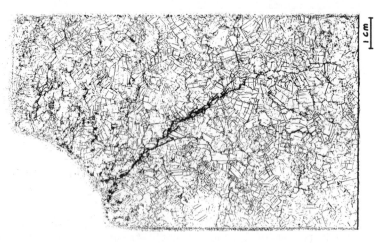

Fig. 5. Thin section shows a radial crack, caused by a combination of fast temperature fall and lowering of
triaxial compressive stress.

444

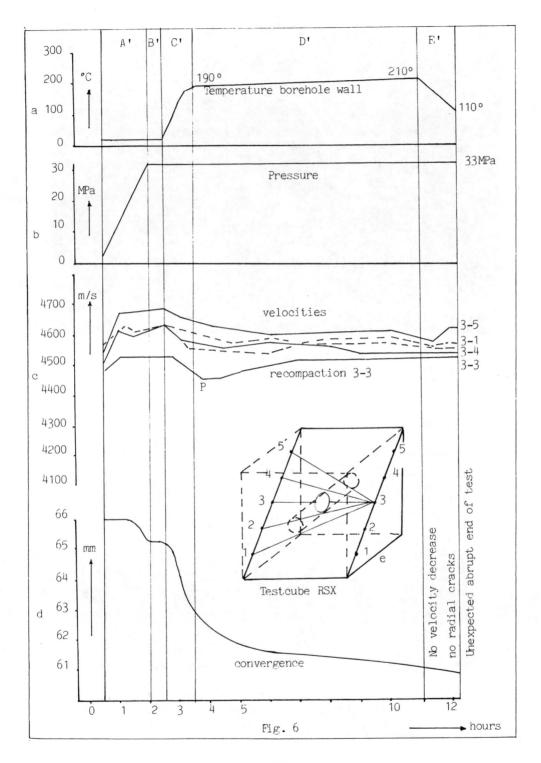

Fig. 6
hours

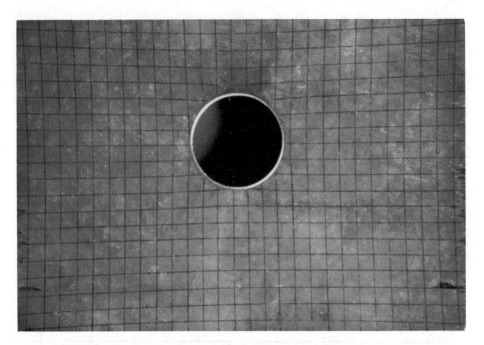

Fig. 7. Flow of rocksalt towards the annular space around a model canister.
Material movement is visible by the deflexion of the 1 cm grid.

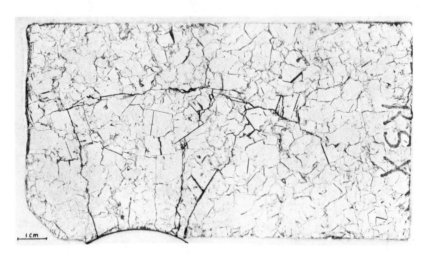

Fig. 8. Thin section showing radial and concentric cracks in the rocksalt around a model canister, caused by a fast temperature fall and abrupt lowering of the triaxial compressive stress.

This indicates the crucial effect of recompaction of the environmental rock salt. This will mean the tightening of the micro cataclasis. It is the expected effect, which has been confirmed rather clearly by this test.

Traject E'. In this case of limited convergence the velocities do not show any fall (fig. 6-c) ; this contrary to the RSV test,with free convergence, fig. 3-c. Obviously the built up radial compressive stress in the wall rock salt, acoustically indicated by the recompaction, is so elevated that it has compensated the shrinkage effect during the temperature fall from 210°C to 110°C during 1¼ hour.

A long time afterwards, however, at room temperature, small radial tensile shrinkage cracks are visible, see fig. 8.

4.   CONCLUSIONS

I.   Heating from the inside of a bore hole activates the micro-cata-
        clastic-plastic flow considerably.

II.   The material supply will be delivered mainly, at least in the be-
        ginning, by a limited zone of the environmental rock salt as a
        result of local decompaction, i.e. by loosening of the structure
        by micro-cataclasis.

III.   In the case of free convergence, temperature fall will cause almost
        instantly open tensile radial (and concentric) cracks as a result
        of shrinkage ; release of the compressive loading stress will in-
        tensify this effect considerably.
        N.B. These circumstances will never occur in the case of the deep
        bore hole H.L.W. reposition technique.

IV.   Limiting of the material flow towards the bore hole by application
        of a rigid lining (canister) will cause recompaction, i.e. re-
        tightening of the previously formed micro-cataclastic structure.

V.   In the case of rigid lining, lowering of temperature will, in the
        first stage of the temperature decrease, not cause tensile cracks;
        only in the final stage of temperature decrease small tensile ra-
        dial and also concentric cracks may occur.

Summary of discussion

Under the chairmanship of Mr. Röthemeyer (PTB), the discussion was
organised around three main topics:
  (a) liberation of gas and/or fluids from heated salt;
  (b) fracturing of salt induced by heat;
  (c) Overall design features of underground experiments in salt.

Liberation of gas and/or fluid components from heated salt

Mr. Müller requested additional details about Mr. Jockwer's presentation. It was made clear that the results of the experiments conducted at ASSE may not be representative of all kinds of salt, but
nevertheless the test methodologies can be transferred. The results
should be applicable to Gorleben salt immediately (Mr. Röthemeyer),
although some measurements might have to be repeated to check that it
is actually the case (Mr. Jockwer).

The tests at Asse show that heating must reach 450°C so that all water
is liberated from salt. But even below 200°C, water of hydration from
polyhalite and kieserite, or water adsorbed to the crystal boundaries,
will be liberated.

These results can be used in the perspective of site safety analysis,
because liberated fluids contacting waste canister may influence the
corrosion rate of metallic envelopes and of waste glass. Additionally, the liberations of gas may generate high pressures in the disposal
holes. Thus, there are at least two influences of these phenomena on
repository design (Mr. Kühn):
  (a) choice of container materials, and
  (b) date of disposal hole closure to avoid excessive gas pressu-
       re build-up. It must be borne in mind, however, that the
       liberated fluid quantities at ASSE are very low.

Mr. Plodinec wondered whether organics, which led to gas release under
heating, were identified, and whether they could produce complexing
agents. It turns out (Mr. Jockwer) that results in this field are not
yet available.

Heat-induced fracturing in salt

It was made clear from the Temperature Test 5 in Asse that no fracturing was induced by heating (Mr. Rothfuchs); the "Hexagon Test" will

448

aim at inducing, on purpose, fractures in salt by excessive heating (Mr. Kühn). Mr. Liedtke pointed out that cracks could be induced at some distance from hot canisters, where the temperature rise, hence the ability of salt to creep, is low. Therefore, this question deserves attention. According to Mr. Kopietz, the critical stresses (i.e. fracture-inducing) should not be generated during normal operation of a repository in salt. In-situ heating experiments producing thermally-induced fracturing should be considered as an overtesting method with regard to the evaluation and demonstration of limits of thermal loadings for repository design.

Overall design features of experiments in salt

The design features of the WIPP were specified by Mr. Tyler: the in-floor disposal holes are spaced by 5.5 m, each one hosting 2 canisters; these latter have an initial heat output of 470 W. This leads to a thermal load of 18 W/m^2, resulting in a maximum salt temperature of 77°C.

Mr. Côme wondered whether the dodecahedral arrangement of AWID flat-jacks to measure stresses in salt would not lead to excessive disturbance of this state of stresses. According to Mr. Kessels, the perturbation caused by each flat-jack is minimum due to the reduced thickness of the device.

Session 6/Séance 6
Overviews and conclusions
Résumés et conclusions

Animateur:
L.TYLER
Sandia National Laboratory, Albuquerque, NM, USA

Overviews by the session chairmen

General

For this conclusive session, the five session chairmen (for sessions 1 to 5) were invited to summarize their views as regards the presentation and discussions held, each for his own session. At the same time, comments from the audience were addressed. A general discussion could take place before the concluding remarks were made. Mr. Tyler (Sandia) kindly accepted the sensitive role of "animateur" for this session.

Overview of Session no. 1/Résumé de la Séance no. 1 (Mr. Barthoux, ANDRA)

Il s'agissait, dans cette séance d'introduction, de faire le point sur l'expérience acquise en matière de laboratoires souterrains à des fins "non-nucléaires", et d'identifier les objectifs des laboratoires souterrains proprement orientés vers l'évacuation des déchets radioactifs.

Les présentations des différents orateurs, et la discussion qui les a suivies, ont provoqué un certain nombre de commentaires que je vous livre en même temps que mes impressions propres.

On a fait remarquer que la mécanique des roches "classique" peut et doit contribuer au succès des travaux en laboratoires souterrains "nucléaires". C'est à la fois souhaitable et nécessaire, car les méthodes d'essais et les analyses théoriques développées dans ce domaine ont été largement évoquées tout au long de cette réunion. Il serait même souhaitable d'aller plus loin, et de réaliser une revue extensive de toutes les expériences acquises en souterrain, à des fins expérimentales, dans des laboratoires "non-nucléaires", étendant ainsi ce qu'a esquissé M. Duffaut.

Une recommandation importante concerne la nécessité d'envisager les phénomènes dans le milieu souterrain, non plus indépendamment les uns des autres, mais de manière couplée. Ceci est vrai pour les expériences proprement dites, mais aussi et surtout pour les personnes impliqués dans ces programmes expérimentaux. Il est maintenant nécessaire d'intensifier ou même de promouvoir le dialogue entre les expérimentateurs, les physiciens chargés de l'interprétation des expériences, les "naturalistes" qui vont décrire les particularités du milieu naturel, et aussi bien sûr ceux qui vont utiliser les résultats des expériences en vue d'optimiser les futurs projets de dépôts.

Compte-tenu des dépenses importantes en temps, personnel et moyens financiers, nécessaires à la conduite de ces programmes d'essais en souterrain, il est de plus en plus nécessaire de s'interroger, avant de lancer des programmes d'essais, sur l'utilité des résultats qui seront acquis. Dans ce domaine, une planification soigneuse des expériences à réaliser doit être faite en fonction des résultats à obtenir, lesquels doivent être identifiés au préalable.

En ce qui concerne l'objet même de cette réunion, un certain "flou" subsiste sur les objectifs à atteindre par les différents types de laboratoires souterrains. Il semble qu'on n'ait pas clairement distingué ce qui doit être fait, du point de vue expérimental, soit en laboratoire "de surface", soit en laboratoire souterrain sur un site non destiné à l'évacuation de déchets, soit en cavité expérimentale souterraine sur un site d'évacuation, soit encore dans une installation de démonstration.

Enfin, un certain nombre de questions sont restés ouvertes, dont une des plus importantes serait: "Faut-il, dans une installation souterraine, réaliser des essais avec des déchets réels, et si oui, avec ou sans conteneur?" La question reste posée.

Translation of Mr. Barthoux's overview.

This introductory session aimed at presenting the status of expertise gained in the field of underground laboratories for "non-nuclear" purposes, and at identifying the objectives of underground laboratories for radioactive waste disposal.

The presentation by the various speakers, and the ensuing discussion, provoked some comments, which I herewith summarize together with my own reactions.

It was stated that "conventional" rock mechanics can and must contribute to the development of works in radwaste underground laboratories. This is advisable and necessary, for mention was made frequently of the experimental procedures and of the theoretical analyses available in this field. It could even be suggested to go one step beyond, and to carry out an extensive review of all the expertise gained underground, for experimental purposes, in "non-nuclear" laboratories; this would be an extension of Mr. Duffaut's review.

It is now recommended to consider underground phenomena as coupled processes, instead of independent ones. This is valid for the experiments themselves, and also -and mainly- for the specialists involved in these experimental programmes. It is now necessary to increase and/or to promote exchange between the experimentalists, the scientists in charge of interpreting the experiments, the "naturalists" who have to describe the particular features of the environment, and, last but not least, those who will have to use the results of the experiments with a view to optimize future repository concepts.

Large expenses in time, manpower and budget, are necessary for the implementation of these underground research programmes. It is therefore necessary to determine, prior to launching a test programme, whether or not the result will actually be useful. The experiments to be carried out must be carefully planned taking account of the results to be obtained; these latter must have been identified in a previous step of the work.

Now, as regards the aim of our present workshop, there seems to remain some ambiguity about the objectives to be obtained by the various types of underground laboratories. The experimental activities to be carried out either (a) in surface laboratories, or (b) in underground laboratories on non-disposal sites, or (c) in underground laboratories on future disposal sites, were not distinguished quite satisfactorily.

Finally, some questions remained open. One of these could be formulated as follows : "Have we got to perform underground tests with real radioactive waste, and, if affirmative, must these waste be overpacked or not ?".
The answer is still to come.

Overview of Session no. 2/Résumé de la Séance no. 2 (Mr. Feates, UK-DOE)

This meeting has drawn attention to at least seven underground laboratories in granite, three in clay and three in halite which are either planned or operating. Experiments in the first of these cavities started about 30 years ago. Therefore, a large amount of experiences are now available in various fields.

Two main purposes for these excavations can be found:

    (i)   testing of hypotheses and equipment and
    (ii) testing of total systems.

and the work has generally been carried out in locations not intended for actual disposal operations.

Now the next step must be made: to satisfy regulators, it will be necessary to show that tests presently carried out at non-disposal sites are applicable to actual disposal sites. This involves a degree of duplication with consequently additional expenses and longer development times. A notable exception is the case of Mol, where all activities are fully integrated on the same clay site, which will eventually be used for disposal purposes.

The question "to backfill or not to backfill" was evoked. In my opinion, backfilling and sealing a repository is a necessary condition for the safety of any disposal system. Leaving a repository open could allow it to become water-filled in the long term, thus transforming the galleries and shafts into direct paths for radionuclides to the biosphere.

Retrievability of the waste canisters should be considered only for demonstration purposes, as was the case in the Climax facility for spent fuel disposal. One argument for retrievability in an actual repository would be to allow access in future if there was a case for re-use some substances from the waste. However, this possibility should not be allowed to inhibit effective disposal operations.

The development of underground laboratories seems presently to show some duplications. In order to avoid too large overlappings, these laboratories should be widely open to international co-operation. Good examples of such co-operative attitude is the German facility at Asse, the Swedish facility at Stripa and the proposed facility in Canada.

To conclude, experiments must be designed and carried out in such a way that regulating bodies, and also the public opinion, are satisfied by the results.

Overview of Session 3/Résumé de la Séance 3 (Mr. Robinson)

From the presentation and discussions, it is quite clear that "useful" experiments are those which contribute to safety analysis and licensing procedures.

For fractured rocks, the key factor to be determined by analysis and experimentation is the measurement of radionuclide migration in fractures. Also, there are concerns as to the validity of DARCY's law to predict groundwater flows in such fractured, low-permeability media. Therefore, further attention must be given to both theoretical and experimental approaches to fracture-flow and radionuclide migration in granitic rocks.

The principal areas of concern were :

(1) Radionuclide migration in fractures
(2) Hydrologic characterization of fractured granites
(3) Non applicability of Darcy's Law for analysis of fracture flow
(4) In situ stress instrumentation and measurement techniques, and
(5) Disturbance caused by measuring instrument emplacement or excavation for the test to measure the quantity of interest (e.g. permeability, fracture flow, or radionuclide migration).

The presentations and subsequent discussions from the floor during Session 3 resulted in the following general areas of agreement :

(1) Several instruments are capable of reliably measuring the magnitude and change of in situ stress providing that the instrument is carefully emplaced in the rock and measurement procedures are adhered to vigorously.

(2) Instrumentation and techniques are available for adequately measuring rock temperatures, temperature distributions, and thermal properties.

Technologies requiring additional work :

(1) Hydrologic and geochemical measurements obtained from surface boreholes and underground openings.

(2) Radionuclide transport measurements from surface boreholes and underground openings.

Overview of Session 4/Résumé de la Session 4 (Mr. Baetslé, CEN/SCK)

Two major achievements in clay have been reported:

- The construction of an underground laboratory in the clay formation of Boom underneath the Mol site in Belgium.

455

- The construction and operation of a backfill mock-up in the Stripa mine.

The construction of the underground laboratory and more particularly the successful digging operation in non-frozen clay have shown that the construction of large galleries will be possible without using extraordinary means but more probably using an adapted tunnelling machine.

The use of compacted bentonite as backfill material is of great importance for all geologic formations where some water movement is possible (fissured granite, basalt). However, attention will have to be paid to the accelerated degradation of waste glass when contacted by bentonite.

The use of old mines with very limited water flow (Konrad mine) for disposal of radioactive waste is a typical example of economical use of existing facilities.

A new initiative in clay is the construction of an underground gallery at Pasquasia in Italy.

As regards instrumentation, there seems to be a large consensus on the use of Glötzl pressure cells, electrical piezometers and extensometers in underground excavations in order to measure the geotechnical characteristics.

The measurement techniques of water content and permeability of the formation need further development. Since the values to be measured are very small (10^{-10} to 10^{-12} m/s), more sophisticated methods (ventilation tests, Rn emanation tests) may be required.

Interesting types of in situ equipment for the measurement of corrosion and electrochemical properties (Redox) have been tested in clay quarries and are ready to be used in underground experiments.

Very little is being done or at least reported on the migration of radionuclides in clay.

One of the major advantages of underground laboratories is that real data can be put forward but therefore a large design effort is necessary. The following tests have to be performed:

- heat transfer and associated water migration;
- study of container corrosion under irradiation in presence of a heat source;
- study of radionuclide migration (actinides) under a combined field of heat and radiation.

Each of these experiments, if representative, needs an extensive underground laboratory equipment. International collaboration is in such cases very important.

In order to convince the public opinion of the feasibility of the geological disposal, the in situ laboratory phase has to be followed by a demonstration phase :

- civil engineering of underground galleries with representative equipment (shielded vehicles, manipulation ...);
- handling of intense radiation sources (HLW);
- experimental disposal of typical waste types (alpha waste, HLW) in instrumented galleries with surveillance and retrievability.

Within the third Radioactive Waste R & D Programme of the CEC three demonstrations or in situ projects are scheduled and one of them is the facility at Mol.

Overview of Session 5/Résumé de la Séance 5 (Mr. Röthemeyer, PTB)

This meeting has shown that much advance has been made in the knowledge of salt as a candidate disposal medium.

The needs for in situ testing can be categorized as follows:

- pilot functions, i.e. testing of a general nature, aiming at improving non site-specific R & D;
- site characterisation and development.

The objectives of in situ testing should result from overall performance criteria, still to be defined in a site-specific context. A German example can be given here: it has been shown by safety analyses for the Gorleben site that building tight dams in the repository galleries would slightly improve the overall safety of the system but cost millions of D. Marks. Therefore, quantitative overall performance criteria have to be established quickly in order to help defining areas in which additional research and effort is needed.

The German position is that all techniques involved in disposal must now be tested so that they can be used directly in the future industrial repo-sitory. More specifically, in situ investigation is still necessary for:

- integral testing of high-level waste form including emplacement techniques;
- site-specific data for modelling the groundwater paths and the associated geochemical phenomena;
- plugging and sealing of repositories.

Finally, in situ testing should help "expecting the unexpected in geology". It is for this reason -the impossibility of standardization of possible sites- that quantitative regulation should be reduced to the formulation of the overall performance criteria.

Summary of general discussion

Coupled testing

Mr. Nataraja raised the question of coupled experiments (e.g. thermal-mechanical, or thermal-hydrological) to be performed in underground labo-ratories. Although the necessity of such testing is clear, there remain some uncertainties as to the priorities to establish between these tests.

Mr. de Marsily pointed out that these coupled testings must be supported by research about "analogues" or natural geological systems, mainly to bring some clarification to such large-scale problems as (i) preferential dis-solution of minerals by groundwater, or dissolution-reprecipitation and (ii) triggering of seismic activity by thermal load from a repository.

Based on a suggestion by Mr. Müller, Mr. Tyler proposed that this need for coupled testing be the subject of a future workshop, to be defined sub-sequently.

Backfilling of repositories

Mr. Saari wondered whether more emphasis than can concluded from the contributions and discussion this far should be put on conventional laboratory testing prior to the underground experiments. He suggested that a thorough laboratory testing programme might help to reduce the costs, and unnecessary tests would be avoided.

Mr. Baetslé pointed out that in situ actual conditions are essential in this field and that some of them can hardly be reproduced in the laboratory (e.g. reducing environment and several MPa earth pressure). This is the main incentive to perform such tests underground and at full-scale.

Use of, and needs for, in situ testing

Mr Tyler recalled Mr. Black's comment, according to which it becomes necessary to determine what kind of tests are best suited for underground laboratories, and added that future tests will have to be identified taking account of existing experience in order to avoid too large duplications.

This leads to the question of transferability of data from the underground laboratories. Mr. Barthoux recalled the distinction between "generic" laboratories which may be best used to validate models, and "site-specific" ones, which aim at determining on-site parameters.

Mr. Simmons added another distinction between "laboratories" and "facilities". The former ones (e.g. the URL) aim at developing/testing methodologies which will be transferable to produce site-specific data on other sites.

Instrumentation in salt

Mr. Kalia questioned about the determination of in situ stresses in salt. Invited by Mr. Röthemeyer, Mr. Hunsche pointed out that in situ stresses are frequently calculated, given the constitutive law of the salt and the in situ behaviour of excavations. On the other hand, the measurement methods presented in Session 5 can help cross-checking the results.

Mr. Hunsche pointed out that the durability of instrumentation in salt largely depends upon the moisture content of the salt considered.

Finally, Mr. Jezierski mentioned that long horizontal borehole drilling in salt (up to 1.5 km long in 100 mm dia., the maximum deviation being some meters) has been sucessfully investigated.

List of participants

BELGIUM

L. BAETSLE
CEN/SCK
Boeretang 200
B-2400 Mol
      Tel.: 014/311801
      Tlx.: 31922 Atomol b

A. BONNE
CEN/SCK
Boeretang 200
B-2400 Mol
      Tel.: 014/311801
      Tlx.: 31922 Atomol b

M. BUYENS
CEN/SCK
Boeretang 200
B-2400 Mol
      Tel.: 014/311801
      Tlx.: 31922 Atomol b

F. CASTEELS
CEN/SCK
Boeretang 200
B-2400 Mol
      Tel.: 014/311801
      Tlx.: 31922 Atomol b

R. DE BATIST
CEN/SCK
Boeretang 200
B-2400 Mol
      Tel.: 014/311801
      Tlx.: 31922 Atomol b

R. FUNCKEN
Tractionel
rue de la Science 31
B-1040 Bruxelles
      Tel.: (02) 234-47-10
      Tlx.: 64860 B

BELGIUM (Cont'd)

N. HENRION
CEN/SCK
Boeretang 200
B-2400 Mol
 Tel.: 014/311301
 Tlx.: 31922 Atomol b

H. HEREMANS
Organisme National des Déchets
Radioactifs et des Matières Fissiles
ONDRAF/NIRAS
Bd. du Régent 54 (B5)
B-1000 Bruxelles
 Tel.: 02-5137460
 Tlx.: 65784 NIROND

M. MAYENCE
Tractionel
rue de la Science 31
B-1040 Bruxelles
 Tel.: (02) 234-44-23
 Tlx.: 64860 B

A. MERTENS
Laboratoire de Génie Civil
Université Catholique de Louvain
Place du Levant 1
B-1348 Louvain-La-Neuve
 Tel.: (010) 418181
 Tlx.: B 59037

B. NEERDAEL
CEN/SCK (Geo-Technology)
Boeretang 200
B-2400 Mol
 Tel.: 014/311801 (Ext. 3917)
 Tlx.: 31922 Atomol b

J. THIMUS
Laboratoire de Génie Civil
Université Catholique de Louvain
Place du Levant 1
B-1348 Louvain-La-Neuve
 Tel.: (010) 418181
 Tlx.: B 59037

P. VAN ISEGEM
CEN/SCK
Boeretang 200
B-2400 Mol
 Tel.: 014/311301
 Tlx.: 31922 Atomol b

CANADA	G. SIMMONS Atomic Energy of Canada Limited Whiteshell Nuclear Research Establishment Pinawa, Manitoba, ROE 1L0 Tel.: Tlx.: 757553 AECL WNRE PIN
DENMARK	J. ANDERSEN Geological Survey of Denmark 31 Thoravej DK-2400 Copenhagen NV A. JOSHI ELSAM DK-7000 Fredericia B. SKYTTE JENSEN Riso/ National Laboratories P.O. Box 49 DK-4000 Roskilde Tel.: (02) 37-12-12 Tlx.: 43116
FINLAND	A. ÖHBERG Saanio & Laine Ltd. Annankatu 25 A 41, SF-00100 Helsinki 10 Tel.: 90-601-166 V. PIRHONEN Technical Research Centre of Finland Geotechnical Laboratory VTT/GEO, Lehtisaarentie 2 SF-00340 Helsinki 34 Tel.: 456 61 77 Tlx.: 122972 VTTHASF K. SAARI Technical Research Centre of Finland Geotechnical Laboratory VTT/GEO, Lehtisaarentie 2 SF-00340 Helsinki 34 Tel.: 90-456-61-72 Tlx.: 122972 VTTHASF
FRANCE	R. ANDRE-JEHAN ANDRA (Agence Nationale pour la Gestion des Déchets Radioactifs) 31-33, rue de la Fédération 75015 Paris Tel.: 273-60-00 Ext. 40-49 Tlx.: 200671

FRANCE (Cont'd)

R. ATABEK
CEA-DRDD-SESD
Centre d'Etudes Nucléaires
B.P. No. 6
F-92260 Fontenay-Aux-Roses
            Tel.: (1) 654-81-29
            Tlx.: 204841

A. BARBREAU
CEA-Institut de Protection et de Sûreté
Nucléaire
Département de Protection Technique
IPSN-DPT - CEN-FAR
B.P. N° 6
92260 Fontenay-aux-Roses
            Tel.: 654-70-76

A. BARTHOUX
ANDRA (Agence Nationale pour la
Gestion des Déchets Radioactifs)
31-33, rue de la Fédération
75015 Paris
            Tel.: 273-60-00 Ext. 40-49
            Tlx.: 200671

A. BOULANGER
Geostock
Tour Aurore
Cedex No. 5
F-92080 Paris La Défense
            Tel.: (1) 778-63-47
            Tlx.: 610898

J. CHARPENTIER
Laboratoire de Mécanique des Solides
Ecole Polytechnique
F-91128 Palaiseau Cedex
            Tel.: (6) 941-82-00
            Tlx.: 691596 F

P. DE LAGUERIE
GEOSTOCK
Tour Aurore
Cedex N° 5
92080 Paris-la-Défense
            Tel.: 778-63-80
            Tlx.: 610898

G. DE MARSILY
Directeur du Centre d'Informatique
Géologique
Ecole des Mines de Paris
35, rue Saint Honoré
77305 Fontainebleau
            Tel.: (6) 422 48 21
            Tlx.: 600736

FRANCE (Cont'd)

H. DERLICH
Commissariat à l'Energie Atomique
IPSN/DPT
B.P. N° 6
92260 Fontenay-aux-Roses
            Tel.: 654 73 79

P. DUFFAUT
Pierre Londe et Associés
130 rue de Rennes
F-75006 Paris
            Tel.: (Home: (1) 548 91 39)
            Tlx.: AENPC 201955
                   (Ecole Ponts et Chaussées)

E. DURAND
Bureau de Recherches Géologiques
et Minières
B.P. 6009
F-45060 Orléans Cedex
            Tel.: (38) 63 80 01
            Tlx.: BRGM A. 780258 F

P. ESCALIER DES ORRES
CEA-IPSN-DAS
Centre d'Etudes Nucléaires
B.P. No. 6
F-92260 Fontenay-Aux-Roses
            Tel.: (1) 654-89-29
            Tlx.: 204841

M. GHOREYCHI
Laboratoire de Mécanique des Solides
Ecole Polytechnique
F-91128 Palaiseau Cedex
            Tel.: (6) 9418200
            Tlx.: 691596 F

M. JORDA
Institut de Recherche Technologique
et de Développement Industriel
Centre d'Etudes Nucléaires
B.P. N° 6
F-92260 Fontenay-aux-Roses
            Tel.: 654-81-29
            Tlx.: 204841 FNAYR

G. LEDOUX
Ecole des Mines de Paris
35, rue Saint Honoré
77305 Fontainebleau
            Tel.: (6) 422 48 21
            Tlx.: 600736

FRANCE (Cont'd)

C. LOUIS
SIMECSOL
115, rue St-Dominique
F-75007 Paris
              Tel.: (01) 555-07-11
              Tlx.: 270703

J. OUVRY
BRGM
F-59260 Hellemmes-Lille
              Tel.: (20) 91 38 19

P. PEAUDECERF
Bureau de Recherches Géologiques
et Minières
B.P. 6009
F-45060 Orléans Cedex
              Tel.: (38) 63 80 01
              Tlx.: BRGM A. 780258 F

R. PINCENT
SIMECSOL
115, rue St. Dominique
F-75007 Paris
              Tel.: (1) 555-07-11
              Tlx.: 370703

FEDERAL REPUBLIC
OF GERMANY

W. BECHTHOLD
Kernforschungszentrum Karlsruhe GmbH.
Institut für Nukleare Entsorgungstechnik
(INE)
P.O. Box 3640
D-7500 Karlsruhe 1
              Tel.: 07247-82 22 23
              Tlx.: 782 64 84 (Reaktor Karlsruhe d)

W. BREWITZ
Gesellschaft für Strahlen- und
  Umweltforschung mbH
Institut für Tieflagerung
Wissenschaftliche Abteilung
Theodor-Heuss-Strasse 4
D-3300 Braunschweig
              Tel.: 0531/80121
              Tlx.: 952865

D. FLACH
GSF, Institut für Tieflagerung
Theodor-Heuss-Strasse 4
D-3300 Braunschweig
              Tel.: 0531/8012280
              Tlx.: 952865

B. HENTE
GSF, Institut für Tieflagerung
Theodor-Heuss-Strasse 4
D-3300 Braunschweig
        Tel.:   0531/8012274
        Tlx.:   952865

U. HUNSCHE
Federal Institute for Geosciences
  and Natural Resources
Stilleweg 2
D-3000 Hannover 51
        Tel.:   0511/643-28-75
        Tlx.:   0923730 bgr ha d

H. JEZIERSKI
Deutsche Gesellschaft zum Bau und
  Betrieb von Endlagern für Abfallstoffe mbH.
Postfach 11 69
D-3150 Peine 1
        Tel.:   5171/43-300
        Tlx.:   92646

N. JOCKWER
GSF, Institut für Tieflagerung
Wissenschaftliche Abteilung
Theodor-Heuss-Strasse 4
D-3300 Braunschweig
        Tel.:   0531/80121
        Tlx.:   952865

W. KESSELS
GSF, Institut für Tieflagerung
Wissenschaftliche Abteilung
Theodor-Heuss-Strasse 4
D-3300 Braunschweig
        Tel.:   0531/8012-213
        Tlx.:   952865 iftta

J. KOPIETZ
Bundesanstalt für Geowissenschaften
  und Rohstoffe
Stilleweg 2
D-3000 Hannover 51
        Tel.:   0511/6432878
        Tlx.:   0923730 bgr ha d

K. KÜHN
GSF, Institut für Tieflagerung
Wissenschaftliche Abteilung
Theodor-Heuss-Strasse 4
D-3300 Braunschweig
        Tel.:   0531/80121
        Tlx.:   952865 iftta

FEDERAL REPUBLIC
OF GERMANY (Cont'd)

L. LIEDTKE
Federal Institute for Geosciences
  and Natural Resources
Postfach 51 01 53
D-3000 Hannover 51
          Tel.:   0511/643-2418 or: 643-2422
          Tlx.:   923730 bfb

H. RÖTHEMEYER
Physikalisch-Technische Bundesanstalt
Bundesalle 100
D-3300 Braunschweig
          Tel.:   0531/592 76 00
          Tlx.:   952822 PTB D

T. ROTHFUCHS
GSF, Institut für Tieflagerung
Wissenschaftliche Abteilung
Schachtanlage Asse
D-3346 Remlingen
          Tel.:   05336/19-232
          Tlx.:   95617 asse d

A. SCHNEEFUSS
Gesellschaft für Strahlen- und
  Umweltforschung mbH
Institut für Tieflagerung
Wissenschaftliche Abteilung
Theodor-Heuss-Strasse 4
D-3300 Braunschweig
          Tel.:   0531/80121
          Tlx.:   952865

ITALY

F. GERA
ISMES SpA.
Via Torquato Taramelli 14
I-00197 Roma
          Tel.:   06/804351-5
          Tlx.:   614615 PER ISMES

G. GIROLIMETTI
ENEA -
CRE/Casaccia/Dipartimento PAS
Divisione SCAMB
Laboratorio Geolog.
S. Maria di Galeria
Roma
Via Anguillarese Km. 1,300
          Tel.:   06/69483177
          Tlx.:   613296 ENEACA I

467

ITALY (Cont'd)

E. TASSONI
ENEA
CRE/Casaccia/Dipartimento PAS
Divisione SCAMB
Laboratorio Geolog.
S. Maria di Galeria
Roma
Via Anguillarese Km. 1,300
        Tel.: 06/69483978
        Tlx.: 613296 ENEACA I

JAPAN

K. AMEMIYA
Hazama-gumi Ltd.
5-8 Kita-Aoyama
2 Chome, Minato-ku
Tokyo
        Tel.: 03 405 1111
        Tlx.: J 25898

K. DOI
PNC
9-13 1-Chome
Akasaka, Minato-ku
Tokyo
        Tel.: 03 586 3311
        Tlx.: J 26462

M. KUMATA
Japan Atomic Energy Research Institute
Tokai-mura
Naka-gun
Ibaraki-ken 319 11
        Tel.: 02928-2-6180
        Tlx.: J 3632340 TOKAI

THE NETHERLANDS

J. GRAMBERG
Technical University Delft
Mijnbouwstraat 120
2628 RX Delft
        Tel.: 015 - 786024
        Tlx.: 38070

J. PRIJ
ECN, Netherlands Energy Research
  Foundation
P.O. Box 1
1755 ZG Petten
        Tel.: 2246 4429
        Tlx.: 57211 REACP NL

J. ROEST
Technical University Delft
Mijnbouwstraat 120
2628 RX Delft
        Tel.: 015 786024
        Tlx.: 38070

THE NETHERLANDS (Cont'd)   L. VONS
                            ECN, Netherlands Energy Research
                               Foundation
                            P.O. Box 1
                            1755 ZG Petten
                                       Tel.:   2246 4431
                                       Tlx.:   57211 REACP NL

SPAIN   J. SANCHEZ SANCHEZ
                            Junta de Energia Nuclear
                            Avda. Complutense N° 22
                            Madrid
                                       Tel.:   244 12 00
                                       Tlx.:   1301

SWEDEN   M. ABELIN
                            Department of Chemical Engineering
                            Royal Institute of Technology
                            Roslagsvägen 101
                            Byggnad 2
                            S-104 05 Stockholm
                                       Tel.:   +46 815 74 28
                                       Tlx.:   10389 kthb Stockholm

                            L. BIRGERSSON
                            Department of Chemical Engineering
                            Royal Institute of Technology
                            Roslagsvägen 101
                            Byggnad 2
                            S-104 05 Stockholm
                                       Tel.:   +46 815 74 28
                                       Tlx.:   10389 kthb Stockholm

                            H. CARLSSON
                            SKBF/KBS
                            Box 5864
                            S-102 48 Stockholm
                                       Tel.:   8-679540
                                       Tlx.:   13108 CDL

                            R. PUSCH
                            University of Lulea
                            Swedish Geological
                            Box 1424
                            S-751 44 Uppsala
                                       Tel.:   018-156420

SWITZERLAND   E. FRANK
                            Nuclear Safety Department
                            Swiss Federal Office of Energy
                            CH-5303 Würenlingen
                                       Tel.:   056/99 39 45 or 99 38 11

469

SWITZERLAND (Cont'd)

H. KEUSEN
Geotest A.G.
Birkenstr. 15
CH-3052 Zollikofen
                Tel.: 031/57 20 74
                Tlx.: 33740

R. LIEB
NAGRA (Swiss Cooperative for the
Disposal of Radioactive Waste)
Parkstrasse 23
CH-5401 Baden
                Tel.: 056/20 52 63 or 20 55 11
                Tlx.: 57333

R. ROMETSCH
NAGRA (Swiss Cooperative for the
Disposal of Radioactive Waste)
Parkstrasse 23
CH-5401 Baden
                Tel.: 056/22 01 82
                Tlx.: 57333

M. SCHWEINGRUBER
EIR (Federal Institute for Reactor
  Research)
CH-5303 Würenlingen
                Tel.: 056/99 24 15
                Tlx.: 53714 eir ch

UNITED KINGDOM

J. BLACK
British Geological Survey
Fluid Processes Research Group
Nicker Hill, Keyworth
Nottingham NG12 5GG
                Tel.: PLUMTREE (06077) 6111
                Tlx.: 378173 igs keyg

C. COOLING
Building Research Establishment
Bucknalls Lane
Garston, Watford Herts. WD2 7JR
                Tel.: 0923 674040
                Tlx.: 923220

F. FEATES
Department of the Environment
Romney House
43 Marsham Street
London SW1P 3PY
                Tel.: 212 88 04
                Tlx.: 22221

UNITED KINGDOM (Cont'd)   M. HEATH
                                 UKAEA, AERE Harwell
                                 Elcon (Western) Ltd.
                                 Trewirgie Rd.
                                 Redruth, Cornwall
                                           Tel.: (0209) 8317 54
                                           Tlx.: 45315

                                 L. LAKE
                                 Mott, Hay & Anderson
                                 St Anne House,
                                 Wellesley Road
                                 Croydon CR9 2UL
                                           Tel.: (01) 686-50-41
                                           Tlx.: 917241

UNITED STATES   H. DIETZ
                                 Basalt Waste Isolation Project
                                 Rockwell Hanford Operations
                                 Richland, WA. 99352
                                           Tel.: 509-373-3992
                                           Tlx.: 5107705108

                                 M. FREI
                                 Office of Civilian Radioactive Waste
                                   Management
                                 Office of Geologic Repository Deployment
                                 US Department of Energy
                                 RW-23 GTN
                                 Washington, D.C. 20545
                                           Tel.: 301 353 3013
                                           Tlx.: 7108280475

                                 A. HUNT
                                 US Department of Energy
                                 Albuquerque Operations Office
                                 P.O. Box 5400
                                 Albuquerque, NM. 87115

                                 H. KALIA
                                 Office of Nuclear Waste Isolation
                                 Battelle Project Management Division
                                 505 King Ave.,
                                 Columbus, OH. 43201
                                           Tel.: 614/424-5015

                                 V. MONTENYOHL
                                 Roy F. Weston Inc.
                                 2301 Research Blvd.
                                 Rockville, MD. 20850
                                           Tel.: 301/963-6852
                                           Tlx.: 835348

UNITED STATES (Cont'd)

M. NATARAJA
Rock Mechanics Section
Nuclear Regulatory Commission
Mail Stop 623-SS
Washington, D.C. 20555
        Tel.: 301/427-4678
        Tlx.: 4274298

W. PATRICK
Spent Fuel Test, Climax
Lawrence Livermore National Laboratory
Post Office Box 808;
M/S L-204
Livermore, Ca. 94550
        Tel.: 415/422-6495
        Tlx.: 9103868339

M. PLODINEC
E.I. Du Pont de Nemours & Co. Inc.
Bldg. 773-A,
Savannah River Lab.,
Aiken, SC 29808
        Tel.: 803/725-2170
        Tlx.: 2392170 FTS

B. ROBINSON
Battelle Project Management Division
Office of Crystalline Repository Development
505 King Avenue
Columbus, OH. 43200
        Tel.: 614/242-4172

P. STEVENS
US Geological Survey
  Water Resources Division
410 National Center
12201 Sunrise Valley Drive
Reston, Virginia 22092
        Tel.: 703/860-6976
        Tlx.: 899153

L. TYLER
Sandia National Laboratory
P.O. Box 5800
Albuquerque, NM. 87158
        Tel.: 505/844-8174

NEA

P. JOHNSTON
(not present)
current address:
Radioactive Waste (Professional) Division
Department of the Environment
Romney House
43 Marsham Street
London SW1P 3PY
UK
            Tel: 01-212-5676

A. MULLER
Division of Radiation Protection
and Waste Management
38 Bd. Suchet
F-75016 Paris
France
            Tel.:   524-96-78
            Tlx.:   630668

CEC

B. COME
Commission of the European Communities
200 rue de la Loi
B-1040 Brussels
Belgium
            Tel.:   02/235 11 11
            Tlx.:   COMEU B 21877

S. FINZI
Commission of the European Communities
200, rue de La Loi
B-1049 Brussels
Belgium
            Tel.:   (02) 235 9155
            Tlx.:   COMEU B 21877

B. HAIJTINK
Commission of the European Communities
200, rue de La Loi
B-1049 Brussels
Belgium
            Tel.:   (02) 235 3695
            Tlx.:   COMEU B 21877

S. ORLOWSKI
Commission of the European Communities
200, rue de La Loi
B-1049 Brussels
Belgium
            Tel.:   (02) 235 4063
            Tlx.:   COMEU B 21877

CEC (Cont'd)

P. VENET
Nuclear Fuel Cycle Division
Commission of the European Communities
200 rue de la Loi
B-1040 Brussels
Belgium
                 Tel.: (02) 235 5936
                 Tlx.: COMEU B 21877